Environment-Friendly Construction Materials

Environment-Friendly Construction Materials

Volume 1

Special Issue Editors

Shaopeng Wu
Inge Hoff
Serji N. Amirkhanian
Yue Xiao

MDPI • Basel • Beijing • Wuhan • Barcelona • Belgrade

MDPI

Special Issue Editors

Shaopeng Wu
Wuhan University of Technology
(WUT)
China

Inge Hoff
Norwegian University of Science and Technology
(NTNU)
Norway

Serji N. Amirkhanian
University of Alabama
USA

Yue Xiao
Wuhan University of Technology (WUT)
China

Editorial Office
MDPI
St. Alban-Anlage 66
4052 Basel, Switzerland

This is a reprint of articles from the Special Issue published online in the open access journal *Materials* (ISSN 1996-1944) from 2018 to 2019 (available at: https://www.mdpi.com/journal/materials/special_issues/EFCM).

For citation purposes, cite each article independently as indicated on the article page online and as indicated below:

LastName, A.A.; LastName, B.B.; LastName, C.C. Article Title. *Journal Name* **Year**, *Article Number, Page Range.*

Volume 1
ISBN 978-3-03921-012-1 (Pbk)
ISBN 978-3-03921-013-8 (PDF)

Volume 1-3
ISBN 978-3-03897-418-5 (Pbk)
ISBN 978-3-03897-419-2 (PDF)

Contents

About the Special Issue Editors vii

Shaopeng Wu, Inge Hoff, Serji Amirkhanian and Yue Xiao
Special Issue of Environment-Friendly Construction Materials
Reprinted from: *Materials* **2019**, *12*, 1101, doi:10.3390/ma12071101 1

Yazhen Sun, Chenze Fang, Jinchang Wang, Xuezhong Yuan and Dong Fan
Method of Fatigue-Life Prediction for an Asphalt Mixture Based on the Plateau Value of
Permanent Deformation Ratio
Reprinted from: *Materials* **2018**, *11*, 722, doi:10.3390/ma11050722 6

Yazhen Sun, Chenze Fang, Jinchang Wang, Zuoxin Ma and Youlin Ye
Energy-Based Approach to Predict Fatigue Life of Asphalt Mixture Using Three-Point Bending
Fatigue Test
Reprinted from: *Materials* **2018**, *11*, 1696, doi:10.3390/ma11091696 22

Zhenjun Wang, Linlin Cai, Xiaofeng Wang, Chuang Xu, Bo Yang and Jingjing Xiao
Fatigue Performance of Different Thickness Structure Combinations of Hot Mix Asphalt and
Cement Emulsified Asphalt Mixtures
Reprinted from: *Materials* **2018**, *11*, 1145, doi:10.3390/ma11071145 34

Peide Cui, Yue Xiao, Mingjing Fang, Zongwu Chen, Mingwei Yi and Mingliang Li
Residual Fatigue Properties of Asphalt Pavement after Long-Term Field Service
Reprinted from: *Materials* **2018**, *11*, 892, doi:10.3390/ma11060892 52

Yongchun Cheng, Di Yu, Guojin Tan and Chunfeng Zhu
Low-Temperature Performance and Damage Constitutive Model of Eco-Friendly Basalt
Fiber–Diatomite-Modified Asphalt Mixture under Freeze–Thaw Cycles
Reprinted from: *Materials* **2018**, *11*, 2148, doi:10.3390/ma11112148 65

**Yongchun Cheng, Wensheng Wang, Yafeng Gong, Shurong Wang, Shuting Yang and
Xun Sun**
Comparative Study on the Damage Characteristics of Asphalt Mixtures Reinforced with an
Eco-Friendly Basalt Fiber under Freeze-thaw Cycles
Reprinted from: *Materials* **2018**, *11*, 2488, doi:10.3390/ma11122488 86

Yazhen Sun, Bincheng Gu, Lin Gao, Linjiang Li, Rui Guo, Qingqing Yue and Jinchang Wang
Viscoelastic Mechanical Responses of HMAP under Moving Load
Reprinted from: *Materials* **2018**, *11*, 2490, doi:10.3390/ma11122490 103

**Jinxuan Hu, Shaopeng Wu, Quantao Liu, Maria Inmaculada García Hernández, Wenbo Zeng,
Shuai Nie, Jiuming Wan, Dong Zhang and Yuanyuan Li**
The Effect of Ultraviolet Radiation on Bitumen Aging Depth
Reprinted from: *Materials* **2018**, *11*, 747, doi:10.3390/ma11050747 122

**Ning Tang, Yu-Li Yang, Mei-Ling Yu, Wen-Li Wang, Shi-Yue Cao, Qing Wang and
Wen-Hao Pan**
Investigation of Ageing in Bitumen Using Fluorescence Spectrum
Reprinted from: *Materials* **2018**, *11*, 1325, doi:10.3390/ma11081325 136

Rui He, Shuhua Wu, Xiaofeng Wang, Zhenjun Wang and Huaxin Chen
Temperature Sensitivity Characteristics of SBS/CRP-Modified Bitumen after Different Aging
Processes
Reprinted from: *Materials* **2018**, *11*, 2136, doi:10.3390/ma11112136 . . . . **147**

Ling Pang, Xuemei Zhang, Shaopeng Wu, Yong Ye and Yuanyuan Li
Influence of Water Solute Exposure on the Chemical Evolution and Rheological Properties of
Asphalt
Reprinted from: *Materials* **2018**, *11*, 983, doi:10.3390/ma11060983 . . . . **162**

Wensheng Wang, Yongchun Cheng and Guojin Tan
Design Optimization of SBS-Modified Asphalt Mixture Reinforced with Eco-Friendly Basalt
Fiber Based on Response Surface Methodology
Reprinted from: *Materials* **2018**, *11*, 1311, doi:10.3390/ma11081311 . . . . **179**

Yongchun Cheng, Di Yu, Yafeng Gong, Chunfeng Zhu, Jinglin Tao and Wensheng Wang
Laboratory Evaluation on Performance of Eco-Friendly Basalt Fiber and Diatomite Compound
Modified Asphalt Mixture
Reprinted from: *Materials* **2018**, *11*, 2400, doi:10.3390/ma11122400 . . . . **201**

Chao Yang, Jun Xie, Xiaojun Zhou, Quantao Liu and Ling Pang
Performance Evaluation and Improving Mechanisms of Diatomite-Modified Asphalt Mixture
Reprinted from: *Materials* **2018**, *11*, 686, doi:10.3390/ma11050686 . . . . **217**

Menglin Li, Ling Pang, Meizhu Chen, Jun Xie and Quantao Liu
Effects of Aluminum Hydroxide and Layered Double Hydroxide on Asphalt Fire Resistance
Reprinted from: *Materials* **2018**, *11*, 1939, doi:10.3390/ma11101939 . . . . **232**

Wei Guo, Xuedong Guo, Mengyuan Chang and Wenting Dai
Evaluating the Effect of Hydrophobic Nanosilica on the Viscoelasticity Property of Asphalt and
Asphalt Mixture
Reprinted from: *Materials* **2018**, *11*, 2328, doi:10.3390/ma11112328 . . . . **243**

About the Special Issue Editors

Shaopeng Wu is a chief professor of materials science and engineering at Wuhan University of Technology. Dr. Wu is a prominent researcher in the field of bituminous materials and asphalt pavement. He has completed research projects on subjects including electrically conductive asphalt pavement, rubberized asphalt binder, recycling asphalt materials, self-healing asphalt, and asphalt preventive maintenance technologies. Dr. Wu is an editor of "Journal of Testing and Evaluation" and "International Journal of Pavement Research and Technology", and a member of the International Society of Asphalt Pavement. He has received 14 provincial awards, including 3 first-prize Science and Technology Progress Awards. He has already supervised more than 30 research projects and published more than 300 SCI peer-reviewed journal papers.

Inge Hoff started his research career in the independent research organization SINTEF (Norway) doing contracted research projects for 10 years. Most of the projects were laboratory-based projects financed by the Norwegian Public Roads Administration or producers of different types of materials. Dr. Hoff was appointed to professor in 2009 and has been working at NTNU since then. He was a supervisor of nine completed Ph.D. projects and is currently supervising six Ph.D. students. Dr. Hoff has authored more than 80 scientific and popular scientific papers and is an active reviewer for several international scientific journals. In addition, he has authored several SINTEF reports. Dr. Hoff is the leader of the only research laboratory for materials for transport infrastructure in Norway. The laboratory is used for education of master's and Ph.D. students, project-based testing for industry, and research purposes.

Serji N. Amirkhanian was the Mays Professor of Transportation and the director of the Asphalt Rubber Technology Services (ARTS) in the Department of Civil Engineering at Clemson University until June of 2010, before becoming a professor of civil engineering at the University of Alabama, Tuscaloosa, USA. He is also a professor of civil engineering at Wuhan University of Technology (Wuhan, China), in addition to being an adjunct professor of materials at Norwegian University of Science and Technology (NTNU), Norway.

His research has resulted in over 300 refereed journal papers, conference papers, and research reports. He has also published several book chapters. In addition, he has given over 300 presentations, presenting his research findings in the US and internationally. He has supported over 100 graduate students and over 15 post-doctoral scholars, conducting research for many agencies (e.g., the Federal Highway Administration (FHWA)). He has consulted for many companies or agencies, such as the World Bank, United Nations (UNIDO), BMW, Owens Corning, Michelin, Honeywell International, Ontario Ministry of Transportation, and Honduras Ministry of Transportation, among many others.

Yue Xiao has been an associate research professor at State Key Lab of Silicate Materials for Architectures in the Wuhan University of Technology since 2014. He was named the Fok Ying Tung Outstanding Young Teacher by the Ministry of Education of China in 2018. He received the title of CHUTIAN Scholar in material science and engineering from the Hubei provincial department of education in 2014. Dr. Xiao obtained his PhD degree in road and railway engineering from Delft University of Technology, the Netherlands. Subsequently, he joined Wuhan University of Technology. His research interests include eco-efficient pavement materials and construction material recycling.

Dr. Xiao is now conducting three innovative projects founded by the National Natural Science Foundation of China (NSFC), as well as projects supported by provincial departments. Since 2011, Dr. Xiao has published 49 SCI peer-reviewed journal papers.

materials

MDPI

Editorial

Special Issue of Environment-Friendly Construction Materials

Shaopeng Wu [1], Inge Hoff [2], Serji Amirkhanian [3] and Yue Xiao [1,*]

[1] State Key Laboratory of Silicate Materials for Architectures, Wuhan University of Technology, Wuhan 430070, China; wusp@whut.edu.cn
[2] Department of Civil and Environmental Engineering, Norwegian University of Science and Technology, Hogskoleringen 7A, NO-7491 Trondheim, Norway; Inge.Hoff@ntnu.no
[3] Department of Civil Construction and Environmental Engineering, University of Alabama, Tuscaloosa, AL 35487, USA; samirkhanian@eng.ua.edu
* Correspondence: xiaoy@whut.edu.cn; Tel.: +86-1817-108-9165

Received: 18 March 2019; Accepted: 27 March 2019; Published: 3 April 2019

Abstract: This special issue, "Environment-Friendly Construction Materials", has been proposed and organized as a means to present recent developments in the field of construction materials. For this reason, the articles highlighted in this editorial relate to different aspects of construction materials, from pavement materials to building materials, from material design to structural design, from self-healing to cold recycling, from asphalt mixture to cement concrete.

Keywords: construction materials; fatigue life; ageing resistance; modified asphalt materials; rejuvenator; self-healing asphalt; recycling; cold recycled asphalt mixture; ultra-high performance concrete

Construction materials are the most widely used materials for civil infrastructures in our daily life. However, from an environmental point of view, they consume a huge amount of natural resources and generate the majority of greenhouse gasses. Therefore, many new and novel technologies for designing environment-friendly construction materials have been developed recently. This special issue, "Environment-Friendly Construction Materials", has been proposed and organized as a means to present recent developments in the field of construction materials. It covers a wide range of selected topics on construction materials. A brief summary of the articles is given in this editorial.

Service life prediction is essentially important in designing construction materials. Researchers all over the world are devoting themselves to life prediction analysis. Sun et al. [1,2] used a plateau value and permanent deformation ratio from three-point bending fatigue tests with cyclic loading to predict the fatigue life of as asphalt mixture. The fatigue equation based on a plateau value can well predict the fatigue life. Wang et al. [3] studied the fatigue performance of combined structures with hot mix asphalt and cement emulsified asphalt mixtures. An artificial neural network was used and fatigue equations were established for fatigue life prediction. Residual fatigue properties of asphalt pavement after long-term field service [4], low-temperature performance [5] and damage characteristics [6] were reported. Eco-friendly fiber was used to improve the performance of mixtures. Sun et al. [7] studied the viscoelastic mechanical responses of high-modulus asphalt pavement by numerical simulation with a moving load. In three articles, ageing resistances of asphalt were reported, including ageing depth resulting from ultraviolet radiation [8], fluorescence spectrum ageing analysis [9], and ageing improvement by SBS/CRP (Styrene–butadiene–styrene polymer/crumb rubber powder) modification [10]. One research article focused on the chemical evolution and rheological properties of asphalt under water solute exposure [11]. Saturates and aromatics were partly dissolved in water and then moved out.

Modification on construction materials are being conducted in many research institutes to design durable civil infrastructures. Fiber is a widely used strengthening additive in asphalt mixtures. Eco–friendly basalt fiber was incorporated with SBS and diatomite by Wang et al. [12] and Cheng et al. [13]. Another article by Yang et al. [14] presented improving mechanisms of diatomite modified asphalt mixtures, by means of permanent deformation resistance and moisture resistance. Aluminum hydroxide and layered double hydroxide were proposed by Li et al. [15] to improve the fire resistance of asphalt. Another nanomaterial, named nanosilica, was evaluated by Guo et al. [16].

Improving the aggregate morphology characteristics is another effective way to get durable asphalt mixtures. Xiao et al. [17] established the relationship between fine aggregate morphology and skid-resistance of micro-surfacing, while Cheng et al. [18] and Wang et al. [19] reported the influence of aggregate morphological characteristics on asphalt mixtures. The studied aggregate morphological characteristics include roundness, perimeter index, erosion-dilation area ratio, angularity, and surface texture. Influence of aggregate characteristics on the demulsification speed of asphalt emulsion was presented by Tang et al. [20]. Furthermore, Liu et al. [21] proposed to use ash byproduct to improve the asphalt-aggregate adhesion properties.

Rejuvenator, a healing agent to recover aged asphalt binder, is a widely used material in pavement preventive maintenance. There are many different rejuvenators. For instance, soybean oil based [22], dodecyl benzene sulfonic acid based [23], bio-oil based [24], petroleum based [25], isocyanate and epoxy substances based [26] were detailed in this special issue. The interesting rejuvenation enhancement was investigated and reported by these articles. Healing behavior of asphalt materials is another key issue in the pavement preventive maintenance. Calcium alginate capsules were designed by both Xu et al. [27] and Shu et al. [28]. The former article investigated the healing capacity of asphalt mixture when calcium alginate capsules were used, while the second article presented a preparation process for calcium alginate capsules with a multinuclear structure. In the study by Wan et al. [29], self-healing properties of steel fiber and steel slag based ultra-thin wearing course were studied by a semi-circular bending test under induction heating. Other researches focused on the healing agent effect [30], induced healing efficiency of induction heating and microwave heating [31], and initial self-healing temperature [32].

Andrzejuk et al. [33] and Ogrodnik et al. [34] reported their research on reusing the wastes of sanitary ceramics as aggregates for asphalt mixture and cement concrete, respectively. Waste concrete powder [35], low-grade aggregate [36], crumb rubber waste [37], and recycled concrete aggregate [38] were also successfully reused as construction materials. In the study by Li et al. [39], the reclaimed asphalt pavement was reused 100% in cold recycled asphalt mixtures. Asphalt emulsion and cement were used to improve the interfacial bonding between binders and fillers, aiming to enhance the moisture resistance and high temperature stability.

Several other studies involved the evaluation of eco-friendly railway concrete sleepers [40] and engineered cementitious composites [41]. In the research of the former article, waste rubber was reused for high-strength rubberized concrete. It was found that a decrease of compressive strength can be expected when rubber content increased, and 10% was recommended as the optimal reuse content. In the latter article, modified polyvinyl alcohol fiber was added into engineered cementitious composites to enhance the mechanical performance. Research on self-compacting concrete [42], ultra-high performance concrete [43], cement paste plasticized by polycarboxylate superplasticizer [44], and pozzolanic additive in cement [45] were also discussed in this special issue.

Last, but not least, there are two articles focusing on functional construction materials, like phase change materials for building energy conservation [46], and graphene-modulated removal performance of nitrogen and phosphorus pollutants [47].

Funding: This research was funded by the National Natural Science Foundation of China (grant number U1733121, 51878526 and 51778515).

Conflicts of Interest: The authors declare no conflict of interest.

References

1. Sun, Y.; Fang, C.; Wang, J.; Yuan, X.; Fan, D. Method of Fatigue-Life Prediction for an Asphalt Mixture Based on the Plateau Value of Permanent Deformation Ratio. *Materials* **2018**, *11*, 722. [CrossRef] [PubMed]
2. Sun, Y.; Fang, C.; Wang, J.; Ma, Z.; Ye, Y. Energy-Based Approach to Predict Fatigue Life of Asphalt Mixture Using Three-Point Bending Fatigue Test. *Materials* **2018**, *11*, 1696. [CrossRef] [PubMed]
3. Wang, Z.; Cai, L.; Wang, X.; Xu, C.; Yang, B.; Xiao, J. Fatigue Performance of Different Thickness Structure Combinations of Hot Mix Asphalt and Cement Emulsified Asphalt Mixtures. *Materials* **2018**, *11*, 1145. [CrossRef] [PubMed]
4. Cui, P.; Xiao, Y.; Fang, M.; Chen, Z.; Yi, M.; Li, M. Residual Fatigue Properties of Asphalt Pavement after Long-Term Field Service. *Materials* **2018**, *11*, 892. [CrossRef] [PubMed]
5. Cheng, Y.; Yu, D.; Tan, G.; Zhu, C. Low-Temperature Performance and Damage Constitutive Model of Eco-Friendly Basalt Fiber–Diatomite-Modified Asphalt Mixture under Freeze–Thaw Cycles. *Materials* **2018**, *11*, 2148. [CrossRef] [PubMed]
6. Cheng, Y.; Wang, W.; Gong, Y.; Wang, S.; Yang, S.; Sun, X. Comparative Study on the Damage Characteristics of Asphalt Mixtures Reinforced with an Eco-Friendly Basalt Fiber under Freeze-thaw Cycles. *Materials* **2018**, *11*, 2488. [CrossRef] [PubMed]
7. Sun, Y.; Gu, B.; Gao, L.; Li, L.; Guo, R.; Yue, Q.; Wang, J. Viscoelastic Mechanical Responses of HMAP under Moving Load. *Materials* **2018**, *11*, 2490. [CrossRef] [PubMed]
8. Hu, J.; Wu, S.; Liu, Q.; García Hernández, M.I.; Zeng, W.; Nie, S.; Wan, J.; Zhang, D.; Li, Y. The Effect of Ultraviolet Radiation on Bitumen Aging Depth. *Materials* **2018**, *11*, 747. [CrossRef] [PubMed]
9. Tang, N.; Yang, Y.-L.; Yu, M.-L.; Wang, W.-L.; Cao, S.-Y.; Wang, Q.; Pan, W.-H. Investigation of Ageing in Bitumen Using Fluorescence Spectrum. *Materials* **2018**, *11*, 1325. [CrossRef]
10. He, R.; Wu, S.; Wang, X.; Wang, Z.; Chen, H. Temperature Sensitivity Characteristics of SBS/CRP-Modified Bitumen after Different Aging Processes. *Materials* **2018**, *11*, 2136. [CrossRef]
11. Pang, L.; Zhang, X.; Wu, S.; Ye, Y.; Li, Y. Influence of Water Solute Exposure on the Chemical Evolution and Rheological Properties of Asphalt. *Materials* **2018**, *11*, 983. [CrossRef]
12. Wang, W.; Cheng, Y.; Tan, G. Design Optimization of SBS-Modified Asphalt Mixture Reinforced with Eco-Friendly Basalt Fiber Based on Response Surface Methodology. *Materials* **2018**, *11*, 1311. [CrossRef] [PubMed]
13. Cheng, Y.; Yu, D.; Gong, Y.; Zhu, C.; Tao, J.; Wang, W. Laboratory Evaluation on Performance of Eco-Friendly Basalt Fiber and Diatomite Compound Modified Asphalt Mixture. *Materials* **2018**, *11*, 2400. [CrossRef] [PubMed]
14. Yang, C.; Xie, J.; Zhou, X.; Liu, Q.; Pang, L. Performance Evaluation and Improving Mechanisms of Diatomite-Modified Asphalt Mixture. *Materials* **2018**, *11*, 686. [CrossRef]
15. Li, M.; Pang, L.; Chen, M.; Xie, J.; Liu, Q. Effects of Aluminum Hydroxide and Layered Double Hydroxide on Asphalt Fire Resistance. *Materials* **2018**, *11*, 1939. [CrossRef]
16. Guo, W.; Guo, X.; Chang, M.; Dai, W. Evaluating the Effect of Hydrophobic Nanosilica on the Viscoelasticity Property of Asphalt and Asphalt Mixture. *Materials* **2018**, *11*, 2328. [CrossRef]
17. Xiao, Y.; Wang, F.; Cui, P.; Lei, L.; Lin, J.; Yi, M. Evaluation of Fine Aggregate Morphology by Image Method and Its Effect on Skid-Resistance of Micro-Surfacing. *Materials* **2018**, *11*, 920. [CrossRef]
18. Cheng, Y.; Wang, W.; Tao, J.; Xu, M.; Xu, X.; Ma, G.; Wang, S. Influence Analysis and Optimization for Aggregate Morphological Characteristics on High- and Low-Temperature Viscoelasticity of Asphalt Mixtures. *Materials* **2018**, *11*, 2034. [CrossRef]
19. Wang, W.; Cheng, Y.; Tan, G.; Tao, J. Analysis of Aggregate Morphological Characteristics for Viscoelastic Properties of Asphalt Mixes Using Simplex Lattice Design. *Materials* **2018**, *11*, 1908. [CrossRef]
20. Tang, F.; Xu, G.; Ma, T.; Kong, L. Study on the Effect of Demulsification Speed of Emulsified Asphalt based on Surface Characteristics of Aggregates. *Materials* **2018**, *11*, 1488. [CrossRef] [PubMed]
21. Liu, Z.; Huang, X.; Sha, A.; Wang, H.; Chen, J.; Li, C. Improvement of Asphalt-Aggregate Adhesion Using Plant Ash Byproduct. *Materials* **2019**, *12*, 605. [CrossRef] [PubMed]
22. Kuang, D.; Jiao, Y.; Ye, Z.; Lu, Z.; Chen, H.; Yu, J.; Liu, N. Diffusibility Enhancement of Rejuvenator by Epoxidized Soybean Oil and Its Influence on the Performance of Recycled Hot Mix Asphalt Mixtures. *Materials* **2018**, *11*, 833. [CrossRef] [PubMed]

23. Kuang, D.; Ye, Z.; Yang, L.; Liu, N.; Lu, Z.; Chen, H. Effect of Rejuvenator Containing Dodecyl Benzene Sulfonic Acid (DBSA) on Physical Properties, Chemical Components, Colloidal Structure and Micro-Morphology of Aged Bitumen. *Materials* **2018**, *11*, 1476. [CrossRef] [PubMed]

24. Yang, T.; Chen, M.; Zhou, X.; Xie, J. Evaluation of Thermal-Mechanical Properties of Bio-Oil Regenerated Aged Asphalt. *Materials* **2018**, *11*, 2224. [CrossRef]

25. Pan, P.; Kuang, Y.; Hu, X.; Zhang, X. A Comprehensive Evaluation of Rejuvenator on Mechanical Properties, Durability, and Dynamic Characteristics of Artificially Aged Asphalt Mixture. *Materials* **2018**, *11*, 1554. [CrossRef] [PubMed]

26. Li, Z.; Xu, X.; Yu, J.; Wu, S. Assessment on Physical and Rheological Properties of Aged SBS Modified Bitumen Containing Rejuvenating Systems of Isocyanate and Epoxy Substances. *Materials* **2019**, *12*, 618. [CrossRef] [PubMed]

27. Xu, S.; Liu, X.; Tabaković, A.; Schlangen, E. Investigation of the Potential Use of Calcium Alginate Capsules for Self-Healing in Porous Asphalt Concrete. *Materials* **2019**, *12*, 168. [CrossRef]

28. Shu, B.; Wu, S.; Dong, L.; Wang, Q.; Liu, Q. Microfluidic Synthesis of Ca-Alginate Microcapsules for Self-Healing of Bituminous Binder. *Materials* **2018**, *11*, 630. [CrossRef] [PubMed]

29. Wan, J.; Xiao, Y.; Song, W.; Chen, C.; Pan, P.; Zhang, D. Self-Healing Property of Ultra-Thin Wearing Courses by Induction Heating. *Materials* **2018**, *11*, 1392. [CrossRef] [PubMed]

30. Pan, C.; Tang, P.; Riara, M.; Mo, L.; Li, M.; Guo, M. Effect of Healing Agents on Crack Healing of Asphalt and Asphalt Mortar. *Materials* **2018**, *11*, 1373. [CrossRef] [PubMed]

31. Liu, Q.; Chen, C.; Li, B.; Sun, Y.; Li, H. Heating Characteristics and Induced Healing Efficiencies of Asphalt Mixture via Induction and Microwave Heating. *Materials* **2018**, *11*, 913. [CrossRef] [PubMed]

32. Li, C.; Wu, S.; Tao, G.; Xiao, Y. Initial Self-Healing Temperatures of Asphalt Mastics Based on Flow Behavior Index. *Materials* **2018**, *11*, 917. [CrossRef] [PubMed]

33. Andrzejuk, W.; Barnat-Hunek, D.; Siddique, R.; Zegardło, B.; ągód, G. Application of Recycled Ceramic Aggregates for the Production of Mineral-Asphalt Mixtures. *Materials* **2018**, *11*, 658. [CrossRef] [PubMed]

34. Ogrodnik, P.; Szulej, J.; Franus, W. The Wastes of Sanitary Ceramics as Recycling Aggregate to Special Concretes. *Materials* **2018**, *11*, 1275. [CrossRef] [PubMed]

35. Wang, K.; Ren, L.; Yang, L. Excellent Carbonation Behavior of Rankinite Prepared by Calcining the C-S-H: Potential Recycling of Waste Concrete Powders for Prefabricated Building Products. *Materials* **2018**, *11*, 1474. [CrossRef] [PubMed]

36. Zhang, X.; Zhang, B.; Chen, H.; Kuang, D. Feasibility Evaluation of Preparing Asphalt Mixture with Low-Grade Aggregate, Rubber Asphalt and Desulphurization Gypsum Residues. *Materials* **2018**, *11*, 1481. [CrossRef]

37. Li, H.; Jiang, H.; Zhang, W.; Liu, P.; Wang, S.; Wang, F.; Zhang, J.; Yao, Z. Laboratory and Field Investigation of the Feasibility of Crumb Rubber Waste Application to Improve the Flexibility of Anti-Rutting Performance of Asphalt Pavement. *Materials* **2018**, *11*, 1738. [CrossRef]

38. Hou, Y.; Ji, X.; Li, J.; Li, X. Adhesion between Asphalt and Recycled Concrete Aggregate and Its Impact on the Properties of Asphalt Mixture. *Materials* **2018**, *11*, 2528. [CrossRef]

39. Li, Y.; Lyv, Y.; Fan, L.; Zhang, Y. Effects of Cement and Emulsified Asphalt on Properties of Mastics and 100% Cold Recycled Asphalt Mixtures. *Materials* **2019**, *12*, 754. [CrossRef]

40. Kaewunruen, S.; Li, D.; Chen, Y.; Xiang, Z. Enhancement of Dynamic Damping in Eco-Friendly Railway Concrete Sleepers Using Waste-Tyre Crumb Rubber. *Materials* **2018**, *11*, 1169. [CrossRef]

41. Sun, M.; Chen, Y.; Zhu, J.; Sun, T.; Shui, Z.; Ling, G.; Zhong, H.; Zheng, Y. Effect of Modified Polyvinyl Alcohol Fibers on the Mechanical Behavior of Engineered Cementitious Composites. *Materials* **2018**, *12*, 37. [CrossRef] [PubMed]

42. Ling, G.; Shui, Z.; Sun, T.; Gao, X.; Wang, Y.; Sun, Y.; Wang, G.; Li, Z. Rheological Behavior and Microstructure Characteristics of SCC Incorporating Metakaolin and Silica Fume. *Materials* **2018**, *11*, 2576. [CrossRef] [PubMed]

43. Liu, K.; Yu, R.; Shui, Z.; Li, X.; Ling, X.; He, W.; Yi, S.; Wu, S. Effects of Pumice-Based Porous Material on Hydration Characteristics and Persistent Shrinkage of Ultra-High Performance Concrete (UHPC). *Materials* **2018**, *12*, 11. [CrossRef] [PubMed]

44. Ma, B.; Peng, Y.; Tan, H.; Lv, Z.; Deng, X. Effect of Polyacrylic Acid on Rheology of Cement Paste Plasticized by Polycarboxylate Superplasticizer. *Materials* **2018**, *11*, 1081. [CrossRef]

45. Xu, W.; Wei, J.; Chen, J.; Zhang, B.; Xu, P.; Ren, J.; Yu, Q. Comparative Study of Water-Leaching and Acid-Leaching Pretreatment on the Thermal Stability and Reactivity of Biomass Silica for Viability as a Pozzolanic Additive in Cement. *Materials* **2018**, *11*, 1697. [CrossRef]

46. Zhang, D.; Chen, M.; Liu, Q.; Wan, J.; Hu, J. Preparation and Thermal Properties of Molecular-Bridged Expanded Graphite/Polyethylene Glycol Composite Phase Change Materials for Building Energy Conservation. *Materials* **2018**, *11*, 818. [CrossRef]

47. Xia, G.; Xu, W.; Fang, Q.; Mou, Z.; Pan, Z. Graphene-Modulated Removal Performance of Nitrogen and Phosphorus Pollutants in a Sequencing Batch Chlorella Reactor. *Materials* **2018**, *11*, 2181. [CrossRef] [PubMed]

materials

MDPI

Article

Method of Fatigue-Life Prediction for an Asphalt Mixture Based on the Plateau Value of Permanent Deformation Ratio

Yazhen Sun [1], Chenze Fang [1], Jinchang Wang [2], Xuezhong Yuan [3,*] and Dong Fan [3]

[1]   School of Transportation Engineering, Shenyang Jianzhu University, Shenyang 110168, China;
      syz16888@126.com (Y.S.); fangchenze@126.com (C.F.)
[2]   Institute of Transportation Engineering, Zhejiang University, Hangzhou 310058, China; wjc501@zju.edu.cn
[3]   School of Science, Shenyang Jianzhu University, Shenyang 110168, China; xiaodaxia1122@126.com
*   Correspondence: yuanxuezhong1@163.com; Tel.: +86-136-1404-0298

Received: 11 April 2018; Accepted: 25 April 2018; Published: 3 May 2018

Abstract: Laboratory predictions for the fatigue life of an asphalt mixture under cyclic loading based on the plateau value (PV) of the permanent deformation ratio (PDR) were carried out by three-point bending fatigue tests. The influence of test conditions on the recovery ratio of elastic deformation (RRED), the permanent deformation (PD) and PDR, and the trends of RRED, PD, and PDR were studied. The damage variable was defined by using PDR, and the relation of the fatigue life to PDR was determined by analyzing the damage evolution process. The fatigue equation was established based on the PV of PDR and the fatigue life was predicted by analyzing the relation of the fatigue life to the PV. The results show that the RRED decreases with the increase of the number of loading cycles, and the elastic recovery ability of the asphalt mixture gradually decreases. The two mathematical models proposed are based on the change laws of the RRED, and the PD can well describe the change laws. The RRED or the PD cannot well predict the fatigue life because they do not change monotonously with the fatigue life, and one part of the deformation causes the damage and the other part causes the viscoelastic deformation. The fatigue life decreases with the increase of the PDR. The average PDR in the second stage is taken as the PV, and the fatigue life decreases in a power law with the increase of the PV. The average relative error of the fatigue life predicted by the fatigue equation to the test fatigue life is 5.77%. The fatigue equation based on PV can well predict the fatigue life.

Keywords: road engineering; fatigue life; three-point bending fatigue test; asphalt mixture; plateau value of permanent deformation ratio; damage evolution; fatigue equation

---

1. Introduction

In recent years, the road construction industry has made great progress with the development of technologies, but the qualities of existing roads are mixed. The asphalt concrete pavement, as the surface course of the road structure, is subjected to repeated actions of vehicle loads and the environmental influence of seasonal changes. Under cyclic loading, the stresses and strains in the material change continuously, which results in the reduction of the strength. Pavement under cyclic loading is prone to fatigue failure. The fatigue life of pavement material has become the focus of research for an increasing number of researchers [1–8]. Three approaches are usually used to study fatigue life: the phenomenological approach, the fracture-mechanics-based approach, and the energy (damage) approach [9–15].

The fracture mechanics approach was used to study fatigue by monitoring crack development in its early stages. Researchers carried out many studies on the fatigue properties of asphalt concrete with rubber grains based on fracture mechanics [16,17]. Principles of fracture mechanics were applied to data obtained by monitoring the size and the length of the crack opening to discover the stress intensity factors at the crack tip [18,19]. The Paris equation can describe the relation of the stress intensity factors to the crack propagation and was used to describe the growth process of fatigue cracks and predict the fatigue life of the asphalt mixture [20–22]. Limitations of this approach include the need for a large amount of experimental data, considering only the crack propagation, and the stress intensity factor K_I being not a material constant at higher temperatures [23,24]. Also, fracture mechanics cannot accurately describe the viscoelastic plastic mechanical properties of the asphalt mixture.

Researchers carried out many studies on the fatigue properties of bitumen and the asphalt mixture based on the dissipated energy concept. These studies assumed that fatigue life depends on the accumulation of dissipated energy in each loading cycle. The fatigue equation established based on this assumption was used to predict fatigue life [25–28]. Later studies demonstrated that the damage was related to the rate of change in dissipated energy from one loading cycle to the next and only part of the dissipated energy can cause damage to the material. Therefore, it is inaccurate to use the total dissipated energy to predict the fatigue life [29–31].

The phenomenological approach provides an important idea for the early study of fatigue properties. This approach assumes that the stress or the strain in the asphalt layer is related to the number of load repetitions to failure. Some fatigue equations established based on this assumption were used to predict fatigue life [32,33]. There was a large discreteness in the fatigue life predicted by these equations. Also, the phenomenological approach cannot reveal the mechanism of damage evolution [34–37].

The fatigue equations in the above studies, established from different perspectives, were used to predict the fatigue life of the asphalt mixture. These equations, established after processing a large amount of experimental data, were complex and inconvenient to use. Permanent deformation (PD) is accumulated during the damage evolution process of the asphalt mixture, and since the PD is easy to obtain and deal with, establishing the fatigue equation based on the analysis of the PD has important implications for researchers in predicting the fatigue life of the asphalt mixture.

To address the shortcomings of the above studies, understand the change law of deformation of the asphalt mixture under cyclic loading, and predict the fatigue life from the perspective of deformation, the AC-13I asphalt mixture was used under different experimental conditions to carry out the three-point bending fatigue tests in this paper. By analyzing the change laws of the recovery ratio of elastic deformation (RRED), the conclusions drawn from this are that the elastic recovery ability of the asphalt mixture under cyclic loading gradually decreases, and two mathematical models describing the change laws of the RRED and the PD are proposed. By analyzing the influence of experimental factors on the RRED and the PD, the conclusions also drawn are that the RRED or the PD does not change monotonously with fatigue life, and the RRED or the PD cannot be directly used to predict the fatigue life of the asphalt mixture under cyclic loading. The PD of the asphalt mixture does not change monotonously with fatigue life, indicating that only part of the PD causes the damage of the asphalt mixture, and that the other part causes the viscoelastic deformation. By analyzing the relationship between the damage evolution speed and the value of the permanent deformation ratio (PDR), it was found that the faster the damage evolution speed of the asphalt mixture under cyclic loading was, the earlier the damage was close to the threshold of failure, and the fatigue life of the asphalt mixture under cyclic loading decreased with the increase of the damage evolution speed. It was also found that the power function could be used to describe the mathematical relationship between fatigue life and the plateau value of the PDR of the asphalt mixture. The fatigue equation was established based on the power-function relationship between the plateau value of the PDR and the fatigue life. The predicted fatigue life of the asphalt mixture under different test conditions was calculated by the fatigue equation and compared with the test fatigue life. The relative error between the fatigue life predicted by the fatigue equation and the test fatigue life was small, which indicates that the proposed equation can be used to accurately predict the fatigue life of an asphalt mixture.

2. Materials and Experimental Procedures

2.1. Test Materials

A 70-penetration bitumen was used as asphalt binder for preparations of the specimens, with its specifications listed in Table 1. Limestone was used as the aggregate. The ratio of binder to aggregate is 8.1% by weight. The continuous aggregate gradation, having the nominal maximum size of 16 mm, is listed in Table 2.

Table 1. Properties of asphalt rubber.

Material Property	Used Standard	Value
Softening point	T0606—2000	57 (°C)
Viscosity	T0625—2000	1.5–4.0 (Pa·s)
Elastic recovery	T0662—2000	30 (%)

Table 2. Aggregate gradation.

Sieve Size (mm)	16.0	13.2	9.5	4.75	2.36	1.18	0.6	0.3	0.15	0.075
Passing percentage	100.0	91.1	80.2	54.0	33.2	22.5	16.0	12.1	8.7	5.5

2.2. Preparation of Specimen

A rut board of 400 mm × 400 mm × 70 mm was made by a hydraulic sample forming machine [38], as shown in Figure 1. By cutting the rut board, specimen beams of 250 mm × 30 mm × 35 mm were obtained, as shown in Figure 2. The average density of the beam specimens is 2.445 g/cm^3.

Figure 1. Rut board forming machine.

Figure 2. Specimen beams.

2.3. Test Conditions and Methods

The fatigue test is shown in Figure 3. The flexural tensile strengths of specimens under different test conditions were measured. To better analyze the influence of experimental factors on test results, three groups of contrast tests were arranged:

Group 1. For this group, the temperature was 25 °C, the loading rate was 10 mm/min, and the stress-strength ratios (SSR) were 0.6, 0.7, and 0.8 respectively.

Group 2. The temperatures was 5 °C, 15 °C, and 25 °C, respectively, the loading rate was 10 mm/min, and the SSR was 0.6.

Group 3. The temperature was 25 °C, the loading rates were 10 mm/min and 20 mm/min, respectively, and the SSR was 0.6.

Figure 3. Fatigue test.

2.4. Test Results

The fatigue tests were carried out according to test scheme and the fatigue lives are listed in Table 3. It can be seen that fatigue life decreases with the increase of the stress-strength ratio (SSR) and the loading rate and increases with the increase of the temperature.

Table 3. Fatigue lives.

SSR-Temperature-Loading Rate	Value
0.6–25 °C-10 mm/min	313
0.7–25 °C-10 mm/min	161
0.8–25 °C-10 mm/min	139
0.6–15 °C-10 mm/min	109
0.6–5 °C-10 mm/min	97
0.6–25 °C-20 mm/min	113

3. Analysis of Recovery Ratio of Elastic Deformation

The deformation time curve for the three-point bending fatigue test is shown in Figure 4. As shown in the figure, during each loading cycle, the deformation of the asphalt mixture increases linearly to the peak value at the set loading rate in the loading stage and decreases linearly to the point where the load is zero in the unloading stage. In this fashion, the asphalt mixture specimen is subjected to cyclic loading until it is broken. The elastic recovery ability of the asphalt mixture under cyclic loading gradually decreases and the RRED represents the elastic recovery ability of the asphalt mixture [31]. The RRED is given as

$$RRED = \frac{d_{max,N} - d_{min,N}}{d_{max,N}} \tag{1}$$

where $d_{max,N}$ is the peak deformation of loading cycle N and $d_{min,N}$ is the minimum deformation of loading cycle N.

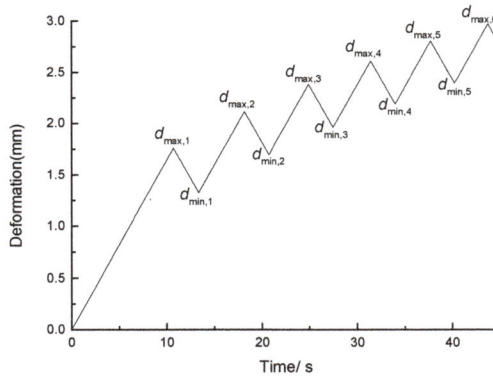

Figure 4. Deformation time.

3.1. Construction of RRED Mathematical Model

The curve of the RRED vs. the number of loading cycles (RRED-N curve) of the three-point bending fatigue test calculated by Equation (1) is shown in Figure 5. As shown in the figure, the RRED decreases with the increase of the number of loading cycles, and the process can be divided into two stages. In the first stage, the RRED lasts a short time but decreases rapidly, and at the end of this stage, it drops to a small value. In the second stage, the RRED lasts a long time and decreases slowly. The RRED of the asphalt mixture decreases with the increase of the number of loading cycles, which indicates that both the elastic recovery ability and the integrity of the asphalt mixture decrease [39]. The RRED mathematical model was proposed to accurately describe the change law of the elastic recovery ability and the integrity of the asphalt mixture under cyclic loading. The RRED mathematical model is defined as

$$RRED = aN^b \tag{2}$$

where a and b are fitting parameters and N is the number of loading cycles.

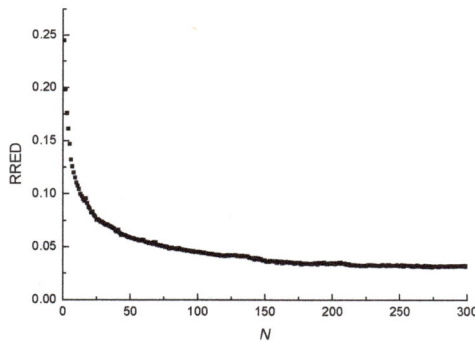

Figure 5. RRED-N curve.

The proposed RRED mathematical model given by Equation (2) was used to fit the test data and the correlation coefficient is greater than 0.99, which indicates that the proposed model describes very well the elastic recovery ability and the integrity of the asphalt mixture under cyclic loading. The fitting results at the temperature of 25 °C, the stress-strength ratio of 0.6, and the loading rate of 10 mm/min are listed in Table 4, and the fitting data and test data are shown in Figure 6.

Table 4. Fitting results of RRED mathematical model.

Fitting Parameters	*a*	*b*	R^2
Value	0.25937	−0.37801	0.99456

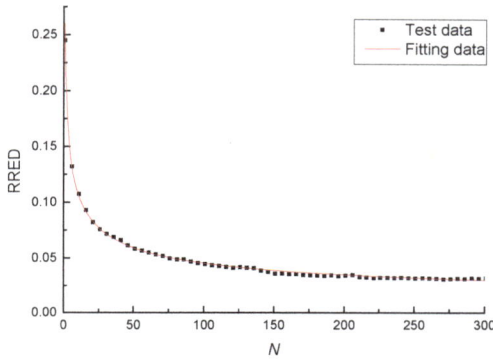

Figure 6. Test data and fitting data.

3.2. Influence of Experimental Factors on RRED

The results in different test conditions are shown in Figure 7. As shown in Figure 7a, the RRED calculated by Equation (1) at the stress-strength ratio of 0.7 is very close to that at the stress-strength ratio of 0.8, and they are greater than that at the stress-strength ratio of 0.6. However, the sequence of the magnitude of the fatigue life at different stress-strength ratios corresponds to the stress-strength ratios of 0.6, 0.7, and 0.8, respectively. As shown in Figure 7b, the RRED is influenced by the temperature, and the sequence of the magnitude of the recovery ratio at different temperatures corresponds to the temperatures of 15 °C, 5 °C, and 25 °C, respectively. However, the sequence of fatigue life corresponds to the temperatures of 25 °C, 15 °C, and 5 °C, respectively. As shown in Figure 7c, the RRED at the loading rate of 20 mm/min is greater than that at the loading rate of 10 mm/min, but the fatigue life at 10 mm/min is greater.

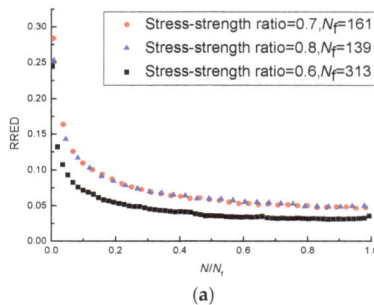

(a)

Figure 7. *Cont.*

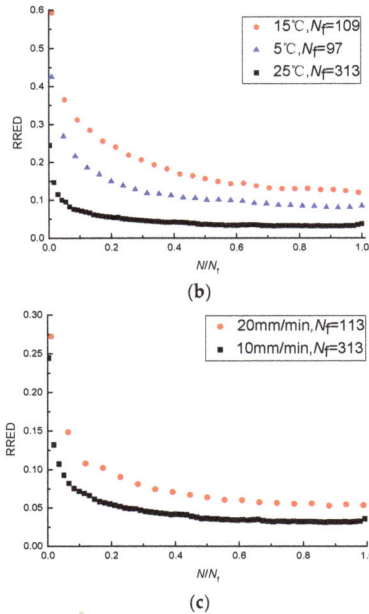

Figure 7. RREDs under different test conditions: (**a**) RREDs at different stress-strength ratios; (**b**) RREDs at different temperatures; (**c**) RREDs at different loading rates.

The above analyses indicate that the RRED does not change monotonously with the stress-strength ratio, the temperature, and the loading rate. It can be seen from Table 3 that there is a negative correlation between the fatigue life and the SSR and loading rate, and a positive correlation between the fatigue life and the temperature. So, the RRED does not change monotonously with the fatigue life and the RRED cannot well predict the fatigue life of the asphalt mixture.

4. Analysis of PD

4.1. Construction of PD Mathematical Model

The asphalt mixture, a viscoelastic plastic material subjected to cyclic loading, undergoes three different types of deformation: elastic deformation, viscous-flow deformation, and delayed deformation. After the unloading stage of each cycle, the elastic deformation of the asphalt mixture can fully recover, but the viscous-flow deformation of the asphalt mixture cannot recover. One part of the delayed deformation can recover, but the other part cannot recover [32]. The PD of the asphalt mixture under cyclic loading is composed of all the viscous-flow deformation and the delayed deformation that cannot recover. The fatigue failure of the beam specimen is mainly caused by the tensile stress at the bottom of the specimen, because the ultimate tensile stress of the asphalt mixture is far less than its ultimate compressive stress. When a stress produced by a cyclic loading is below the ultimate stress, the fatigue failure of the beam specimen is due to the stress concentration at the bottom, and the fatigue damage is mainly manifested by fatigue cracks. The growth process of a fatigue crack consists of two stages: the crack initiation stage and the crack propagation stage. Damage gradually accumulates with the increase of the number of loading cycles. Fatigue cracks occur at the bottom of the beam specimen when the damage is accumulated to a certain extent. Fatigue cracks expand due to the increase of the number of loading cycles. The cumulative result of the expansion is the fracture of the beam specimen [40]. The whole process of the damage evolution of the asphalt mixture is accompanied by the accumulation of the PD, and the accumulation of the PD will further the evolution of damage. Thus, the accumulation of the PD is related to the damage evolution and fatigue life of the asphalt mixture.

The change of the PD of the asphalt mixture under cyclic loading (vs. the number of loading cycles) is a monotonous accumulation process. The permanent deformation and the number of loading cycles (PD-N) curve of the three-point bending fatigue test is shown in Figure 8. As shown in the figure, the PD increases with the increase in the number of loading cycles, which can be divided into three stages. The first stage lasts a short time and the PD increases rapidly, but the growth rate gradually decreases. At the end of the first stage, the PD reaches a large value. The second stage lasts a long time and the PD increases stably. The third stage lasts a short time, but the PD increases rapidly and the growth rate gradually increases.

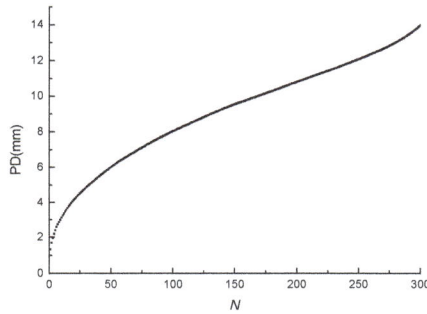

Figure 8. PD-N curve.

In order to accurately describe the change law of the PD of the asphalt mixture under cyclic loading (vs. the number of loading cycles), a PD mathematical model was proposed:

$$PD = cN^d \qquad (3)$$

where c and d are fitting parameters and N is the number of loading cycles.

The model proposed was used to fit the experimental data of the PD and the correlation coefficient is greater than 0.98, which indicates that the proposed model can accurately describe the change law of the PD of the asphalt mixture under cyclic loading (vs. the number of loading cycles). The fitting results of the PD mathematical model are listed in Table 5.

Table 5. Fitting results of PD mathematical model.

Parameter	c	d	R^2
Parameter value	0.85777	0.56389	0.98372

4.2. Influence of Experimental Factors on PD

The results under different test conditions are shown in Figure 9. It can be seen from Figure 9a that the PD is obviously influenced by the stress-strength ratio. The sequence of the magnitude of the PD at different stress-strength ratios corresponds to the stress-strength ratios of 0.6, 0.8, and 0.7, respectively. However, the sequence of the magnitude of the fatigue life at different stress-strength ratios corresponds to the stress-strength ratios of 0.6, 0.7, and 0.8, respectively. As shown in Figure 9b, the PD is obviously influenced by the temperature. The sequence of the magnitude of the PD at different temperatures corresponds to the temperatures of 25 °C, 5 °C, and 15 °C, respectively. However, the sequence of the fatigue life at different temperatures corresponds to the temperatures 25 °C, 15 °C, and

5 °C, respectively. As shown in Figure 9c, the PD is obviously influenced by the loading rate. The PD at the loading rate of 10 mm/min is greater than that at 20 mm/min, and the fatigue life at the loading rate of 10 mm/min is also greater. It is evident that the PD of the asphalt mixture does not change monotonously with the fatigue life. Therefore, the PD of the asphalt mixture cannot well predict the fatigue life.

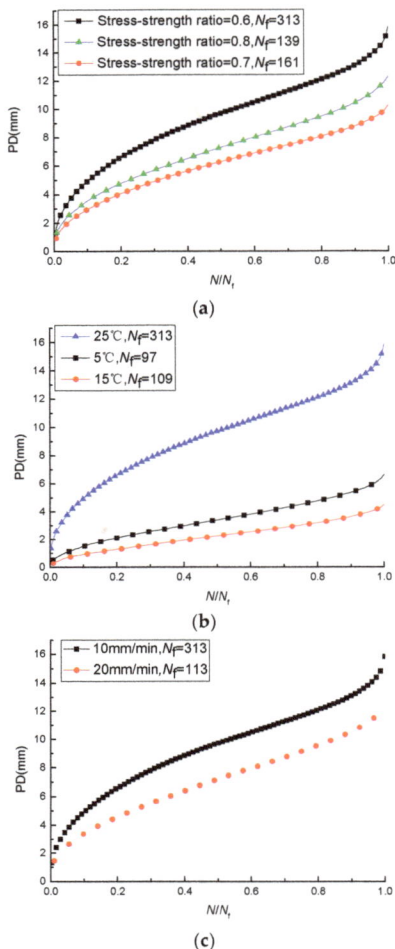

(a)

(b)

(c)

Figure 9. PDs under different test conditions: (**a**) PDs at different stress-strength ratios; (**b**) PDs at different temperatures; (**c**) PDs at different loading rates.

The process of damage evolution is accompanied by the accumulation of PD, and the accumulation of PD will further the evolution of damage. Therefore, the accumulation of the PD is related to the damage evolution speed. The faster the damage evolution speed, the sooner the damage is close to the threshold of the failure of the asphalt mixture, and the shorter the fatigue life is. Therefore, the fatigue life of the asphalt mixture is related to the PD. The PD of the asphalt mixtures does not change monotonously with the fatigue life, which indicates that all of the PD is not used to damage the asphalt mixture. One part of the PD causes the damage to the asphalt mixture under cyclic loading, and the other part causes the viscoelastic deformation of the asphalt mixture [41].

5. Analysis of PDR

The PD is related to damage evolution and fatigue life, but the PD cannot be directly used to predict the fatigue life of the asphalt mixture. To reveal the relationship between the PD and the fatigue life, the definition of PDR is proposed in this paper as

$$PDR = \frac{|PD_{N+1} - PD_N|}{PD_N} \tag{4}$$

where PD_{N+1} and PD_N are the PDs of the loading cycle (N+1) and the loading cycle N. Under certain conditions, the PD of the viscoelastic deformation part of each loading cycle is a fixed value [41]. The PDR reflects the proportion of the PD that produces the viscoelastic deformation to the total PD of each loading cycle. The value of PDR is only related to the damage deformation.

5.1. Influence of Experimental Factors on PDR

The PDRs at different loading cycles were calculated using Equation (4). To better analyze the PDR under different experimental factors, the number of loading cycles is converted into dimensionless quantities. As shown in Figure 10, at first, the PDR decreases with the increase of the number of loading cycles, then the ratio increases, and the permanent-deformation-ratio plot can be divided into three stages. The trends of the PDRs (vs. the number of loading cycles) under different test conditions are the same. The PDR decreases with the increase of the stress-strength ratio and the loading rate and increases with the decrease of the temperature, so there is a negative correlation between fatigue life and the PDR.

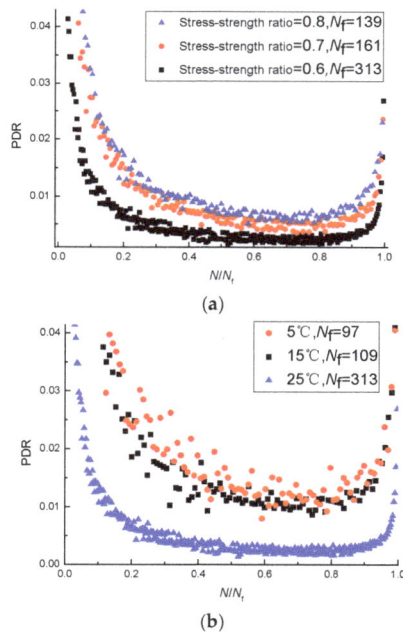

Figure 10. *Cont.*

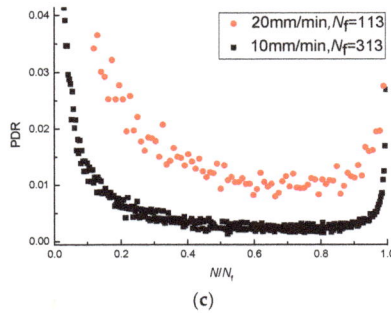

(c)

Figure 10. PDRs under different test conditions: (**a**) PDRs at different stress-strength ratios; (**b**) PDRs at different temperatures; (**c**) PDRs at different loading rates.

5.2. Analysis of Damage Evolution

Under certain conditions, the PD of the viscoelastic deformation part of each loading cycle is a fixed value, and the value of the PDR is only related to the damage deformation. Therefore, the damage factor can be expressed based on the PDR as

$$D = \frac{\sum\limits_{k=1}^{N} PDR_k}{\sum\limits_{k=1}^{N_f} PDR_k} \tag{5}$$

where D is the damage factor of the loading cycle N and PDR_k is the PDR at the loading cycle k.

In order to study the damage evolution process of the asphalt mixture from the perspective of deformation, the damage evolution curve (the D-N/N_f curve) and the PDR evolution scatter plots (PDR-N/N_f scatter plot) for the fatigue test at the temperature of 25 °C, at the stress-strength ratio of 0.4, and at the loading rate of 10 mm/min are shown in Figures 11 and 12. To better analyze the damage evolution process, the number of loading cycles is converted into dimensionless quantities.

As shown in Figure 11, the damage increases nonlinearly during the fatigue test, and the process can be divided into three stages. In the first stage, the damage lasts a short time but increases rapidly, and at the end of this stage, it reaches a large value. In the second stage, the damage lasts a long time and increases stably. In the third stage, the damage increases sharply and the beam specimen fractures in the end [42].

As shown in Figure 12, at first, the PDR decreases with the increase of the number of loading cycles, then the ratio increases, and the permanent-deformation-ratio plot can be divided into three stages. The first stage lasts a short time and the PDR decreases rapidly, but the speed of decreasing gradually slows down. At the end of the first stage, the PDR drops to a small value. The second stage lasts a long time and the PDR decreases slowly. The third stage lasts a short time, but the PDR increases sharply and the growth rate of the PDR increases gradually. The rapid decline in the first stage of the PDR shows that a considerable portion of the PD contributes to material damage. After the material damages to a certain extent, the PDR moves into the second stage of the low stable value, indicating that the PD mainly contributes to the viscoelastic deformation in the material. The PD of this part tends to be stable, which is far greater than what the PD contributed to material damage, which makes up a large proportion of the total deformation. In the third stage, the PD reaches the failure threshold and the specimen breaks quickly because of the deformation in the material accumulated in the first and second stage.

As shown in Figures 11 and 12, the three stages of the PDR and the three stages of the damage correspond to each other, which indicates that there exists a connection between the damage evolution and the value of PDR. It can be seen from the growth rate of the damage and the value of PDR that the damage of the asphalt mixture increases rapidly in the first and third stage and the growth rates of damage in the first and third stage are greater than that in the second stage. In the same way, the values of the PDR in the first and third stage are greater than that in the second stage. Therefore, the PDR is an energy parameter reflecting the speed of damage evolution of the asphalt mixture, and the speed of damage evolution increases with the PDR of the asphalt mixture.

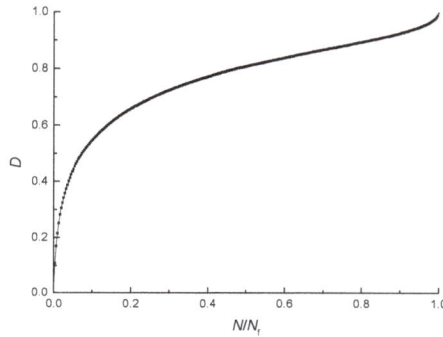

Figure 11. D-N/N_f curve.

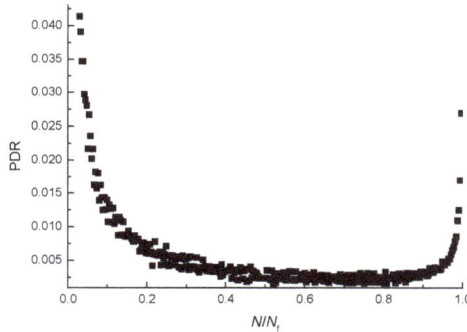

Figure 12. PDR-N/N_f scatter plot.

The faster the damage evolution speed of the asphalt mixture under cyclic loading is, the earlier the damage is close to the threshold of failure and the shorter the corresponding fatigue life is. Therefore, the fatigue life decreases with the increase of the PDR.

5.3. Establishment of Fatigue Equation Based on Plateau Value (PV) of PDR

The damage evolution analysis in Section 5.2 showed that fatigue life decreases with the increase of the PDR. At first, the PDR decreases with the increase in the number of loading cycles, then the ratio increases, and the process can be divided into three stages. Although the PDR in the first and third stages is greater than that in the second stage, the second stage accounts for the major portion of the entire PD in the loading process. Therefore, the PDR in the second stage reflects the overall PDR of the three stages. In this paper, the average PDR in the second stage is taken as the plateau value (PV), and the PDR of the three stages is represented by the PV.

The results of the fatigue life (N_f) and the PV of the three-point bending fatigue tests, as listed in Table 6, indicate that fatigue life decreases with the increase of the plateau value (PV). The N_f-PV scatter plot is shown in Figure 13, which indicates that fatigue life can be expressed in a power function with the increase of PV. The power function, simple and easy to use, is established as

$$N_f = A(PV)^B \tag{6}$$

where N_f is the fatigue life and A and B are fitting parameters. The fitting results are listed in Table 7. The correlation coefficient is greater than 0.98, so the observed data are well represented by the equation.

According to the fatigue equation for the asphalt mixture, the predicted fatigue lives of the asphalt mixture under different test conditions were calculated and compared with the test fatigue lives, as listed in Table 8. The average relative error of the fatigue life predicted by the fatigue equation to the test fatigue life is 11.03%. Therefore, the fatigue equation based on the plateau value of the PDR can well predict the fatigue life of the asphalt mixture under cyclic loading.

Table 6. Results of N_f and PV.

Test Fatigue Life (N_f)	PV
313	0.00303819
161	0.007641095
139	0.008407475
113	0.012586557
109	0.014762424
97	0.012631972

Table 7. Fitting parameters of fatigue equation.

Parameter	A	B	R^2
Parameter value	4.24479	−0.74172	0.98653

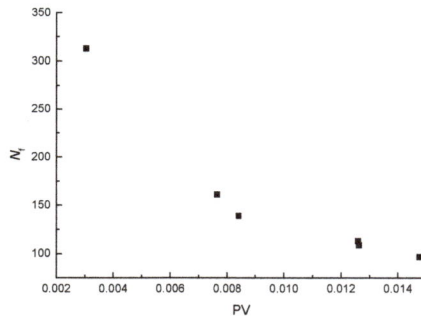

Figure 13. Scatter plots of N_f-PV.

Table 8. Contrast results of fatigue life.

SSR-Temperature-Loading Rate	Test Fatigue Life	Predicted Fatigue Life	Relative Error
0.6–25 °C-10mm/min	313	312	0.11%
0.7–25 °C-10mm/min	161	158	2.02%
0.8–25 °C-10mm/min	139	147	5.72%
0.6–25 °C-20mm/min	113	109	3.59%
0.6–15 °C-10mm/min	109	97	11.20%
0.6–5 °C-10mm/min	97	109	12.00%

6. Conclusions

Laboratory investigations of the change law of deformation of an asphalt mixture under cyclic loading and prediction of the fatigue life based on the PV of PDR were carried out by three-point bending fatigue tests. With the study above, the following conclusions can be drawn:

(1) The recovery ratio of the elastic deformation of the asphalt mixture under cyclic loading decreases with the increase of the number of loading cycles, and the elastic recovery ability of the asphalt mixture under cyclic loading decreases gradually. The proposed RRED mathematical model in this paper can well describe the change law of the RRED.

(2) The process of damage evolution of the asphalt mixture is accompanied by the accumulation of PD, and the accumulation of PD will further the evolution of damage. One part of the PD causes the damage of the asphalt mixture under cyclic loading, and the other part causes the viscoelastic deformation of the asphalt mixture. The proposed PD mathematical model can well describe the change law of RRED.

(3) The RRED or the PD cannot well predict the fatigue life of the asphalt mixture because the RRED and the PD of the asphalt mixture do not change monotonously with the fatigue life.

(4) The definition of PDR is proposed in this paper. The influence of experimental factors on the PDR indicates that the PDR increases with the stress-strength ratio and the loading rate, and increases with the decrease of the temperature, so there exists a negative correlation between fatigue life and the PDR.

(5) The PDR reflects the proportion of the PD that produces the viscoelastic deformation to the total PD of each loading cycle, and the value of the PDR is only related to the damage deformation. The damage evolution based on the PDR indicates that the faster the damage evolution speed is, the earlier the damage is close to the threshold of failure, and the fatigue life decreases with the increase of PDR.

(6) The average PDR in the second stage is taken as the PV, and the overall status of the PDR of the three stages and the resistance to fatigue damage are represented by the PV. The fatigue life decreases in power law with the increase of the PV. The average relative error of the fatigue life predicted by the fatigue equation to the test fatigue life is 5.77%. The fatigue equation based on the PV can well predict the fatigue life of the asphalt mixture.

Author Contributions: Yazhen Sun organized the research; Dong Fan and Chenze Fang performed the three-point bending fatigue tests; Yazhen Sun and Chenze Fang wrote the manuscript; Xuezhong Yuan checked the manuscript.

Acknowledgments: This research was performed at the Shenyang Jianzhu University and Institute of Transportation Engineering of Zhejiang University. The research is funded by the National Natural Science Fund (51478276), the Natural Science Foundation of Liaoning Province (20170540770).

Conflicts of Interest: The authors declare no conflict of interest.

References

1. Sas, W.; Głuchowski, A.; Gabrys, K.; Sobol, E.; Alojzy, S. Deformation Behavior of Recycled Concrete Aggregate during Cyclic and Dynamic Loading Laboratory Tests. *Materials* **2016**, *9*, 780. [CrossRef] [PubMed]

2. Król, J.B.; Niczke, Ł.; Kowalski, K.J. Towards Understanding the Polymerization Process in Bitumen Bio-Fluxes. *Materials* **2017**, *10*, 1058.

3. Wu, S.P.; Zhang, G.; Han, J.; Liu, G.; Zhou, J. Fatigue Performance of Bridge Deck Pavement Materials. *Trans. Nonferrous Met. Soc. China* **2006**, *24*, 318–320. [CrossRef]

4. Wu, S.P.; Liu, C.; Cao, T.; Pang, L. Investigation of the Low-temperature Performance of Asphalt Mixtures via Fatigue and Linear Contraction Test. *J. Huazhong Univ. Sci. Technol.* **2008**, *25*, 30–33.

5. Wu, S.P.; Jiang, C.J.; Lin, J.T.; Kim, J.; Zhu, Z.H. Research on Fatigue Characteristic of Aged Asphalt Binder. *J. Wuhan Univ. Technol. (Transp. Sci. Eng.)* **2013**, *37*, 451–455.

6. Pasetto, M.; Baldo, N. Unified approach to fatigue study of high performance recycled asphalt concretes. *Mater. Struct.* **2017**, *50*. [CrossRef]

7. Kavussi, A.; Qazizadeh, M.J. Fatigue characterization of asphalt mixes containing electric arc furnace (EAF) steel slag subjected to long term aging. *Constr. Build. Mater.* **2014**, *72*, 158–166.

8. Shu, X.; Huang, B.; Vukosavljevic, D. Laboratory evaluation of fatigue characteristics of recycled asphalt mixture. *Constr. Build. Mater.* **2008**, *22*, 1323–1330.

9. Li, L.B. Fatigue Life Prediction of Fiber-Reinforced Ceramic-Matrix Composites with Different Fiber Preforms at Room and Elevated Temperatures. *Materials* **2016**, *9*, 207. [CrossRef] [PubMed]

10. Yu, Z.Y.; Zhu, S.P.; Liu, Q.; Liu, Y. Multiaxial Fatigue Damage Parameter and Life Prediction without Any Additional Material Constants. *Materials* **2017**, *10*, 923. [CrossRef] [PubMed]

11. Khalid, A.G.; Samuel, H.C. Fatigue damage analysis in asphalt concrete mixtures using the dissipated energy approach. *Can. J. Civ. Eng.* **2006**, *33*, 890–901.

12. Li, C. The Experimental Study of Asphalt Mixture Fatigue Property under Different Impact Factor. Master's Thesis, Dalian University of Technology, Dalian, China, 2009.

13. Maggiore, C.; Grenfell, J.; Airey, G.; Collop, A.C. Evaluation of fatigue life using dissipated Energy methods. In *7th RILEM International Conference on Cracking in Pavements, RILEM Bookseries*; Springer: Dordrecht, The Netherlands, 2012; Volume 4, pp. 643–652.

14. Pasetto, M.; Baldo, N. Fatigue performance of recycled hot mix asphalt: A laboratory study. *Adv. Mater. Sci. Eng.* **2017**. [CrossRef]

15. Shen, S.; Carpenter, S.H. Application of the dissipated Energy concept in fatigue endurance limit testing. *Transp. Res. Rec.* **2005**, *1929*, 165–173.

16. D'Andrea, A.; Fiore, N. Fatigue life of asphalt concrete with rubber grains. *Adv. Damage Mech.* **2003**, *40*, 65–74.

17. Fiore, N.; Caro, S.; D'Andrea, A.; Scarsella, M. Evaluation of bitumen modification with crumb rubber obtained through a high pressure water jet (HPWJ) process. *Constr. Build. Mater.* **2017**, *151*, 682–691. [CrossRef]

18. Okazaki, Y. Comparison of Fatigue Properties and Fatigue Crack Growth Rates of Various Implantable Metals. *Materials* **2012**, *5*, 2981–3005. [CrossRef]

19. Wang, Q.; Zhang, W.; Jiang, S. Fatigue Life Prediction Based on Crack Closure and Equivalent Initial Flaw Size. *Materials* **2015**, *8*, 7145–7160. [CrossRef] [PubMed]

20. Wang, H.; Zhang, W.F.; Sun, F.Q.; Zhang, W. A Comparison Study of Machine Learning Based Algorithms for Fatigue Crack Growth Calculation. *Materials* **2017**, *10*, 543.

21. Zhang, W.; Liu, H.L.; Wang, Q.; He, J.J. A Fatigue Life Prediction Method Based on Strain Intensity Factor. *Materials* **2017**, *10*, 689. [CrossRef] [PubMed]

22. Deng, H.L.; Li, W.; Zhao, H.Q.; Sakai, T. Multiple Fatigue Failure Behaviors and Long-Life Prediction Approach of Carburized Cr-Ni Steel with Variable Stress Ratio. *Materials* **2017**, *10*, 1084. [CrossRef] [PubMed]

23. Paris, P.; Erodgan, F. A critical analysis of crack propagation laws. Transactions of the ASME. *J. Basic Eng.* **1963**, *85*, 528–534. [CrossRef]

24. Majidzadeh, K.; Kauffmann, E.M.; Rasmsamooj, D.V. Application of fracture mechanics in the analysis of pavement fatigue. *J. Assoc. Asphalt Paving Technol.* **1971**, *40*, 227–246.

25. Li, L.B. Comparison of Cyclic Hysteresis Behavior between Cross-Ply C/SiC and SiC/SiC Ceramic-Matrix Composites. *Materials* **2016**, *9*, 62. [CrossRef] [PubMed]

26. Ayman, A.A.; Ruan, D.; Lu, G.X.; Arafat, A.B. Finite Element Analysis of Aluminum Honeycombs Subjected to Dynamic Indentation and Compression Loads. *Materials* **2016**, *9*, 162.

27. Li, L.B. Fatigue Damage and Lifetime of SiC/SiC Ceramic-Matrix Composite under Cyclic Loading at Elevated Temperatures. *Materials* **2017**, *10*, 371. [CrossRef] [PubMed]

28. Tayebali, A.A.; Rowe, G.M.; Sousa, J.B. Fatigue response of asphalt aggregate mixtures. *J. Assoc. Asphalt Paving Technol.* **1992**, *61*, 333–360.

29. Yan, H.; Zhu, H.Z.; Tang, B.M. Study on Fatigue Dissipated Energy Model of AC-13 Asphalt Mixture. *J. Chongqing Jiaotong Univ. (Nat. Sci.)* **2010**, *29*, 559–562.

30. Li, X.; Liang, N.X. Research on Splitting Fatigue Fest of Asphalt Mixture Based on Strain Control. *J. China Foreign Highw.* **2013**, *33*, 276–280.

31. Luo, X.; Luo, R.; Lytton, R.L. Characterization of Asphalt Mixtures Using Controlled-Strain Repeated Direct Tension Test. *Am. Soc. Civ. Eng.* **2013**, *25*, 194–207. [CrossRef]

32. Chen, W.Q. *Research on Fatigue Failure Mechanism of Asphalt Mixture Based on Dissipated Energy*; South China University of Technology: Guangzhou, China, 2014.

33. Pell, P.S.; Cooper, K.E. The effect of testing and mix variables on the fatigue performance of bituminous materials. *Proc. Paving Technol.* **1975**, *44*, 1–37.

34. Gao, L.; Xie, L.G.; Jia, S.C.; Zhu, Z.A. Virtual test of type I fatigue crack of asphalt mixture based on discrete element method. *J. Huazhong Univ. Sci. Technol. (Nat. Sci. Ed.)* **2018**, *46*, 92–97.

35. Ye, Q. Evaluation Based on the Viscoelastic Properties of Asphalt Mixture Fatigue. Master's Thesis, Harbin Institute of Technology, Harbin, China, 2016.

36. Pasetto, M.; Baldo, N. Fatigue performance of asphalt concretes with rap aggregates and steel slags. In *7th RILEM International Conference on Cracking in Pavements, RILEM Bookseries*; Springer: Dordrecht, The Netherlands, 2012; Volume 4, pp. 719–727.

37. Artamendi, I.; Khalid, H. Characterization of fatigue damage for paving asphaltic materials. *Fatigue Fract. Eng. Mater. Struct.* **2005**, *28*, 1113–1118.

38. Highway Research Institute of the Transportation Department. *Standard Test Methods of Bitumen and Bituminous Mixtures for Highway Engineering, JTG E20-2011*; China Communications Press: Beijing, China, 2011.

39. Luo, X.; Luo, R. Energy-Based Mechanistic Approach to Characterize Crack Growth of Asphalt Mixtures. *Am. Soc. Civ. Eng.* **2013**, *25*, 1198–1208. [CrossRef]

40. Zhang, Z.; Roque, R.; Birgisson, B.; Sangpetgnam, B. Identification and verification of a suitable crack growth law for asphalt mixtures. *Asphalt Paving Technol.* **2001**, *70*, 206–241.

41. Zhang, J.P.; Huang, X.M.; Li, H. Permanent deformation of asphalt mixture under repeated load. *J. Southeast Univ. (Nat. Sci. Ed.)* **2008**, *38*, 511–515.

42. Luan, L.Q.; Tian, X.G. Nonlinear Analysis of Fatigue Damage of Asphalt Mixture. *J. Build. Mater.* **2012**, *15*, 508–512.

materials

MDPI

Article

Energy-Based Approach to Predict Fatigue Life of Asphalt Mixture Using Three-Point Bending Fatigue Test

Yazhen Sun [1,*], Chenze Fang [1], Jinchang Wang [2,*], Zuoxin Ma [1] and Youlin Ye [1]

[1] School of Transportation Engineering, Shenyang Jianzhu University, Shenyang 110168, China; fangchenze@126.com (C.F.); 15541947325@163.com (Z.M.); yeyoulin1985@126.com (Y.Y.)

[2] Institute of Transportation Engineering, Zhejiang University, Hangzhou 310058, China

* Correspondence: syz16888@126.com (Y.S.); wjc501@zju.edu.cn (J.W.); Tel.: +86-139-4002-1867 (Y.S.); +86-135-8803-0136 (J.W.)

Received: 4 August 2018; Accepted: 10 September 2018; Published: 12 September 2018

Abstract: The three-point bending fatigue tests were carried out in order to accurately predict the fatigue life of an asphalt mixture based on the plateau value (PV) of the dissipated strain energy ratio (DSER). The relations of the dissipated strain energy (DSE) to the stress-strength ratio, temperature and loading rate were studied, and the constructions of the mathematical models of DSE and DSER were completed based on the change laws of the DSE. The relation of the fatigue life to the PV was determined based on the analysis of damage evolution, based on which the fatigue equation was established and used to predict the fatigue life. The results show that the change laws of DSE and DSER can be well described by the proposed mathematical models. The PV is defined as the average value of the DSER in the second stage and the fatigue life decreases in power function with the increase of PV, based on which the fatigue equation of $N_f = A(PV)^B$ was established, and the established fatigue equation is very close to that is used in the MEPDG. The fatigue equation can well predict the fatigue life asphalt mixture.

Keywords: energy-based approach; dissipated strain energy; plateau value of dissipated strain energy ratio; fatigue life; three-point bending fatigue test

---

1. Introduction

In recent years, more and more asphalt pavements have been built, but the qualities of part of them are not guaranteed. The asphalt pavement, subjected to repeated actions of different kinds of loads, is prone to fatigue failure, because of the strength reduction and the fatigue damage of materials [1]. The strength of asphalt mixture decreases gradually with the increase of the cyclic loading times, and the attenuated material strength can be used to define the damage. Some damage models were proposed to study the fatigue performance of asphalt mixture based on the change laws of the strength reduction [2,3]. The bearing capacity of asphalt mixture decreases with the increase of damage. The strength reduction will lead to the damage and the damage evolution, and the damage evolution accelerates the strength reduction [4,5]. When the damage evolves to the threshold of the failure, the bearing capacity will be less than the applied load, and the fatigue failure will occur to the material structure [6,7].

Many researches have been carried out to study the influence of experimental factors on the fatigue resistance of asphalt mixture. Furthermore, many mathematical models (such as the mathematical models of recovery ratio of elastic deformation, permanent deformation and permanent deformation ratio, etc.) were proposed to study the fatigue resistance of asphalt mixture, and it was found that the permanent deformation ratio reflects the damage evolution speed and can be well used to predict the fatigue life [8]. Logarithmic fatigue life linearly decreases with the increase of stress-strength ratio, and the fatigue life at low frequency is much less than that at high frequency [9]. There are many researches about the effects of the asphalt type on the fatigue life were carried out by experts. Three kinds of asphalt were used to study the effects of the asphalt type on the fatigue life through the fatigue test, and the results show that the rubber asphalt can well improve the fatigue performance of asphalt mixture compared with the base asphalt, styrene butadiene styrene (SBS) modified asphalt [9]. When the bitumen aggregate ratio is between 7.5% and 9%, the fatigue life of the rubber asphalt mixture increases with the increase of asphalt content [10]. The fatigue life of the rubber asphalt mixture increases with the decrease of the air voids in the reasonable range [11]. The high fatigue life of the rubber asphalt mixture was found to be closely related to the gradation [12]. Three kinds of gradation were used to study the effects of the asphalt mixture type on the fatigue life through the fatigue test, and the results of the research show that when the stress-strength ratio is relatively low (0.3–0.5), the gradation has a significant effect on the fatigue life of the asphalt mixture, and the increasing sequence of the fatigue life of different asphalt mixture types corresponds to the asphalt mixture types of AC-13, stone mastic asphalt (SMA), and gap gradation, respectively, which was caused by the reason that the internal structure and the air voids are different when the gradation changes. However, when the stress-strength ratio is relatively high (0.6–0.8), the gradation has no obvious effect on the fatigue life of the asphalt mixture [9].

The energy (damage) approach is widely used in the fatigue-life prediction, because of its simple principle and convenient operation [13–16]. The energy dissipation law of asphalt mixture was studied and the viewpoint that the dissipated strain energy (DSE) of material is related to their fatigue properties was proposed by Heukelom [17]. The total amount of DSE before the fracture of specimen was analyzed and the viewpoint that the maximum number of loading cycles can be affected by the cumulative DSE was proposed by Chomton and Valayer [18]. The viewpoint that there is a power function relation between the fatigue life and the cumulative DSE of the material structure before its fatigue failure was proposed based on the assumption that all the DSE before failure can cause the damage of material by Van Dijk [19]. The conclusion that temperature and loading mode can affect the mathematical relationship between the fatigue life and cumulative DSE of asphalt mixture was proved by Tayebali [20]. The damage evolution process was studied from the angle of energy by using the change laws of DSE and recoverable strain energy based on the different viscoelastic properties of asphalt mixture in the tensile and compressive portion by Xue Luo [21].

Most of the above researches on the fatigue-life prediction for asphalt mixture, based on the assumption that all the DSE before failure can cause the damage of material, cannot accurately reveal the damage evolution mechanism. In fact, only partial DSE can cause the fatigue damage to asphalt mixture, and the change of DSE is the real cause of material damage [22–26]. Therefore, the damage variable, defined by the cumulative total amount of dissipative energy before the fatigue failure, cannot accurately reveal the damage evolution process of asphalt mixture. The fatigue equation based on the damage variable cannot be used to correctly predict the fatigue life before fatigue failure.

In order to accurately reveal the damage evolution process and accurately predict the fatigue life of asphalt mixture, three-point bending fatigue tests were carried out. The conclusion that the DSE becomes smaller, with the increase of fatigue life, was obtained by analyzing the influences of temperature, stress ratio and rate on the calculation results of DSE. The mathematical models of DSE and dissipated strain energy ratio (DSER) was established based on the change laws of the DSE. The conclusion that the fatigue life decreases with the increase of DSER was obtained by defining the damage variable using DSER and studying the damage evolution process. The fatigue equation, established based on the relation of the fatigue life to the plateau value (PV) of DSER, was used to predict the fatigue life, and the results show that it can accurately predict the fatigue life of asphalt mixture.

The applied main research methodology was that the fatigue test was used for the theoretical analysis of the relation of the fatigue life to the PV, and the fatigue equation was established based on the relation.

2. Materials and Experimental Procedures

The clamping structure of four-point bending fatigue test is complex, and the three-point bending fatigue test has the strong applicability and the simple loading mode. Therefore, the three-point bending fatigue test was used to study the fatigue life of asphalt mixture in this paper [27].

2.1. Test Materials

The 70# rubber modified asphalt (marking in accordance with the Chinese standard of the Technical Specifications for Construction of Highway Asphalt Pavements (JTGF40-2004)) with a penetration of 70 produced from Ningbo was used as asphalt binder and its specifications provided by the manufacturer are listed in Table 1 [28]. The ratio of binder to aggregate is 8.8% by weight. The limestone was used as the aggregate, and the continuous aggregate gradation type is AC-13 as listed in Table 2, has the nominal maximum aggregate size of 13.2 mm. The AC-13 asphalt mixture is widely built in the surface course of highway with the standard axle load of 100 kN designed by the specifications for design of highway asphalt pavement (JTGD50-2017) [29].

Table 1. Properties of asphalt rubber.

Properties	Standard	Value
Penetration (25 °C,100 g, 5 s)	T0604-2011	70 (0.1 mm)
Softening point	T0606-2011	57 (°C)
Viscosity (177 °C)	T0625-2011	3.8 (Pa·s)
Elastic recovery (25 °C)	T0662-2011	72 (%)

Table 2. Aggregate gradation.

Sieve Size (mm)	16.0	13.2	9.5	4.75	2.36	1.18	0.6	0.3	0.15	0.075
Passing percentage	100.0	91.1	80.2	54.0	33.2	22.5	16.0	12.1	8.7	5.5

2.2. Preparation of Specimen

The temperature of aggregates and asphalt was 160 °C, and the temperature of compaction was 100 °C. The sample forming machine was used to form the rut board (400 mm × 400 mm × 70 mm). The rut board was cut to obtain the specimen beams (250 mm × 30 mm × 35 mm) with the average density of 2.445 g/cm^3 and the size error of the specimen beams should be controlled within 2 mm [30].

2.3. Test Conditions and Methods

The ratio of the peak stress of each cycle to the ultimate material strength is called the stress-strength ratio (SSR). The fatigue tests had three groups of contrast tests in order to study the relationship between the fatigue life and the experimental factors. In order to reduce the test error, three parallel specimens, that is, a total of 18 specimens were selected for the fatigue test:

Group 1. For this group, the temperature was 25 °C, the loading rates were 10 mm/min and 20 mm/min, respectively, and the SSR was 0.6.

Group 2. For this group, the temperature was 5 °C, 15 °C, and 25 °C, respectively, the loading rate was 10 mm/min, and the SSR was 0.6.

Group 3. For this group, the temperature was 25 °C, the loading rate was 10 mm/min, and the SSRs were 0.6, 0.7, and 0.8 respectively.

2.4. Test Results

During each loading cycle, the specimen was loaded linearly until the deformation reached the peak and was unloaded linearly until the stress reached zero. The fatigue life was determined by the loading number corresponding to the apparent fracture of the specimen which loses the carrying capacity.

The 18 specimens were tested according to the test scheme and the average fatigue life of the 3 parallel specimens was taken as the final fatigue life. The statistical results of the fatigue lives are listed in Table 3, from which it can be seen that each coefficient of variation is less than 8.15%. The fatigue life decreases with the increase of the SSR and the loading rate, and increases with the increase of the temperature, form which the spread conclusions that the fatigue resistance of asphalt pavement can be improved by properly reducing the axle load and the driving speed can be drawn.

Table 3. Fatigue lives.

Stress-Strength Ratio (SSR)-Temperature-Loading Rate	Sample 1	Sample 2	Sample 3	Average Value	Coefficient of Variation (%)	Standard Deviation
0.6–25 °C-10 mm/min	329	312	298	313	4.05	12.68
0.7–25 °C-10 mm/min	168	150	165	161	4.89	7.87
0.8–25 °C-10 mm/min	140	128	149	139	6.19	8.60
0.6–15 °C-10 mm/min	105	113	109	109	3.00	3.27
0.6–5 °C-10 mm/min	91	101	99	97	4.45	4.32
0.6–25 °C-20 mm/min	120	119	100	113	8.14	9.20

3. Construction of DSE Mathematical Model

Applying a stress to a material will induce a strain. The energy being input into the material is represented by the area under the stress-strain curve of three-point bending fatigue test. The strain will recover when the stress is removed from the material, as shown in Figure 1. If the loading and unloading curves coincide, all the energy put into the material is recovered or returned after the load is removed. If the two curves do not coincide, there is energy lost in the material, energy that was dissipated through mechanical work, heat generation, or damage to the material in such a manner that it could not be used to return the material to its original shape. This energy difference is the dissipated energy of the material caused by the load cycle.

3.1. Influence of Experimental Factors on DSE

During each loading cycle, the asphalt mixture specimen was loaded linearly at the given rate until the deformation reached the peak in the loading stage; and was unloaded linearly at the given rate until the deformation reached zero in the unloading stage. The curves of stress-time and strain-time of each loading cycle are shown in Figure 2, from which it can be seen that the time corresponding to the strain peak of asphalt mixture is later than it to the stress peak, because of the viscoelastic hysteresis characteristics of asphalt mixture. As shown in Figure 1, the stress-strain curve of each loading cycle before the fatigue fracture are hysteretic closed curve, because of the viscoelastic hysteresis characteristics and the internal area of the curve is equal to the DSE in each loading cycle. The DSE of the loading cycle N (DSE_N), caused by the irrevocable deformation, is composed of the DSE_N^θ and the DSE_N^ε. The DSE_N^ε, used to produce the damage deformation of the asphalt mixture, is closely related to the fatigue life, and DSE_N^θ, used to produce the viscoelastic deformation, is independent of the fatigue life.

$$DSE_N = DSE_N^\theta + DSE_N^\varepsilon \tag{1}$$

where DSE_N^θ is the DSE used to produce the viscoelastic deformation of the loading cycle N, and DSE_N^ε is the DSE used to produce the damage deformation of the loading cycle N. The DSE_N^θ of the loading cycle N is a fixed value under certain condition which contains temperature, stress ratio and loading rate [22].

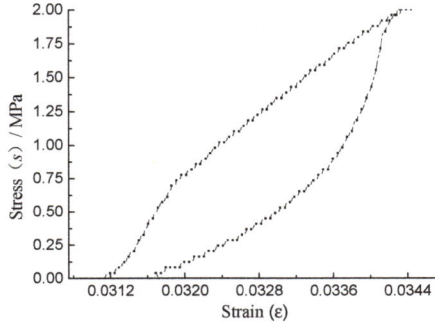

Figure 1. Stress-strain curve.

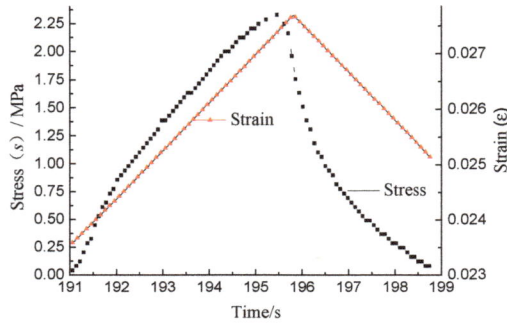

Figure 2. Change curves of stress and strain verses time.

The DSE_N is obtained by calculating the internal area of the closed curve using the tool of ORIGIN. The abscissa is converted into dimensionless quantities, and the DSE-N/N_f curves are shown as Figures 3–5, from which it the can be seen that the DSE-N/N_f curves can be divided into three stages and the overall shape looks like the shape of "U". The DSE of the first stage decreases rapidly, but lasts a short time, and the value of the DSE is small at the end of the first stage. The DSE of the second stage changes stably, but lasts a long time. The DSE of the third stage increases rapidly. As shown in Figures 3–5, the DSE increases with the increase of SSR and loading rate and decreases with the increase of temperature. Therefore, the DSE decreases with the increase of the fatigue life. Because the stress amplitude and deformation of the single cycle specimen increases when the SSR increases, and the specimen with the same force needs greater deformation when the temperature decreases or the loading rate increases. This causes the internal structure of asphalt mixture dissipate more energy to complete the recompositing.

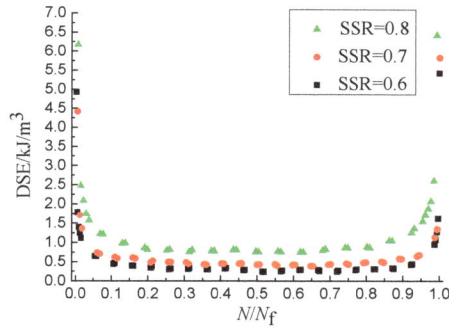

Figure 3. The influence of SSR on dissipated strain energy (DSE).

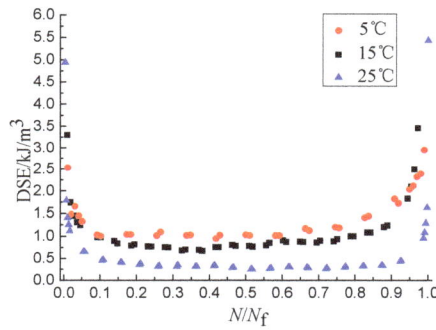

Figure 4. The influence of temperature on DSE.

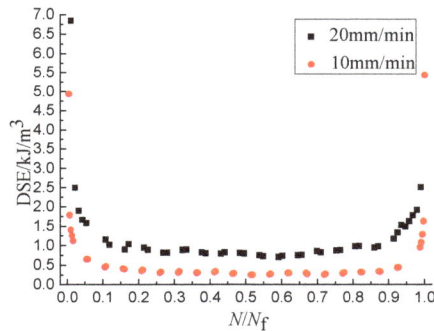

Figure 5. The influence of loading rate on DSE.

3.2. Construction of DSE Mathematical Model

In order to describe the change law of DSE accurately, the DSE mathematical model, shown in Equation (2), was proposed by studying the change of the DSE verses the number of loading cycles. The DSE mathematical model was used to fit the DSE-*N* curve, and the fitting effect is shown in Figure 6, with the correlation coefficient greater than 0.95, which indicates that the change law of DSE can be well described by the model. The fitting results of DSE-*N* curves of different test conditions are listed in Table 4, from which it can be seen that the value ranges of c_1–c_4 are 0.110–0.554, 0.013–0.122, 0.269–1.134 and 0.118–1.878, respectively.

$$DSE(N) = c_1 + c_2 |(N - c_3)^{c_4}|,\qquad(2)$$

where c_1, c_2, c_3 and c_4 are fitting parameters.

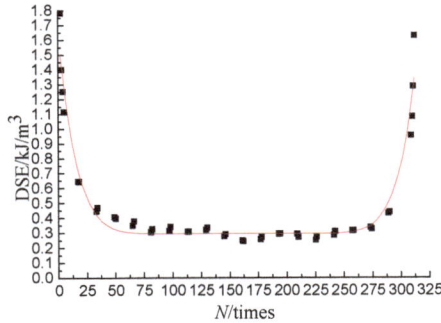

Figure 6. Fitting effect of DSE-*N* curve.

Table 4. Fitting parameters of DSE-*N* curve.

SSR-Temperature-Loading Rate	c_1	c_2	c_3	c_4	R^2
0.6–25 °C-10 mm/min	0.29359	7.7304×10^{-23}	158.1178	10.1179	0.9512
0.7–25 °C-10 mm/min	0.44862	2.0989×10^{-19}	83.74164	9.79295	0.9606
0.8–25 °C-10 mm/min	0.79304	5.6107×10^{-16}	65.24516	8.55627	0.9685
0.6–15 °C-10 mm/min	1.01016	3.1166×10^{-7}	42.66207	3.83088	0.9059
0.6–5 °C-10 mm/min	0.8327	2.6332×10^{-25}	52.42931	14.5094	0.9093
0.6–25 °C-20 mm/min	0.83581	3.9988×10^{-15}	46.86074	8.81419	0.9622

Obviously, the overall calculation result of DSE and the left and right translation of graph can be affected by the c_1 and c_3, respectively. This section takes the parameters obtained at the SSR of 0.6, the loading rate of 10 mm/min, and the temperature of 25 °C as an example to analysis the sensitivity of c_2 and c_4. The c_1 and c_3 keeping invariant, set the values for the c_1 as 7.7304×10^{-23}, 9.7304×10^{-23} and 1.177304×10^{-23}, respectively, and the c_3 as 158.11784, 162.11784, 166.11784, by which the DSE-*N* curves were drawn, as shown in Figures 7 and 8, from which it can be seen that the change of DSE in the second stage is not affected and the DSE in the first and third stage are affected by the c_2 and c_4.

The parameters c_1–c_4 are obtained by fitting the DSE mathematical model, and each parameter has the specific physical meaning. From the above analysis, we can draw the conclusion that the c_1 is the major parameter that affects the overall calculation result of DSE, and the c_3 is the major parameter that affects the left and right translation of graph. The c_2 and c_4 are the main parameters affecting the change of DSE in the first and the third stage.

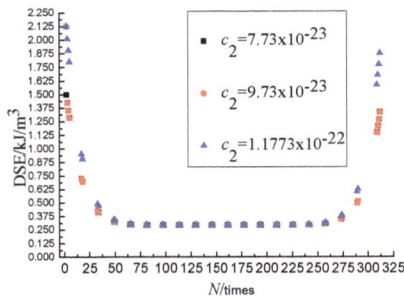

Figure 7. Sensitivity analysis of parameter c_2.

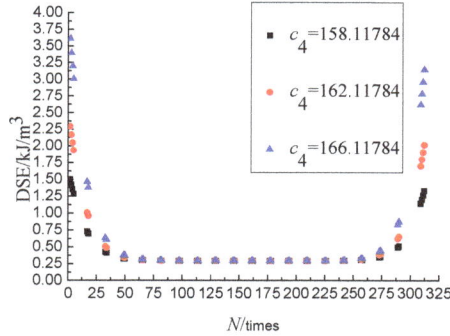

Figure 8. Sensitivity analysis of parameter c_4.

4. Construction of DSER Mathematical Model

The DSE is gradually accumulated during the damage evolution. At present, the study of fatigue damage evolution based on DSE is mostly based on the assumption that all the DSE causes damage to asphalt mixture. The assumption has been proved to be inaccurate and damage to the asphalt mixture will be caused by the change of DSE [31]. The concept of a dissipated strain energy ratio (DSER) was proposed based on the absolute value of the change in dissipated strain energy between load cycle N and load cycle $N + 1$ divided by the dissipated strain energy in load cycle N [22,32]:

$$\text{DSER} = \frac{|\text{DSE}_{N+1} - \text{DSE}_N|}{\text{DSE}_N}, \tag{3}$$

where DSE_{N+1} and DSE_N are the DSE of the loading cycle $N + 1$ and N. The DSER represents the percentage of dissipated energy causing damage to the material.

The DSE_N^ϑ of the loading cycle N is a fixed value under certain conditions, therefore, the proportion of the DSE_N^ε to the DSE_N is reflected by the DSER that is only related to the damage deformation [22,23]. Mastering the change law of DSER can provides a theoretical basis for the study of damage evolution. In order to accurately describe the change law of DSER in the process of damage evolution, the DSER mathematical model of the three-point bending fatigue test, obtained by substituting the Equation (2) into the Equation (3), is shown in Equation (4). The DSER results of the mathematical model were used to compare with that obtained by tests, and the comparison results are shown in Figure 9, from which it can be seen that the mathematical model well predict the overall change trend of DSER.

$$\text{DSER}(N) = \frac{|\text{DSE}_{N+1} - \text{DSE}_N|}{\text{DSE}_N} = \frac{c_2 \left| |(N + 1 - c_3)^{c_4}| - |(N - c_3)^{c_4}| \right|}{c_1 + c_2 |(N - c_3)^{c_4}|} \tag{4}$$

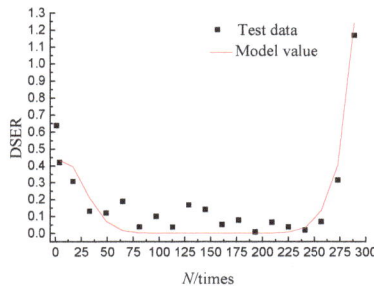

Figure 9. Contrast result of dissipated strain energy ratio (DSER).

5. Fatigue-Life Prediction Based on the Energy Approach

5.1. Analysis of Damage Evolution Based on the DSER

DSE is used to produce damage and viscoelastic deformation, and the DSER reflects the proportion of the DSE that produces the damage deformation to the total DSE. The DSER is only related to the damage deformation, so it was used to analyze the damage evolution. The damage factor can be expressed based on the DSER as,

$$D = \frac{\sum\limits_{k=1}^{N} DSER_k}{\sum\limits_{k=1}^{N_f} DSER_k} \tag{5}$$

where $DSER_k$ is the DSER at the loading cycle k, and D is the damage factor.

In order to study the relationship between the DSER and the damage evolution, the damage evolution process reflected by the D-N/N_f curve (Figure 10) and the DSER changing process reflected by the $DSER$-N/N_f scatter plot (Figure 11) was analyzed, respectively.

The nonlinear damage evolution process can be divided into three stages, as shown in Figure 10, and the proportion of the duration of the first stage to that of the whole process is less than 5%, but the damage value at the end of the first stage is greater than 0.4. The damage of the second stage increases stably, but the proportion of the duration of the first stage to that of the whole process is around 90%, and the damage of the third stage increases sharply until the specimen fractures.

The DSER changing process can be also divided into three stages and the proportion of the duration of each stage to that of the whole changing process is close to that of the damage evolution process, as shown in Figure 11. In the first stage, the DSER value is greater than 0.3, but the DSER decreases gradually and the slope of the descent curve gradually decreases. The DSER value of the second stage is less than 0.2, but the whole trend approaches a stable value. The DSER of third stage increases sharply in a short period of time.

The damage evolution speed and the DSER values in the first and third stages is far greater than that in the second stage [33], from which it can be seen that the three stages of the DSER evolution and the damage evolution correspond to each other, and there is a positive correlation between the damage evolution speed and the DSER. The DSER can be used as the energy parameter to characterize the damage evolution speed. The speed of the damage closing to failure threshold increases with the increase of the damage evolution speed. Therefore, the fatigue life decreases with the increase of the DSER.

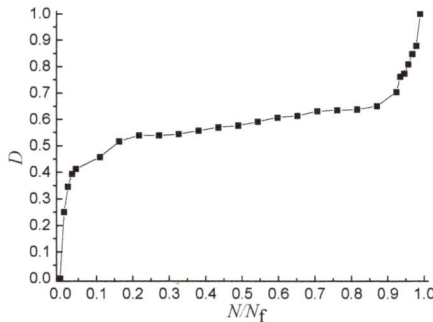

Figure 10. D-N/N_f curve.

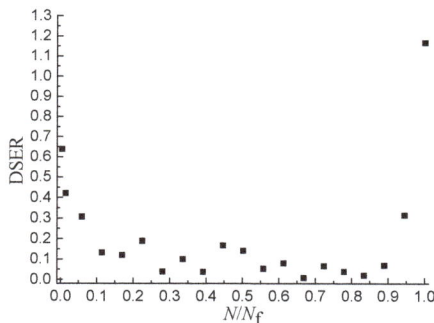

Figure 11. Scatter plots of DSER-N/N_f.

5.2. Establishment of Fatigue Equation Based on the PV of DSER

The DSER changing process can be divided into three stages. Although the calculation results of DSER in the first and third stages are greater than that of the second stage, the second stage is the main part of the whole process which accounts for more than 80% of the whole process. Therefore, the DSER value of second stage reflects the overall DSER value of the three stages. The plateau value (PV) of DER represents a period during which there is a constant percentage of input energy being turned into damage, and as such may prove to be a material property useful in design [32]. Therefore, in this paper, the plateau value (PV) is defined as the average value of DSER in the second stage, and PV reflects the overall DSER value of the three stages and the resistance to fatigue damage.

The results of the PV and the fatigue life (N_f) of the asphalt mixture are listed in Table 5, which indicates that N_f decreases with the increase of PV. The power function with the advantages of simple form and easy generalization was used to establish the fatigue equation as shown below, in Formula (6). The fatigue equation was used to fit the results in Table 5, and the fitting results are shown in Table 6. The correlation coefficient is greater than 0.95, which proves the reliability of the fatigue equation used to calculate the predicted fatigue lives. The parameter B is shown as -3.29 in the Table 6, this is very close to -3.95 that is used in the MEPDG (AASHTO Ware ME Design), thus the research is consistent with known results [34]. The comparison between the predicted life and the test life is listed in Table 7, which indicates that the fatigue life can be well predicted by the fatigue equation, based on the PV of DSER.

$$N_f = A(PV)^B, \tag{6}$$

where A and B are fitting parameters.

Table 5. Statistical Results of N_f and plateau value (PV).

N_f	PV
313	0.0209
164	0.02484
129	0.02592
106	0.02829
98	0.03065
92	0.0315

Table 6. Fitting Parameters of Fatigue Equation.

Fitting Parameters	A	B	R^2
Results	8.91785×10^{-4}	-3.29402	0.95553

Table 7. Contrast results of fatigue life.

Stress Strength Ratio-Temperature-Loading Rate	Test Results of N_f	Predictive Results of N_f	Relative Error
0.6–25 °C-10 mm/min	313	305	2.5%
0.7–25 °C-10 mm/min	164	172	4.9%
0.8–25 °C-10 mm/min	129	150	16.3%
0.6–25 °C-20 mm/min	106	112	5.7%
0.6–15 °C-10 mm/min	98	79	19.4%
0.6–5 °C-10 mm/min	92	76	17.4%

6. Conclusions

(1) The DSE increases with the increase of the SSR and the loading rate, and the decrease of the temperature. That is, the fatigue life of the asphalt mixture decreases with the increase of the DSE.

(2) The change laws of DSE and DSER verses the number of loading cycles can be well described by the proposed mathematical models.

(3) The DSER is only related to the damage deformation. The speed of the damage closing to failure threshold increase with the increase of the damage evolution speed and the fatigue life decreases with the increase of the DSER.

(4) The fatigue life decreases in power function with the increase of PV, based on which the fatigue equation was established. The established fatigue equation is very close to that is used in the MEPDG (AASHTO Ware ME Design). The fatigue equation can well predict the fatigue life asphalt mixture.

Author Contributions: Y.S. and C.F. wrote the manuscript; C.F., Y.Y. and Z.M. performed the three-point bending fatigue tests; Y.S., C.F. and J.W. checked the manuscript.

Funding: The research is funded by the National Natural Science Fund (51478276), the Natural Science Foundation of Liaoning Province (20170540770).

Acknowledgments: This research was performed at the Shenyang Jianzhu University and Institute of Transportation Engineering of Zhejiang University.

Conflicts of Interest: The authors declare no conflict of interest.

References

1. Sas, W.; Głuchowski, A.; Gabrys, K.; Sobol, E.; Alojzy, S. Deformation behavior of recycled concrete aggregate during cyclic and dynamic loading laboratory tests. *Materials* **2016**, *9*, 780–796. [CrossRef] [PubMed]

2. Li, X.; Liang, N.X. Research on splitting fatigue fest of asphalt mixture based on strain control. *J. China Foreign Highw.* **2013**, *33*, 276–280.

3. Wu, S.P.; Zhang, G.; Han, J.; Liu, G.; Zhou, J. Fatigue performance of bridge deck pavement materials. *Trans. Nonferrous Met. Soc. China* **2006**, *24*, 318–320. [CrossRef]

4. Wu, S.P.; Liu, C.; Cao, T.; Pang, L. Investigation of the low-temperature performance of asphalt mixtures via fatigue and linear contraction test. *J. Huazhong Univ. Sci. Technol.* **2008**, *25*, 30–33.

5. Yan, H.; Zhu, H.Z.; Tang, B.M. Study on fatigue dissipated energy model of AC-13 asphalt mixture. *J. Chongqing Jiaotong Univ. (Nat. Sci.)* **2010**, *29*.

6. Sun, Y.Z.; Fang, C.Z.; Fan, D.; Wang, J.C.; Yuan, X.Z. A research on fatigue damage constitutive equation of asphalt mixture. *Math. Probl. Eng.* **2018**, *2018*, 13. [CrossRef]

7. Sha, J.; Xing, M.L.; Zhang, J.Y. Fatigue fest of asphalt mixture based on strain control. *J. China Foreign Highw.* **2016**, *10*, 217–220.

8. Sun, Y.Z.; Fang, C.Z.; Wang, J.C.; Yuan, X.Z.; Fan, D. Method of fatigue-life prediction for an asphalt mixture based on the plateau value of permanent deformation ratio. *Materials* **2018**, *11*, 722. [CrossRef] [PubMed]

9. Li, C. The Experimental Study of Asphalt Mixture on Fatigue Property under Different Impact Factor. Master's Thesis, Dalian University of Technology, Dalian, China, 2009.

10. Yang, Y. Research on Fatigue Damage of Asphalt Mixtures under Different Loading Frequencies. Master's Thesis, Changsha University of Science & Technology, Changsha, China, 2009.

11. Huang, W.D.; Wang, J.; Gao, C.; Li, K. Research on Influence Factors of Rubber Asphalt Mixture Fatigue Performance. *J. Tongji Univ. (Nat. Sci.)* **2009**, *37*, 1608–1614.

12. Gao, C. Fatigue Properties of Asphalt Rubber Mixtures. Master's Thesis, Tongji University, Shanghai, China, 2011.

13. Gao, F. Research on Fatigue Properties of Rubber Asphalt Mixture For Different Gradations. Master's Thesis, Chongqing Jiaotong University, Chongqing, China, 2015.

14. Xue, L.; Rong, L.; Robert, L.L. Energy-based mechanistic approach to characterize crack growth of asphalt mixtures. *Am. Soc. Civ. Eng.* **2013**, *25*, 1198–1208.

15. Xue, L.; Rong, L.; Robert, L.L. Characterization of Fatigue Damage in Asphalt Mixtures Using Pseudostrain Energy. *Am. Soc. Civ. Eng.* **2013**, *25*, 208–218.

16. Sha, J.; Xing, M.L.; Zhang, J.Y. Experimental study on fatigue damage of asphalt mixture under variable amplitude loading. *J. Mater. Sci. Eng.* **2017**, *35*, 306–315.

17. Heukelom, W. Observations on the rheology and fracture of bitumens and asphalt mixes. *J. Assoc. Asphalt Paving Technol.* **1966**, *35*, 359–399.

18. Chomton, G.; Valayer, P.J. Applied rheology of asphalt mixes practical application. In Proceedings of the 3rd International Conference on the Structural Design of Asphalt Pavements, London, UK, 11–15 September 1972.

19. Van Dijk, W.; Moreaud, H.; Quedeville, A.; Uge, U. The fatigue of bitumen and bituminous mixes. In Proceedings of the 3rd International Conference on the Structural Design of Asphalt Pavements, London, UK, 11–15 September 1972.

20. Tayebali, A.A.; Rowe, G.M.; Sousa, J.B. Fatigue response of asphalt aggregate mixtures. *J. Assoc. Asphalt Paving Technol.* **1992**, *61*, 333–360.

21. Xue, L.; Rong, L.; Robert, L.L. Characterization of Asphalt Mixtures Using Controlled-Strain Repeated Direct Tension Test. *Am. Soc. Civ. Eng.* **2013**, *25*, 194–207.

22. Chen, W.Q. Research on Fatigue Failure Mechanism of Asphalt Mixture Based on Dissipated Energy. Master's Thesis, South China University of Technology, Guangzhou, China, 2014.

23. Li, X.; Liang, N.X. Variation ratio of dissipated energy of fatigue failure for porous asphalt mixture. *J. Changan Univ. (Nat. Sci. Edit.)* **2017**, *37*, 98–104.

24. Khalid, A.G.; Samuel, H.C. Fatigue damage analysis in asphalt concrete mixtures using the dissipated energy approach. *Can. J. Civ. Eng.* **2006**, *33*, 890–901.

25. Wu, Z.Y.; Zhang, X.N.; You, H. Prediction of Fatigue Life of Asphalt Mixture Based on Strain Control. *J. South China Univ. Technol. (Nat. Sci. Edit.)* **2014**, *42*, 139–144.

26. Yu, T.Q.; Wang, X.W.; Liu, Z.H. *Elasticity and Plasticity*; China Architecture & Building Press: Beijing, China, 2004.

27. Fan, D. Research on Anti Cracking Control of Composite Pavement Based on Asphalt Rubber-Stress Absorbing Membranes Interlayer. Master's Thesis, Shenyang Jianzhu University, Shenyang, China.

28. *JTGF40-2004—Technical Specifications for Construction of Highway Asphalt Pavements*; China Communications Press: Beijing, China, 2004.

29. *JTDG50-2017—Specifications for Design of Highway Asphalt Pavement*; China Communications Press: Beijing, China, 2017.

30. *JTGE20-2011—Standard Test Methods of Bitumen and Bituminous Mixtures for Highway Engineering*; China Communications Press: Beijing, China, 2011.

31. Carpenter, S.H.; Jansen, M. Fatigue behavior under new aircraft loading conditions. In *Aircraft/Pavement Technology: In the Midst of Change*; Hermann, F.V., Ed.; American Society of Civil Engineers: Seattle, DC, USA, 1997; pp. 259–271.

32. Ghuzlan, K.; Carpenter, S. Energy-Derived, Damage-Based Failure Criterion for Fatigue Testing. *Transp. Res. Rec. J. Transp. Res. Board* **2000**, *1723*, 141–149. [CrossRef]

33. Luan, L.Q.; Tian, X.G. Nonlinear Analysis of Fatigue Damage of Asphalt Mixture. *J. Build. Mater.* **2012**, *15*, 508–512.

34. Podolsky, J.H.; Buss, A.; Williams, R.C.; Cochran, E. Mechanistic empirical performance of warm-mix asphalt with select bio-derived additives in the Midwestern United States using AASHTOW are pavement ME design. *Road Mater. Pavement Des.* **2017**, *18*, 800–816. [CrossRef]

materials **MDPI**

Article

Fatigue Performance of Different Thickness Structure Combinations of Hot Mix Asphalt and Cement Emulsified Asphalt Mixtures

Zhenjun Wang [1,3,*], Linlin Cai [1,*], Xiaofeng Wang [2], Chuang Xu [1], Bo Yang [2] and Jingjing Xiao [3]

1   School of Materials Science and Engineering, Chang'an University, Xi'an 710061, China;
     xc_chd029cico@163.com
2   Henan Provincial Communications Planning & Design Institute Co., Ltd., Zhengzhou 450052, China;
     wangxf0351@sina.com (X.W.); yangbohnrbi@foxmail.com (B.Y.)
3   Engineering Research Central of Pavement Materials, Ministry of Education of China, Chang'an University,
     Xi'an 710061, China; xiaojj029@sina.com
*   Correspondence: zjwang@chd.edu.cn (Z.W.); linlincaia@163.com (L.C.);
     Tel.: +86-(0)29-8233-7245 (Z.W.); +86-(0)29-8233-9328 (L.C.)

Received: 15 June 2018; Accepted: 2 July 2018; Published: 5 July 2018

Abstract: Cement emulsified asphalt mixture (CEAM) is widely used in asphalt pavement for its environmental virtues. However, a CEAM layer can influence fatigue performance of asphalt pavement because of higher air voids of CEAM in contrast to hot mix asphalt (HMA). Therefore, it is common to use HMA and CEAM structure combinations for improving the fatigue performance. In this work, three different thickness structure combinations of HMA (AC-10) and CEAM (AC-16) were designed, in which HMA and CEAM were used as top layer and bottom layer, respectively. The fatigue performance of the three combinations was studied. The fatigue equations of the combinations were established and the rational combination was recommended. The distributions of the internal voids in the combinations were studied with X-ray computed tomography (X-ray CT); and the correlation between the fatigue life and the void ratios were analyzed. Artificial neural network (ANN) was employed to predict the fatigue life of each combination. The results show that the fatigue life of the combinations is inversely proportional to the stress ratio level and environment temperature. The optimal combination is the structure with 40 mm HMA and 40 mm CEAM. The internal void ratio of CEAM is higher than that of HMA. A thinner HMA and thicker CEAM structure can result in higher void ratios and lower fatigue life of the combinations. The prediction results of ANN are similar to the experimental results. The obtained results can potentially guide the design of cement emulsified asphalt pavement structures.

Keywords: cement emulsified asphalt mixture; fatigue performance; thickness combinations; X-ray computed tomography; artificial neural network

---

1. Introduction

1.1. Background

Cement emulsified asphalt mixture (CEAM) consists of asphalt emulsion, cement, aggregate, and mineral fillers, which is widely used in road construction and maintenance for environmental requirements. CEAM has many advantages such as low pollution, low energy consumption, low cost, and easy construction [1,2]. However, the loss of cohesive force in its internal structure causes the CEAM mixture to be poor in water resistance, forming a pit, crack, and other pavement damage. Therefore, CEAM cannot be applied to the road surface alone [3]. On the contrary, hot mix asphalt (HMA) has been widely used in pavement structure because of its high and low temperature performance and high cohesive force [4]. However, the problems such as high pollution, high energy consumption, and high cost caused by HMA cannot be ignored. Therefore, the advantage combination of cement emulsified asphalt mixture and ordinary hot mix asphalt mixture in asphalt pavement is important for asphalt pavement design.

With the rapid development of transportation, the asphalt pavement is facing more and more severe fatigue damages. Previous studies have concentrated on the fatigue performance materials varied with stress ratios, asphalt contents, void ratios, aggregate types, and so on. For example, Li, et al. [5,6] found that the fatigue life of the mixture increased with the increase of asphalt content at the high stress ratio level, and the cement content can improve the fatigue life at the low stress ratio level. Wang, Cory, and Chiara, et al. [7–9] reported that the factors for the fatigue performance of CEAM can be sorted in their influence as emulsified asphalt content > stress ratio level > cement content. Jiang, et al. [10] pointed out that the fatigue life of the asphalt mixture decreases with the increase of void ratios. Chen, Wu, and Gao, et al. [11–13] indicated that the rougher aggregate surface was beneficial to increase the fatigue life of the asphalt mixture; and the fatigue damage tended to occur in areas where more aggregates were concentrated.

The asphalt pavement structure design is facing more and more fatigue failure challenges [4,14]. To predict the fatigue life of the asphalt mixture, scholars have studied the correlation between the vehicle repeated loading times and the mechanical response of the asphalt pavement [15]. Three main research methods for the fatigue characteristics of the asphalt pavement were developed as follows: (1) phenomenological method—a method for characterizing the fatigue fracture characteristics of materials using fatigue life curves [16]; (2) mechanics approximation method—the stress intensity factor, which was generated at the crack tip of the pavement, was used to study the development law of various cracks when the material undergoes fatigue failure under repeated loading [17]; and (3) energy dissipation method—a method for characterizing the fatigue life of the asphalt mixture with the energy accumulation value when the mixture reached failure under a certain cyclic load. The dissipated energy can not only analyze the damage process of the internal structure in a theoretical basis, but also qualitatively study the fatigue performance of the asphalt mixture [18,19]. However, the existing research mainly focuses on the fatigue performances of the HMA mixture as a single surface layer.

1.2. Objectives

CEAM is often used in the middle or upper layers of asphalt pavement structure. There are evident differences in the properties of CEAM and HMA, such as air void and flexural tensile strength. However, less attention is paid to the fatigue performance of different asphalt pavement thickness structure combinations. That is to say, it is rare to find studies concentrating on the fatigue performance of the asphalt pavement layer with various thickness structure combinations of HMA and CEAM. Therefore, three kinds of thickness structure combinations with HMA and CEAM were prepared in this work, in which the top layer was HMA and the bottom layer was CEAM. The effects of thickness combinations on the fatigue performance of the structures were evaluated under different temperatures and stress levels. Analyses between the internal void distributions and fatigue life were carried out with an X-ray computed tomography (X-ray CT) technique. Based on the experimental results, an artificial neural network (ANN) model was developed to predict the fatigue life of the different thickness structure combinations. The results of this work can contribute to putting CEAM to use reasonably in asphalt pavement structures. The conclusions can also be potentially used to guide the asphalt pavement structure design with CEAM and HMA mixtures.

2. Experimental

2.1. Raw Materials

The binder of HMA was asphalt and its properties are shown in Table 1. Ordinary Portland cement was used in CEAM and its main properties are given in Table 2. Cationic emulsified asphalt was adopted in CEAM and its properties are shown in Table 3. Limestone aggregate and mineral fillers were used in two kinds of the mixtures, and their properties are shown in Tables 4 and 5.

Table 1. Properties of asphalt binder.

Properties	Unit	Specification [20]	Test Results
Penetration (25 °C, 5 s, 100 g)	0.1 mm	60–80	72
Softening point (Ring-and-ball method)	°C	≥46	52.3
Ductility at 15 °C	cm	≥40	51
Solubility	%	≥99.5	99.71
Density at 15 °C	g/cm^3	Measured	1.036
Wax content (distillation method)	%	≤2.2	1.7
Flash Point (COC)	°C	≥260	305

Table 2. Properties of ordinary Portland cement.

Fineness/ (%, 80 μm)	Water Consumption of Normal Consistency/%	Stability (Boiled Method)	Setting Time/min		Compressive Strength/MPa		Flexural Strength/MPa	
			Initial	Final	3 d	28 d	3 d	28 d
2.4	23.2	qualified	143	201	28.7	52.5	6.2	9.3

Table 3. Properties of emulsified asphalt.

Properties		Unit	Specification [20]	Test Results
Particle charge type		-	Cationic	Cationic
The remaining percentage on 1.18 mm sieve		%	≤0.1	0.03
Wrap ratio with coarse aggregate		-	≥2/3	>2/3
	Residue content	%	63	≥55
Evaporation residue	Penetration (5 s, 100 g, 25 °C)	0.1 mm	75	50–300
	Ductility (15 °C)	cm	47	≥40
Storage stability at room temperature	1 d	%	0.63	≤1
	7 d	%	3.24	≤5

Table 4. Properties of aggregate.

Properties	Unit	Specification [21]	Test Results
Apparent density	g/cm^3	≥2.50	2.805
Crushing value	%	≤20	13.2
Wear loss in Los Angeles	%	≤24	18.3

Table 5. Properties of mineral fillers.

Properties		Unit	Specification [21]	Test Results
Apparent density		g/cm^3	≥2.50	2.826
Moisture content		%	≤1	0.45
	<0.6 mm		100	100
Particle size	<0.15 mm	%	90–100	98.6
	<0.075 mm		75–100	83.7
Appearance		-	No lumps	No lumps
Hydrophilic coefficient		-	<1	0.6

2.2. Preparation of the Specimens

In accordance with the specification [22], the aggregate gradations for three different thickness structure combinations of HMA (AC-10) and CEAM (AC-16) are shown in Figure 1. The HMA and CEAM mixtures were prepared in reference to the specifications [23]. The properties of the mixtures were tested and the results are shown in Tables 6 and 7. Each specimen was prepared with two layers, which were the top layer and bottom layer. The total structure thickness of each specimen remained at 80 mm. AC-10 HMA was used in the top layer. Its thickness was designed as 40 mm, 30 mm, and 20 mm. Meanwhile, CEAM was used in the bottom layer. Its thickness was designed as 40 mm, 50 mm, and 60 mm.

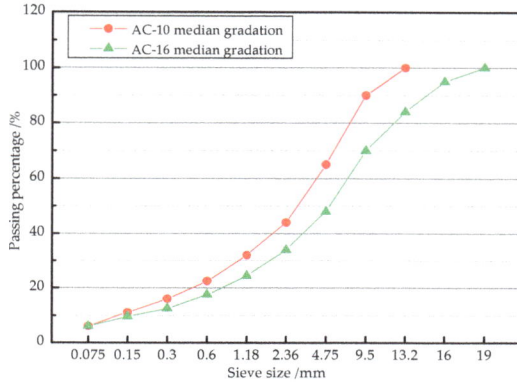

Figure 1. Aggregate gradations of hot mix asphalt (HMA) and cement emulsified asphalt mixture (CEAM).

Table 6. Properties of hot mix asphalt (HMA).

Properties	Unit	Test Results
Optimum asphalt content	%	4.6
Marshall stability	kN	12.2
Flow value	mm	3.28
Dynamic stability	times/mm	942
Residual stability	%	87

Table 7. Properties of cement emulsified asphalt mixture (CEAM).

Properties	Unit	Test Results
Emulsified bitumen content	%	8.0
Cement content	%	2.0
Marshall stability	kN	10.3
Flow value	mm	3.23
Dynamic stability	times/mm	1088
Residual stability	%	83

The bottom layer, CEAM, was compacted by the wheel forming machine (LCX, Beijing, China) with 15 round-trips (30 times) at the temperature of 25 °C. Then, the emulsified asphalt was sprayed as the tack coat layer. The HAM mixture was paved as the upper layer after demulsification of the emulsified asphalt and the HAM mixture was compacted by the wheel machine (LCX, Beijing, China) with 15 round-trips (30 times) at the temperature of 150 °C. Therefore, three thickness structure combinations, with sizes of 300 mm × 200 mm × 80 mm, were prepared as 40 mm (HMA) + 40 mm (CEAM) (marked as C1); 30 mm (HMA) + 50 mm (CEAM) (marked as C2); and 20 mm (HMA) + 60 mm (CEAM) (marked as C3). Finally, the thickness structure combination specimens were cured at the temperature of 25 °C for 28 days, so that the CEAM can possess enough mechanical performance. Afterwards, the specimens were sawed to the trabecular specimens with sizes of 250 mm × 50 mm × 80 mm for the fatigue tests, which are shown in Figure 2.

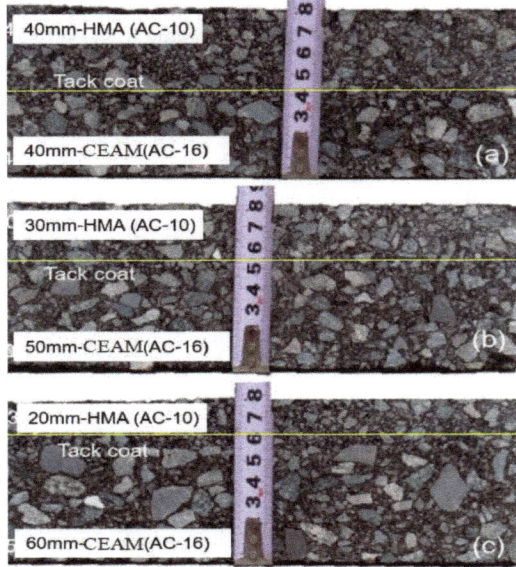

Figure 2. Pictures of different thickness structure combinations: (**a**) C1; (**b**) C2; and (**c**) C3.

2.3. Test Methods

2.3.1. Flexural Tensile Strength Test

The material testing system (MTS-370, MTS Systems Corporation, MN, USA) was used for the flexural tensile strength test at the temperature of 20 °C. The preload was conducted with a cyclic load of approximately 1.0 kN to prevent the poor touch between the indenter and the test piece. The loading speed for the bending strength test was 50 mm/min, which was slowly loaded on the center of the specimen to break it. All of the tests were carried out three times and the average values were used as the testing result. The maximum deflection of the specimen was recorded. The setup for the flexural tensile strength test is shown in Figure 3 and the flexural tensile strength and the flexural stiffness module of the three combinations were calculated by Equations (1) to (3).

$$R_B = \frac{3LP_B}{2bh^2} \tag{1}$$

$$\varepsilon_B = \frac{6hd}{L^2} \tag{2}$$

$$S_B = \frac{R_B}{\varepsilon_B} \tag{3}$$

where, R_B is flexural tensile strength, MPa; L is the distances between two support points, mm; P_B is the load, N; b and h are the width and height of the specimen, mm, respectively; d is the maximum deflection, mm; ε_B is the bending strain, no unit; and S_B is the flexural stiffness modulus, no unit.

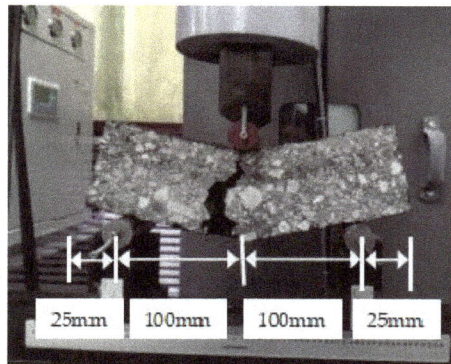

Figure 3. The setup for flexural tensile strength test.

2.3.2. Fatigue Test

The fatigue life N_f of different thickness structure combinations was tested by the three-point bending load fatigue test under the controlled stress loading mode, which is more suitable for the thicker asphalt mixture specimens, with a thickness of 80 mm in this work, and is easy to control in fatigue test. The loading waveform was a continuous full sinusoid wave with a fixed frequency of 10.0 Hz. The material testing system (MTS-370, MTS Systems Corporation, MN, USA) was used to perform the bending fatigue test. The test stress levels were 0.3, 0.4, and 0.5, respectively. The test temperature was 5 °C, 15 °C, and 25 °C, respectively. The axial displacement during the loading process at the temperature of 15 °C was recorded with the increase of the loading time. All of the tests were carried out three times.

2.3.3. X-ray CT Scanning

The industrial X-ray CT machine (YXLON Compact-225, Hamburg, Germany) was adopted to obtain the image of the internal air voids in mixtures to analyze the distribution characteristics of the voids. The parameters of the industrial X-ray CT are shown in Table 8. The scanning space for the combination was 0.01 m. The trabecular specimen with a thickness combination of 40 mm + 40 mm was scanned by CT scanning from the left to the right, with a total of 900 images; and the 180th, 360th, 540th, and 720th images were selected for processing and analyses, which lay at the position of 50 mm, 100 mm, 150 mm, and 200 mm of the specimen. The higher the density, the greater the brightness of the image. The image tended to be white. The lower the density, the smaller the brightness of the image. The image tended to be black. The void ratios were obtained by measuring the grayscale distribution of the images.

Table 8. The parameters of industrial X-ray computed tomography (X-ray CT).

Parameters	Results
Maximum tube voltage	225 kV
Maximum tube current	3.0 mA (D)/1.0 mA (T); 320 W (D)/64 W (T)
Operation mode	Cone beam scanning and digital imaging
Magnification	200 times (D)/100 times (T)
Specimen size	Length: 50 mm–100 mm; Width: 50 mm; Height: 80 mm
Pixel size	1024×1024 (200×200 μm^2)
Filter combination	Al: 1 mm; Cu: 1 mm; Fe: 0.5 mm; Sn: 0.5 mm
Dimension precision	<5 μm

2.3.4. Fatigue Life Prediction of the Combinations Based on ANN

This work developed an ANN for predicting the fatigue lives of different combinations under various conditions, which were not tested by the three-point bending load fatigue test. The main processes for developing an ANN included three steps, namely: (a) pre-treat results of the fatigue tests; (b) training an ANN model using a feed-forward algorithm; and (c) testing the precision of the ANN.

The network structure of the ANN for predicting the fatigue life is shown in Figure 4. As shown in Figure 4, the ANN consists of an input layer, two hidden layers, and an output layer. The size of the input layer is 3 × 1, which means three factors of fatigue life are inputted to the ANN, including the temperature, stress level, and thickness combination. Both of the two hidden layers consist of five neurons. The double-layer structure of the hidden layers and the number of neurons can predict the output precision of the ANN. The size of output layer is 1 × 1, which means that the output of the ANN is the fatigue life.

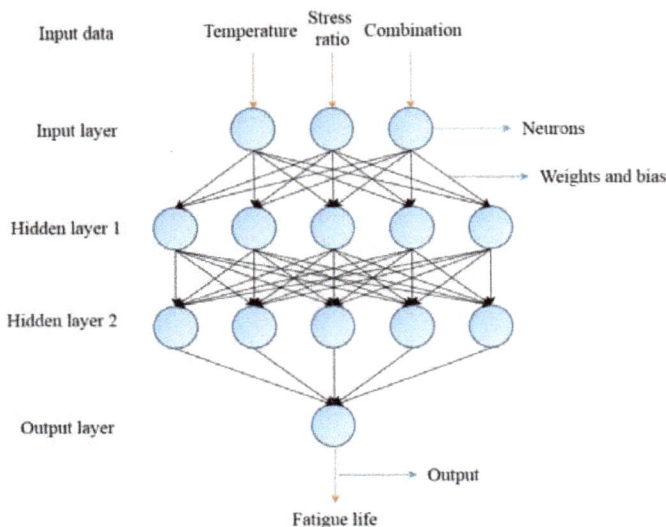

Figure 4. Network structure model of the mixtures.

The first step for developing the ANN to predict the fatigue life is to pretreat the input data and target data. The input data of the ANN contain the temperature, stress level, and thickness combinations. Different thickness combinations of the ANN and the thickness of the top layer are utilized to represent the characters. For example, as a C2 structure combination was 30 mm + 50 mm, the input data of the thickness characteristic of C2 is 50. Then, the temperature data are normalized to eliminate the effect of the dimensions of these three factors and to avoid over fitting. The temperature value is divided by 40. Thus, 5 °C, 15 °C, and 25 °C can be normalized to 0.125, 0.375, and 0.625, respectively. Therefore, the input data of C2 in a 0.3 stress ratio level and 5 °C are (0.125, 0.3, 50). The target data of the ANN are the fatigue life of the combination in different conditions. Then, data of the fatigue life are normalized to eliminate the effect of the dimensions and to avoid over fitting. The fatigue life is divided by 40,000. For example, the fatigue life of C2 at 0.3 stress level and at 5 °C is 28,153 times. Thus, the normalization result is 0.704. Therefore, the target data of C2 at 0.3 stress level and 5 °C is 0.704.

The second step for developing the ANN to predict the fatigue life is to train and to test the ANN using a feed-forward algorithm. Detailed information on the feed-forward algorithm can be found in reference [24]. Training, testing, and target data for developing the ANN are a normalization results of the fatigue test. The last step is to predict fatigue life of different combination under various conditions by the well-trained ANN.

3. Results and Discussion

3.1. Results of Flexural Tensile Strength Test

The test results of flexural tensile strength R_B and the flexural stiffness modulus S_B of three thickness structure combinations are shown in Figure 5. When the thickness of CEAM increases from 40 mm to 60 mm and HMA decreases from 40 mm to 20 mm, the flexural tensile strength R_B and the flexural stiffness modulus S_B are dropped 32.2% and 56.4%, respectively. The values of the flexural tensile strength R_B and the flexural stiffness modulus S_B of the C1 combination are the highest. The bending strain ε_B decreased with the thinner CEAM and the thicker HMA. That is because the AC-10 HMA as the material of the top layer is a viscoelastic material with high ductility and anti-deformation ability. Therefore, the thicker HMA layer and thinner CEAM layer in combination are propitious to enhance the mechanical performance of the thickness structure combinations under the bending force.

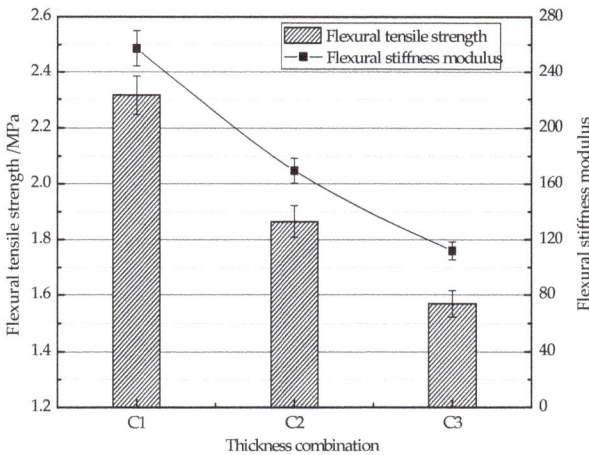

Figure 5. Flexural tensile strength test results of different thickness structure combinations.

3.2. Effects of Test Temperature on Fatigue Performance

Figure 6 shows the effect of temperatures on the fatigue life of each thickness structure combination at different stress levels. The finding is that the fatigue life of all of the combinations increases with the decrease of the temperature under the same stress level. The reason lies in that the stiffness modulus of the asphalt mixtures increase when the temperature decreases within a certain range. In the meantime, strains of the combinations decrease continuously under the same cyclic load stress. The decrement of stiffness at the lower layer and the increment of thicknesses at the top layer can increase the total stiffness modulus. Therefore, the anti-cracking ability of the combinations has been enhanced; thus, the fatigue life of the combinations can be prolonged.

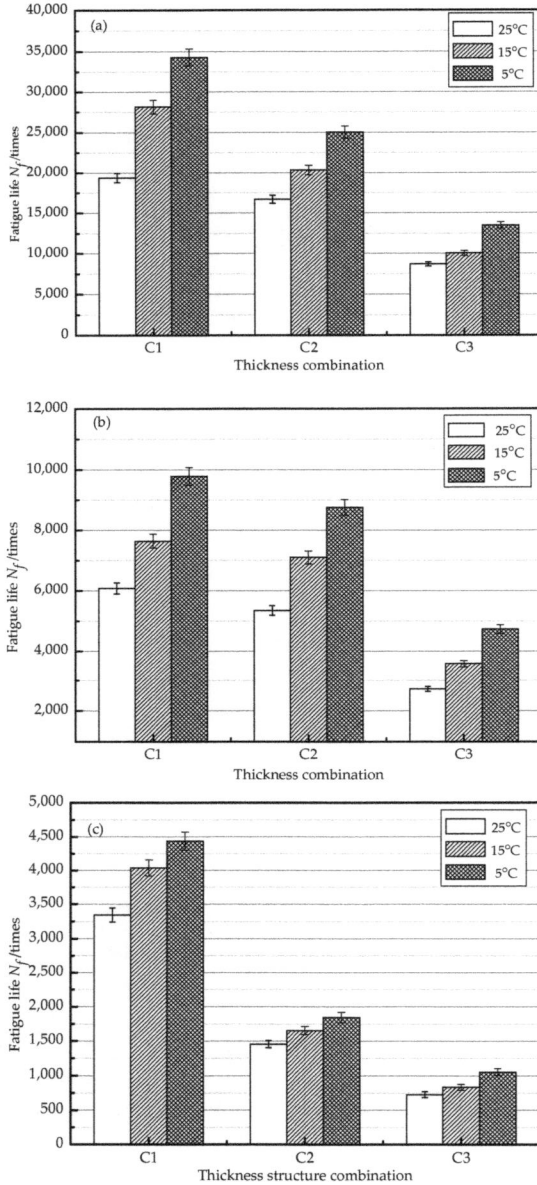

Figure 6. Fatigue life of different thickness structure combinations at different stress levels: (**a**) stress level 0.3; (**b**) stress level 0.4; and (**c**) stress level 0.5.

3.3. *Effect of Stress Level on Fatigue Performance*

Figure 7 shows the relationship between the axial displacement and the loading time during the loading process, at the temperature of 15 °C. The loading time is a total loading time induced by the dynamic load, which is the total time from the start to the end of loading process. The processes of the fatigue fracture of the combinations can be divided into three stages based on loading time, which are the initial stage of cracking, cracking expansion stage, and the final stage of the complete fracture. As shown in Figure 7, the initial stage of cracking for C3 under all of the stress levels is considered to be shorter in contrast to that for the other two combinations. All of the specimens deform quickly and the deformation is not significant enough to be observed in other two stages. Meanwhile, the deformations of all of the specimens present a slow and steady increase in the cracking expansion stage. In the final stage of complete fracture, deformation increases rapidly and the specimens endure complete failure. Furthermore, all of the time–displacement curves are approximately straight in the final stage. In each combination, the loading time for fatigue failure increases as the stress level decreases.

Figure 8 shows the effect of the stress levels on the fatigue life of each thickness structure combination at the temperature of 15 °C. The fatigue life of each combination increases with the decrease of the stress level. The reason is that the stiffness of the combination decreases with the increase of the stress levels. Thus, the fatigue life decreases with the increase of stress levels. Another reason is that the fatigue life is a gradual accumulation process. In the repeated cyclic changes of the long-term maximum axial load and the minimum axial load, the repeated loads exceed the ultimate stresses and strains that the specimen can bear; thus, permanent fracture damages emerge in the specimens. Therefore, the fatigue life of the combinations can decrease with the increase of the stress level.

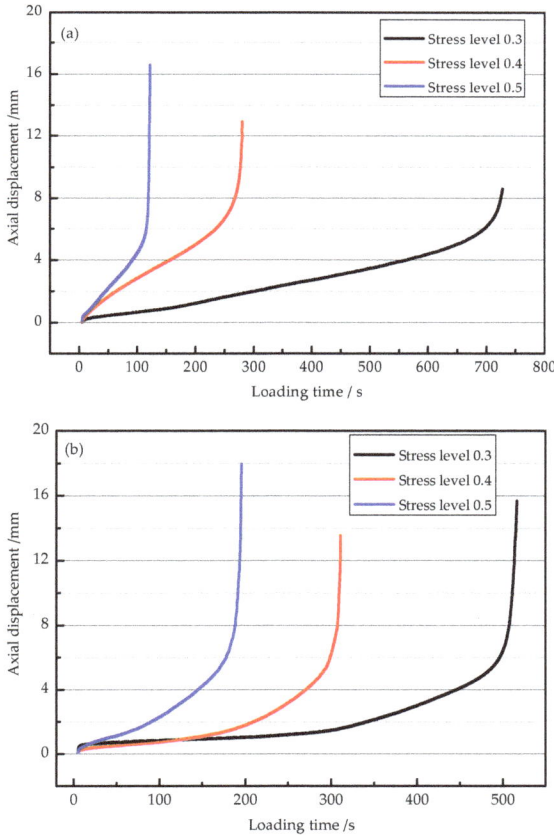

Figure 7. *Cont.*

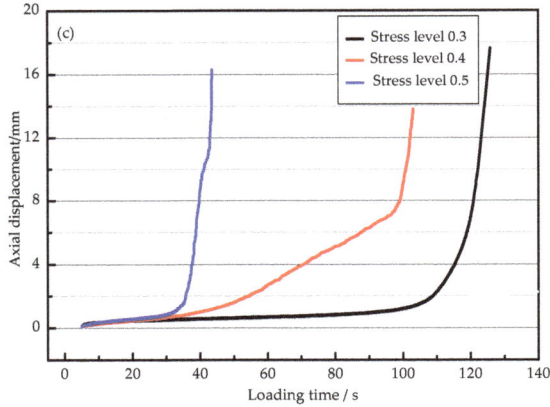

Figure 7. Time–displacement curves of different thickness structure combinations: (**a**) C1; (**b**) C2; and (**c**) C3.

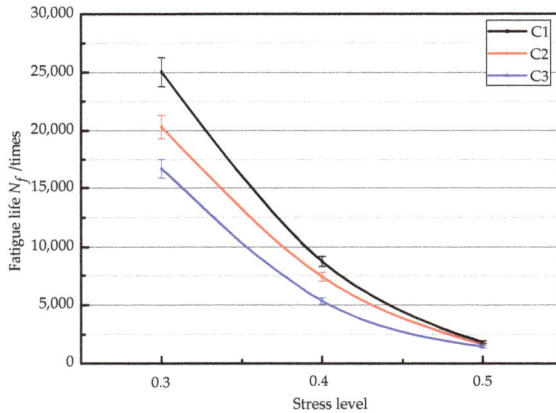

Figure 8. Fatigue life of different thickness structure combinations at 15 °C.

3.4. Establishment of Fatigue Equation

The logarithmic fatigue life of the asphalt mixture at different stress ratios obeys the normal distribution law [25]. The relationship between stress ratio level and fatigue life is shown in Equation (4).

$$\lg N_f = m - n s_i \tag{4}$$

where N_f is the fatigue life of the asphalt mixture; s_i is the stress ratio level; and m and n are the linear regression constant.

According to Equation (4), the single log-linear regression for the fatigue life of the combinations under three stress ratio levels are shown in Figure 9. The regression equations concerning the stress level and fatigue life are shown in Table 9. As shown in Figure 9 and Table 9, the fatigue life of the combinations decreases with the increase of the stress level. In addition, the single log-linear regression lines of the fatigue life of the combinations show the good fittings. At the same stress level, the fatigue life of the combinations increases with the decrease of the temperature. In contrast to the slope and the intercept of the regression line, the fatigue life value of the combinations is the highest and the combinations possess the best fatigue performance at the temperature of 5 °C.

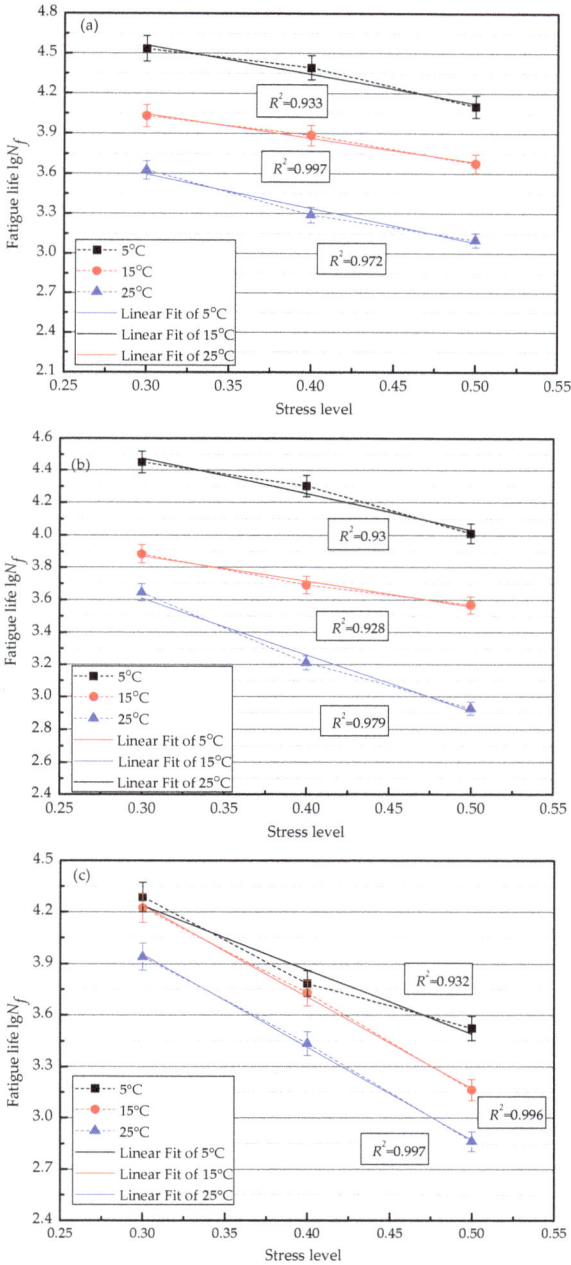

Figure 9. Single logarithmic fatigue curve equation for different thicknesses structure combinations under different stress ratio level: (**a**) C1; (**b**) C2; and (**c**) C3.

Table 9. Stress ratio level—fatigue life regression equation of each combination.

Combinations	Temperature (°C)	Regression	R^2
	5	$\lg N_f = -2.0s_i + 5.153$	0.934
C1 (40 mm + 40 mm)	15	$\lg N_f = -1.6s_i + 4.480$	0.997
	25	$\lg N_f = -3.15s_i + 4.573$	0.972
	5	$\lg N_f = -2.25s_i + 5.153$	0.932
C2 (30 mm + 50 mm)	15	$\lg N_f = -1.65s_i + 4.357$	0.928
	25	$\lg N_f = -3.65s_i + 4.723$	0.979
	5	$\lg N_f = -3.85s_i + 5.403$	0.95
C3 (20 mm + 60 mm)	15	$\lg N_f = -5.3s_i + 5.823$	0.992
	25	$\lg N_f = -5.4s_i + 5.570$	0.997

3.5. Effects of Void Ratios on Fatigue Performance

In order to study the effect of the internal voids on the fatigue performance of the combinations, an X-ray CT scanning machine was used to analyze the internal features of the C1 combination [26], whose fatigue life value is the highest. X-ray CT models were reconstructed in this work. In addition, the voids were marked and distinguished from other objects (aggregate, asphalt mortar) based on the grayscale (0–121). Therefore, software was employed to determine the grayscale threshold of the voids in the combinations. Then, the air voids can be determined by its pixel area. The typical results of the void recognition for C1 are shown in Figure 10, and the results of air voids processed by commercial software are shown in Table 10. It is clear that the voids are mainly distributed at the bottom of the specimens. In addition, the void ratios of HMA and CEAM increase first, and then decrease from 50 mm to 200 mm. The air voids of both the top layer and the bottom layer of the X-ray CT images were calculated and the average was adopted. Previous studies show that the fatigue life of the asphalt mixture increases significantly with the decrease of the air voids of the mixture [27,28]. It is because the greater air voids lead to more micro cracks in the asphalt mixture, which results in the expansion and destruction of the micro cracks under the repeated loads, which then leads to the reduction of the fatigue performance. Therefore, the double layers with thicker HMA and thinner CEAM possess a better fatigue performance. The thickness of HMA decreases with the increase of the CEAM thickness, which results in the decrease of the fatigue life of the combinations. This conclusion can be proven by the fatigue life of the C3 combination, which possesses the lowest fatigue life value.

Table 10. Air voids of the mixtures at different depths.

Different Depths of the Specimens/mm	Air Voids of the Mixtures/%	
	HMA	CEAM
50	1.9	7.7
100	2.4	8.5
150	2.6	8.6
200	2.2	8.3

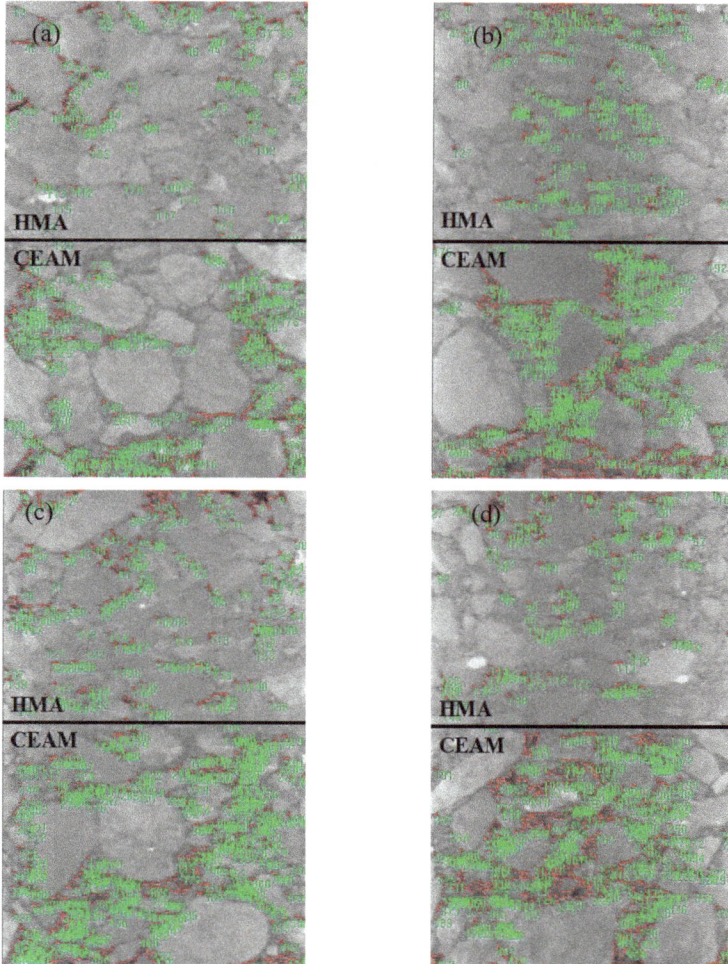

Figure 10. Voids recognition in CT image of C1: (**a**) void recognition in the specimen of 50 mm; (**b**) void recognition in the specimen of 100 mm; (**c**) void recognition in the specimen of 150 mm; and (**d**) void recognition in the specimen of 200 mm.

3.6. Fatigue Life Prediction of Different Combinations Based on ANN

The training performance is the first step for the fatigue life prediction of different combinations, based on ANN. The training results of the ANN are shown in Figure 11. As shown in Figure 11, the mean squared error between the outputs and targets is less than 10^{-1}, which means that the ANN shows good performance in the training after five epochs. Additionally, as shown in Figure 12, the correlations *R* between the outputs and the targets of training and validation are both above 0.900, which means that the errors of the outputs and targets are acceptable. Therefore, the model can be used to predict the fatigue life of other conditions, which are not tested by the three-point bending load fatigue test.

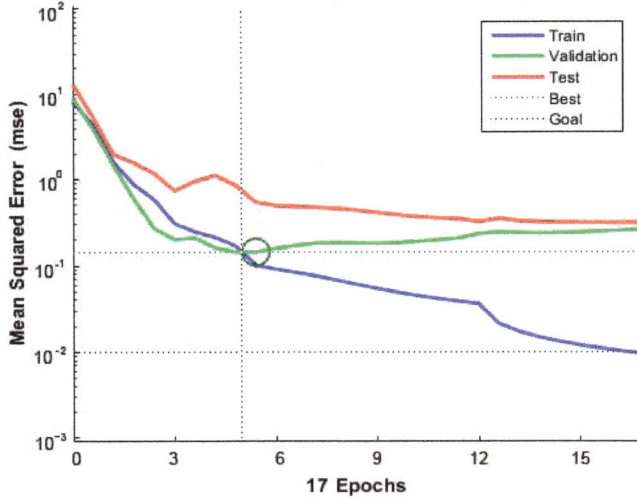

Figure 11. Training results of the artificial neural network (ANN).

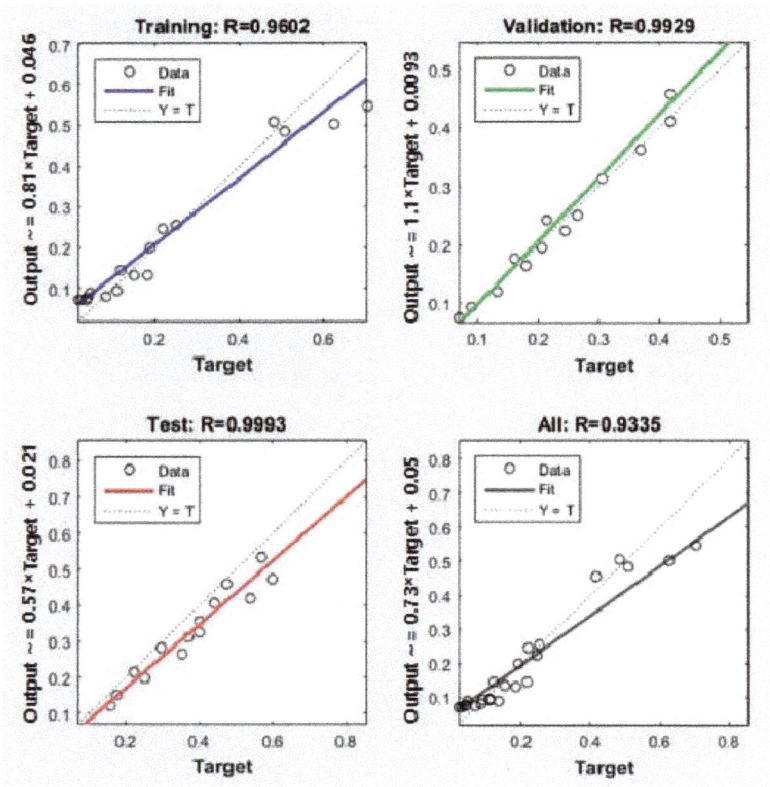

Figure 12. Correlations *R* between outputs and targets.

The testing performance is the second step for the fatigue life prediction of different combinations, based on ANN. The fatigue life of the combinations was predicted by the ANN in the stress levels of 0.20, 0.25, 0.30, 0.35, 0.40, 0.45, and 0.50. The results are shown in Table 11. As shown in Table 11, the fatigue life of the combinations decreases with the increase of temperature. The fatigue life of the combinations decreases with the increase of the stress level. The C1 combination (40 mm + 40 mm) possesses the highest ability to anti fatigue destroy, which is the same as the experimental results. In contrast to the test results, compared to the prediction results in Table 11, the output precision of the ANN is also acceptable. Therefore, the structure of the ANN is reasonable. With the acceptable precision, the prediction of the ANN can be used for analyzing the fatigue life of the HMA and CEAM structure combinations in asphalt pavement engineering.

Table 11. Normalized experimental results and prediction results of the combinations.

Temperature ($\times 40$)/°C	Stress Level	Fatigue Life of C1 ($\times 40,000$) N_f/Times		Fatigue Life of C2 ($\times 40,000$) N_f/Times		Fatigue Life of C3 ($\times 40,000$) N_f/Times	
		Test Data	Prediction	Test Data	Prediction	Test Data	Prediction
	0.20		0.878		0.766		0.560
	0.25		0.862		0.762		0.551
	0.30	0.857	0.849	0.704	0.702	0.484	0.479
0.125	0.35		0.450		0.444		0.383
	0.40	0.219	0.226	0.191	0.196	0.152	0.148
	0.45		0.120		0.106		0.097
	0.50	0.124	0.093	0.111	0.092	0.084	0.091
	0.20		0.705		0.628		0.553
	0.25		0.664		0.516		0.478
	0.30	0.626	0.615	0.508	0.513	0.417	0.416
0.375	0.35		0.441		0.304		0.290
	0.40	0.245	0.234	0.185	0.179	0.134	0.142
	0.45		0.111		0.087		0.074
	0.50	0.046	0.043	0.041	0.043	0.036	0.040
	0.20		0.500		0.458		0.350
	0.25		0.493		0.386		0.283
	0.30	0.337	0.343	0.251	0.254	0.218	0.236
0.625	0.35		0.378		0.145		0.117
	0.40	0.118	0.144	0.089	0.093	0.068	0.074
	0.45		0.082		0.071		0.061
	0.50	0.026	0.024	0.021	0.018	0.018	0.010

4. Conclusions and Recommendation

This work presents the study of the fatigue performance of a double-layer asphalt pavement surface with three thickness combinations. The effect of the stress level and temperature on the fatigue performance was studied to determine the optimal thickness structure combination for the double-layer asphalt pavement surface consisting of HMA and CEAM. Additionally, the effect of the internal voids on the fatigue performance was analyzed by X-ray CT scanning. Finally, the ANN model was established to predict the fatigue life of the combinations. The following conclusions can be drawn:

(1) With the decrease of the CEAM layer thickness, the flexural tensile strength and the maximum load of the combinations increase linearly. The fatigue life of the different thickness combinations decreases with the increase of the stress level and increases with the decrease of the test temperature.

(2) The effect comparison between the stress level and the temperature shows that the 40 mm (HMA) + 40 mm (CEAM) thickness structure combination possesses the best fatigue performance. Under the controlled stress mode, the stiffness of CEAM decreases with the decrease of the temperature; and the fatigue life can be increased with the decrease of the CEAM thickness.

(3)   The results of the X-ray CT test show that the air void of CEAM is greater than that of HMA; and the air voids are mainly distributed at the CEAM layer of the specimens. The air voids of the combinations increase with the increase of the CEAM thickness, which can result in the decrease of fatigue life of the combinations. Therefore, the CEAM layer thickness should be rationally selected in the design of asphalt pavement structure.

(4)   The ANN was established to predict the fatigue life of the combinations. It is approximately equal to the experimental results. Therefore, the outputs of the ANN can be used to predict the fatigue life of different structures under different service conditions, in consideration of the different thickness combinations. Of course, the properties of the asphalt mixture can also influence the fatigue results, and they should be considered in the input layer of the neural network. Therefore, the properties of the mixture itself as the parameters of the ANN are recommended to consider in future research.

(5)   Although CEAM is widely used in asphalt pavement for its environmental virtues, it can influence the fatigue performance of asphalt pavements as to its void characteristics. It is recommended to use HMA and CEAM combinations and to consider their rational thickness structure combination in CEAM asphalt pavement design.

Author Contributions: Conceptualization, Z.W., L.C., and X.W.; data curation, B.Y. and J.X.; formal analysis, Z.W., X.W., and C.X.; investigation, L.C., C.X., B.Y., and J.X.; writing (original draft), Z.W., L.C., and C.X.

Funding: The authors acknowledge the support from the Fundamental Research Funds for the Central Universities of China (No. 300102318402), Scientific Project of Henan Provincial Communications Planning and Design Institute Co., Ltd. (No. 220231180007) and the National Natural Science Foundation of China (No. 51608043).

Acknowledgments: The authors thank Wei Jiang, Zheng Tong, and Jie Gao of Chang'an University for their disinterested help in manuscript corrections and also thank to the reviewers for their valuable comments and suggestions concerning our manuscript.

Conflicts of Interest: There are no conflicts of interest regarding the publication of this paper.

References

1.   Zhang, Y.; Kong, X.; Hou, S. Study on the rheological properties of fresh cement asphalt paste. *Constr. Build. Mater.* **2012**, *27*, 534–544. [CrossRef]

2.   Wang, F.; Liu, Y.; Zhang, Y. Experimental study on the stability of asphalt emulsion for CA mortar by laser diffraction technique. *Constr. Build. Mater.* **2012**, *28*, 117–121. [CrossRef]

3.   Wang, Z.; Shu, X.; Rutherford, T.; Huang, B.; Clarke, D. Effects of asphalt emulsion on properties of fresh cement emulsified asphalt mortar. *Constr. Build. Mater.* **2015**, *75*, 25–30. [CrossRef]

4.   Moreno-Navarro, F.; Rubio-Gámez, M.C. A review of fatigue damage in bituminous mixtures: Understanding the phenomenon from a new perspective. *Constr. Build. Mater.* **2016**, *113*, 927–938. [CrossRef]

5.   Li, L.; Yuan, K.; Wang, T. Study on the fatigue characteristic of mixtures treated by foamed asphalt. *J. Build. Mater.* **2010**, *13*, 687–690.

6.   Ghazi, Al.; Kevin, S.; Walaa, M. Fatigue performance: Asphalt binder versus mixture versus full-scale pavements. *Can. J. Transp.* **2009**, *2*, 131–163.

7.   Wang, D.; Wang, P.; Wei, X. Fatigue performance and influence factor for cold recycling mixture with emulsified asphalt. *J. Beijing Univ. Technol.* **2016**, *42*, 541–546.

8.   Cory, S.; Ali, M.; Lee, H.; Tang, S.; Williams, C.; Schram, S. Effects of high reclaimed asphalt-pavement content on the binder grade, fatigue performance, fractionation process, and mix design. *J. Mater. Civ. Eng.* **2017**, *29*. [CrossRef]

9.   Chiara, R.; Cannone, F.A.; Massimo, L.; Wistubab, M. Back-calculation method for determining the maximum RAP content in stone matrix asphalt mixtures with good fatigue performance based on asphalt mortar tests. *Constr. Build. Mater.* **2016**, *118*, 364–372.

10.   Jiang, W.; Sha, A.; Pei, J. Study on the fatigue characteristic of porous asphalt concrete. *J. Build. Mater.* **2012**, *15*, 513–517.

11.   Chen, J.; Huang, X. Research on influence of distribution characteristics of aggregate on fatigue performance of asphalt mixture. *J. Build. Mater.* **2009**, *12*, 442–447.

12. Wu, S.; Muhunthan, B.; Wen, H. Investigation of effectiveness of prediction of fatigue life for hot mix asphalt blended with recycled concrete aggregate using monotonic fracture testing. *Constr. Build. Mater.* **2017**, *131*, 50–56. [CrossRef]

13. Gao, S.; Zhu, H.; Tang, B. Summary of study on influence factors of fatigue performance of asphalt mixture. *Pet. Asph.* **2008**, *22*, 1–6.

14. Sun, Z.; Yang, G.; Yu, B.; Zhang, M. Study on the fatigue performance of asphalt mixtures for interface layer and underlying surface. *J. Shenyang Jianzhu Univ.* **2011**, *27*, 1093–1098.

15. Delson, B.; Tadeu, L.R.; Laura, M. Research on fatigue cracking growth parameters in asphaltic mixtures using computed tomography. *Nucl. Instrum. Methods Phys. Res.* **2004**, *213*, 498–502.

16. Wasage, T.; Statsna, J.; Zanzotto, L. Repeated loading and unloading tests of asphalt binders and mixes. *Road Mater. Pavement Des.* **2010**, *11*, 725–744. [CrossRef]

17. Deng, S.; Han, X.; Qin, X.; Huang, S. Subsurface crack propagation under rolling contact fatigue in bearing ring. *Sci. China Technol. Sci.* **2013**, *56*, 2422–2432. [CrossRef]

18. Abdulhaq, A.; Abedali, H. Fatigue evaluation of Iraqi asphalt binders based on the dissipated energy and viscoelastic continuum damage (VECD) approaches. *J. Civ. Eng. Constr. Technol.* **2015**, *6*, 27–50.

19. Mahmoud, A.; Reza, S.M.; Massi, A.; Khavandi, K.A. Investigation of fatigue life of asphalt mixtures based on the initial dissipated energy approach. *Liq. Fuels Technol.* **2017**, *35*, 107–112.

20. JTG F40-2017. *Technical Specifications for Construction of Highway Asphalt Pavements*; China Communications Press: Beijing, China, 2017.

21. JTG E42-2015. *Test Methods of Aggregate for Highway Engineering*; China Communications Press: Beijing, China, 2015.

22. JTG D50-2017. *Specifications for Design of Highway Asphalt Pavement*; China Communications Press: Beijing, China, 2017.

23. JTG E20-2011. *Standard Test Methods of Bitumen and Bituminous Mixtures for Highway Engineering*; China Communications Press: Beijing, China, 2011.

24. Tong, Z.; Gao, J.; Han, Z.; Wang, Z. Recognition of asphalt pavement crack length using deep convolutional neural networks. *Road Mater. Pavement Des.* **2018**, *19*, 1334–1349. [CrossRef]

25. Lv, S.; Liu, C.; Chen, D.; Zheng, J.; You, Z.; You, L. Normalization of fatigue characteristics for asphalt mixtures under different stress states. *Constr. Build. Mater.* **2018**, *177*, 33–42. [CrossRef]

26. Wang, H.; Zhang, R.; Chen, Y.; Fang, J. Study on microstructure of rubberized recycled hot mix asphalt based X-ray CT technology. *Constr. Build. Mater.* **2016**, *121*, 177–184. [CrossRef]

27. Al-Khateeb, G.G.; Ghuzlan, K.A. The combined effect of loading frequency, temperature, and stress level on the fatigue life of asphalt paving mixtures using the IDT test configuration. *Int. J. Fatigue* **2014**, *59*, 254–261. [CrossRef]

28. Ma, T.; Zhang, Y.; Zhang, D.; Ye, Q. Influences by air voids on fatigue life of asphalt mixture based on discrete element method. *Constr. Build. Mater.* **2016**, *126*, 785–799. [CrossRef]

Article

Residual Fatigue Properties of Asphalt Pavement after Long-Term Field Service

Peide Cui [1], Yue Xiao [1,*] [iD], Mingjing Fang [2], Zongwu Chen [3,*], Mingwei Yi [4,5] and Mingliang Li [4]

[1] State Key Laboratory of Silicate Materials for Architectures, Wuhan University of Technology,
   Wuhan 430070, China; cuipeide@whut.edu.cn
[2] School of Transportation, Wuhan University of Technology, Wuhan 430070, China;
   mingjingfang@whut.edu.cn
[3] Faculty of Engineering, China University of Geosciences (Wuhan), Wuhan 430074, China
[4] Research Institute of Highway Ministry of Transport, Beijing 100088, China; cwhymw@gmail.com (M.Y.);
   li@rioh.cn (M.L.)
[5] National Engineering Research Center of Road Maintenance Technologies, Beijing 100095, China
* Correspondence: xiaoy@whut.edu.cn (Y.X.); chenzw@cug.edu.cn (Z.C.);
   Tel.: +86-18171089165 (Y.X.); +86-13545894050 (Z.C.)

Received: 17 April 2018; Accepted: 23 May 2018; Published: 25 May 2018

Abstract: Asphalt pavement is widely used for expressways due to its advantages of flexibility, low cost, and easy maintenance. However, pavement failures, including cracking, raveling, and potholes, will appear after long-term service. This research evaluated the residual fatigue properties of asphalt pavement after long-term field service. Fatigue behavior of specimens with different pavement failure types, traffic load, service time, and layers were collected and characterized. Results indicate that after long-term field service, surface layer has a longer fatigue life under small stress levels, but shorter fatigue life under large stress levels. Longer service time results in greater sensitivity to loading stress, while heavier traffic results in shorter fatigue life. Surface and underneath layers present very close fatigue trend lines in some areas, indicating that the fatigue behavior of asphalt mixture in surface and underneath layers are aged to the same extent after eight to ten years of field service.

Keywords: asphalt pavement; fatigue property; pavement failure; long-term field service

---

1. Introduction

Most highways in the world are asphalt pavements. Due to numerous failures of asphalt pavements, which are brought on by heavy vehicles, high volume traffic, and adverse weather, service life of highways is shorter than expected [1–3]. Fatigue fracture is one of the most serious failures pavement can undergo. Fatigue fracture is defined as the phenomenon of deterioration of a material (reduction in stiffness and strength, ending in fracture). Fatigue fracture occurs in areas subjected to repeated traffic loading (wheel paths), and may consist of a series of interconnected cracks in early stages of development [4]. In later stages, fatigue fracture develops into many-sided, sharp-angled pieces, with a characteristic alligator pattern [5,6]. Fatigue fractures decrease the structural capacity of pavements. Furthermore, once fatigue fractures propagate through the entire asphalt thickness, water and aggressive agents have a greater opportunity to infiltrate into the unbound layers, which greatly accelerates the deterioration process [7–9]. Various laboratory methods were used to simulate the actual traffic environment such that fatigue performance of asphalt pavements could be assessed efficiently and professionally.

To extend the life span of asphalt pavement [10,11], many studies have been conducted to understand the fatigue properties from many aspects, such as influence of asphalt, occurrence of fatigue, and how to prevent expanding of the fatigue fractures. Fatigue tests are divided into the simple flexural tests (two-point, three-point, and four-point bending), uniaxial loading test, and diametral loading test (indirect tensile test) [12]. The indirect tensile (IDT) fatigue test is a convenient and practical method for testing fatigue performance of asphalt pavement. IDT is a part of the recommend standard test methods in European standard EN 12697-24. Roque et al. characterize the crack growth rate of asphalt mixtures by using IDT fatigue tests [13]. Kim et al. built a fatigue model for asphalt pavement, which can predict the fatigue life of asphalt mixtures efficiently [14]. Hafeez et al. investigated the fatigue performance of field-laid asphalt mix specimens [15]. Jiang et al. used semi-circular bending (SCB) strength test and fatigue test to investigate the strength and fatigue properties of asphalt mixtures [16]. It is also well known that the study of asphalt aging through field specimens can offer more meaningful information on the performance of asphalt pavements. Praticò et al. model the dependence of the pavement life-cycle cost on asphalt binder quality, and determine the quantitative relationship between bitumen viscosity and the pay adjustment (PA) for a given class of boundary conditions. They found that asphalt binder viscosity can strongly affect the expected pavement life and the PA, and thus needs to be taken into account in contract and construction management [17]. Elkashef et al. used dynamic shear rheometer (DSR), bending beam rheometer (BBR), and disk compact tension (DCT) tests to evaluate rejuvenated RAP (Reclaimed Asphalt Pavement) binder and mixtures. Considerable improvements in fatigue cracking resistance of rejuvenated RAP binder and fracture energy of rejuvenated RAP mixtures were found [18].

Roads can undergo numerous types of failure, so it is necessary to research the fatigue performance of field samples with different failure types. The main objective of this research is to study the residual fatigue characteristics of field specimens collected from expressways that have served for seven, eight, and ten years. Specimens were cored from locations with different failures, like high-severity transverse cracking, longitudinal cracking, alligator cracking, potholes, and raveling. They were detected by IDT fatigue test. The following research studies residual fatigue behavior of specimens after long-term field service, with consideration of different failure types, daily traffic load, service time, and layer positions.

2. Materials and Research Methodologies

2.1. Materials

Expressways located in Hebei Province, China, which have multiple failures, were selected in this study based on the pavement service time and traffic volume characteristics. Hebei province is located in East Asia, which has a temperate continental monsoon climate with an annual average temperature of 9–12.6 °C.

Field specimens used in this study were obtained from pavements using a coring machine. Pavements were cored in the places that have various failures, including high-severity transverse cracking, longitudinal cracking, alligator cracking, potholes, and raveling. Field obtained cores should have relatively smooth, parallel surfaces, and conform to the height and diameter requirements for laboratory IDT fatigue tests. In this study, specimens with full pavement depth were sawed and separated from the layer interface, and tested separately. The information for specimens with different failure types is presented in Table 1.

HD expressway's daily traffic is the smallest of the three highways, and SH expressway's traffic flow is larger than that of HD. Daily traffic of the TJ expressway is the largest, which is two times higher than that of HD expressway. TJ expressway was constructed and available for service in 1994, and was treated with a new surface layer in 2007. Specimens of the TJ expressway were cored in 2014. SH expressway has been in service since 2006 and specimens were collected from field in 2014. The HD expressway had been in service for more than 10 years before it was cored. Both layers used SBS (Styrene-Butadiene-Styrene) modified asphalt binder. Figure 1 summarizes the detailed information of cores used in this research.

Table 1. Information of used cylindrical field specimens.

Expressway No.	Failures	Specimens No.	Field Information
HD expressway	Transversal cracking (TC) Longitudinal cracking (LC) Alligator cracking (AC) Potholes (PH) Raveling (RA)	HD-TC-s and -u HD-LC-s and -u HD-AC-s and -u HD-PH-s and -u HD-RA-s and -u	Built in 2003; Cored in 2013
SH expressway	Longitudinal cracking (LC) Alligator cracking (AC) Raveling (RA)	SH-LC-s and -u SH-AC-s SH-RA-s and -u	Built in 2006; Cored in 2014
TJ expressway	Longitudinal cracking (LC)	TJ-LC-s	Built in 2007; Cored in 2014

Note. "-s" and "-u" stand for specimens from surface and underneath layer, respectively.

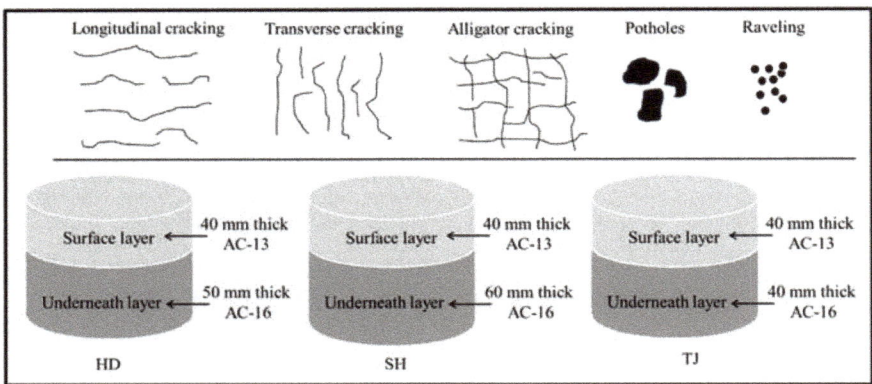

Figure 1. Material information and failure types.

2.2. Research Methodologies

2.2.1. Research Program

Firstly, residual fatigue properties of specimens from one expressway were characterized using IDT fatigue test according to failure types to minimize the influence resulting from material and structural variation. Then, specimens' fatigue characteristics from three expressways were tested and analyzed according to failure types, traffic volume, and service time.

2.2.2. Indirect Tensile Fatigue Test

To get the loading values that were needed in fatigue test, indirect tensile strength (ITS) tests were conducted before fatigue tests. The ITS test was performed at 15 °C with a constant loading rate of 50 mm/min. This loading generates a relatively uniform tensile stress perpendicular to the direction of applied load and along the vertical diametrical plane, which causes the specimen to fail by splitting along the central part of vertical diameter.

Repetitive IDT fatigue tests were used to measure fatigue performance of asphalt mixtures in accordance with the AASHTO-TP31 standard and they were designed to simulate the tensile forces that are generated in asphalt concrete pavements under traffic loading [19–22]. Testing machine used is a universal test machine (UTM-25), which was produced by IPC Global in Australia. In the IDT fatigue test, a cylinder-shaped specimen was exposed to repeated compressive loads with a haversine load signal through the vertical diametrical stripe. A picture of the IDT fatigue testing setup is presented in Figure 2.

The environmental chamber used to store specimens was set at 15 °C. Haversine waveform loading was used with 0.1 s loading and 0.9 s resting time. The sample was placed in the chamber at testing temperature for at least 4 h before testing, so that the samples reached the targeted temperature. The tests were performed under stress-controlled conditions according to various stress levels, which calibrate the ratio between applied stress and ITS. There were three specimens tested under the replicated test conditions.

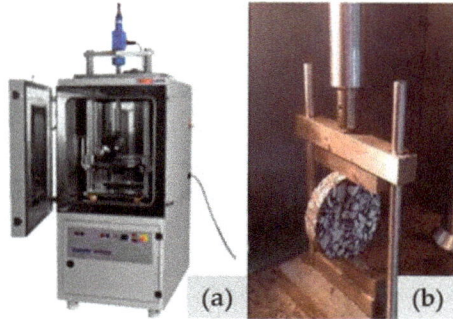

Figure 2. UTM (**a**) and IDT test setup (**b**).

Figure 3 compares the fatigue curves of specimens from H-PH-s with three applied stress levels that are 35%, 45%, and 50% of the ITS values. The vertical displacement is going up along with accumulated stress cycles until specimens fail. Three stages can be discovered during the fatigue tests, including crack initialization, crack propagation, and crack failure. The stage of crack propagation is very short when large stress levels are applied, and the crack propagation stage will last longer when smaller stress levels are applied, since asphalt mixture is a kind of viscoelastic material. These results show linear behavior of crack propagation under minor stress and strain levels, whereas beyond stress and strain levels, crack propagation behavior becomes nonlinear because of accumulation of damage, which may be expressed as the development of micro-cracks [23].

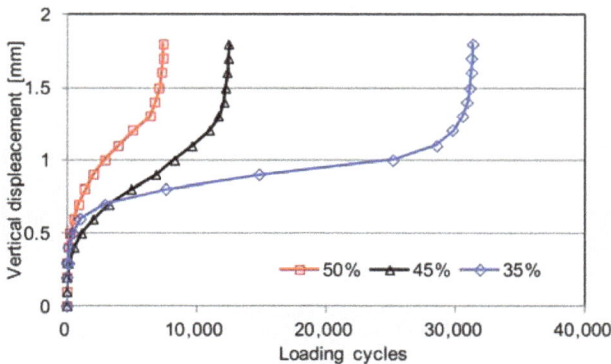

Figure 3. Fatigue curves under three applied stress levels.

There are some different conventional criteria of failure used in the literature. Fatigue failure is normally defined as the moment at which the stiffness has reduced to 50% of its initial value [24]. However, some scholars define fatigue failure as the moment when actual specimen failure was observed [25]. Fatigue failure is also defined using the stress–strain hysteresis loop in each loading cycle of the fatigue test [26,27]. In this research, the fracture fatigue life is determined to be the total number of load applications that caused a complete fracture of the specimen. The vertical deformation detected by UTM's press head will increase sharply when specimen is completely fractured. So, the number of load applications corresponding to a sharp increase in vertical deformation is the fatigue life in this study.

In fatigue testing, fatigue relationships were represented using a line in the double logarithmic coordinate system, which consisted of N_f and σ_0 in the classical fatigue analysis [28].

$$N_f = K\left(\frac{1}{\sigma_0}\right)^n \tag{1}$$

where N_f is the number of cycles to failure, σ_0 is the applied loading level, and K and n are the coefficients related to the material properties.

2.2.3. Indirect Tensile Resilient Modulus

To provide a reference for fatigue performance analysis and assist the description of the aging degree between the surface and underneath layer of asphalt mixture, specimens were also selected for indirect tensile (IT) resilient modulus tests in this study. IT resilient modulus is one of the basic parameters used to describe the viscoelastic properties of asphalt materials. It is determined by the ratio of stress amplitude to strain amplitude at steady state:

$$E^* = \frac{4Ph}{\pi d^2 \Delta L} \tag{2}$$

where E^* is resilient modulus (MPa), P is load (N), h is specimen's height (mm), d is specimen's diameter (mm), and ΔL is resilient deformation.

IT resilient modulus was also found using AASHTO-TP31. Testing machine is UTM-25, which was the same as IT fatigue test. Half-sine intermittent longitudinal loads were applied to the specimens. The system will adjust the applied load according to the target horizontal deformation, and automatically collect the load and displacement data of the last five waveforms. IT resilient modulus is calculated according to the Poisson ratio after obtaining the recoverable vertical deformation of the specimen.

3. Results and Discussions

Firstly, specimens from one expressway were studied to minimize the influence resulting from material and structural variation. Then, specimens from three expressways were tested and compared with each other.

3.1. Fatigue Properties in One Expressway

Firstly, specimens from HD expressway, which have been in service for more than 10 years, were evaluated. No surface dressing, such as micro-surfacing or slurry seal, has been applied during service life. Therefore, specimens from this expressway can be considered as samples that were directly affected by the environment and traffic.

Indirect tensile strength should be conducted and analyzed before fatigue test. The test was replicated three times and average results are listed in Table 2. Firstly, it is clear that ITS values of surface layer samples are all higher than that of corresponding underneath layer. Secondly, specimens from transverse and longitudinal cracking areas have the same ITS values, regardless of whether surface or underneath layer is measured. Thirdly, ITS tests show that the surface and underneath layers from potholes have the same ITS values, and samples from underneath layer of raveling location present the smallest ITS value.

Table 2. Indirect tensile strength of asphalt sample from HD expressway.

Indirect Tensile Strength (MPa)	H-TC	H-LC	H-AC	H-RA	H-PH
Surface layer	4.41	4.64	3.32	3.14	2.82
Underneath layer	3.33	3.56	2.76	2.64	2.97

3.1.1. Cracking

Analysis of fatigue behavior was processed on specimens from cracking areas. Fatigue trend lines and equations for the investigated field specimens from HD are listed in Figure 4 and Table 3, respectively. Simultaneous logarithm on both sides of Equation (1) was conducted before drawing fatigue trend line, therefore log-linear relation can be observed between applied stress and fatigue life. Value of n is the linear gradient of fatigue curve, representing the stress sensitivity of the tested specimen, while K values stand for level of fatigue life for asphalt pavement.

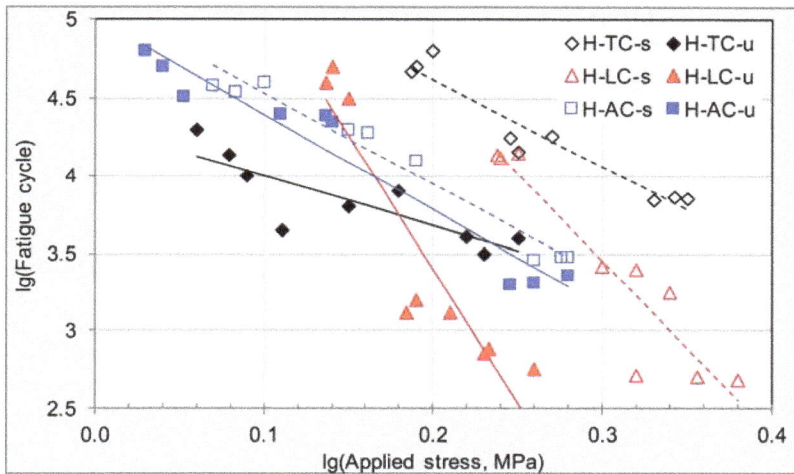

Figure 4. Fatigues trend lines of stress cycles for specimens from cracking areas.

Table 3. Fatigue equations of cracking areas in HD expressway.

Specimen No.	Fatigue Equation	Fatigue Parameters		R^2
		K	n	
H-TC-s	$N_f = 5.58 \times 10^5 \left(\frac{1}{\sigma_0}\right)^{5.6307}$	5.58×10^5	5.631	0.94
H-TC-u	$N_f = 2.12 \times 10^4 \left(\frac{1}{\sigma_0}\right)^{3.2284}$	2.12×10^4	3.228	0.71
H-LC-s	$N_f = 6.33 \times 10^6 \left(\frac{1}{\sigma_0}\right)^{11.173}$	6.33×10^6	11.173	0.87
H-LC-u	$N_f = 8.78 \times 10^6 \left(\frac{1}{\sigma_0}\right)^{18.061}$	8.78×10^6	18.061	0.84
H-AC-s	$N_f = 2.16 \times 10^5 \left(\frac{1}{\sigma_0}\right)^{6.7652}$	2.16×10^5	6.765	0.96
H-AC-u	$N_f = 1.02 \times 10^5 \left(\frac{1}{\sigma_0}\right)^{6.1724}$	1.02×10^5	6.172	0.94

Samples obtained in different cracking areas vary greatly in their fatigue behavior. Firstly, trend lines state that longitudinal cracking has the largest stress sensitivity, as their slope of fatigue trend lines are the biggest. Secondly, the fatigue life of surface layer is significantly higher than that of corresponding underneath layer for transverse cracking and alligator cracking. The n values of the surface layer are similar to the underneath layer in the alligator areas, indicating that surface and underneath layers get aged to the same extent in alligator areas after more than ten years' field service. Thirdly, K and n values of the LC area samples are significantly higher than those of other areas. Moreover, the n value of the underneath layer is even higher than that of surface layer, indicating that the protective effect of surface layer is lost when the longitudinal cracking occurred.

The IT resilient modulus test was conducted using specimens obtained from alligator cracking areas. Test results are shown in Figure 5, which indicates that the resilient modulus increases with increasing frequency and decreasing experimental temperature. The resilient modulus of specimens from surface and underneath layers is close to each other. This also illustrates that more than 10 years' field service will result in serious aging in the location of alligator cracking from both surface and underneath layers, which is identical to the fatigue test.

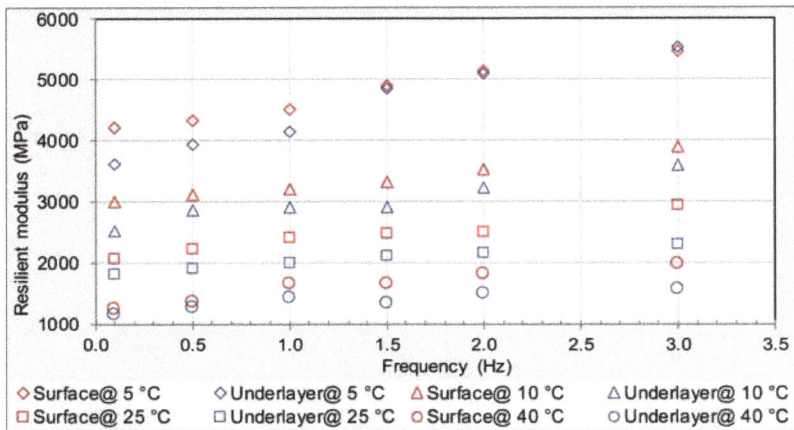

Figure 5. Resilient modulus of specimens from alligator cracking areas.

3.1.2. Raveling and Potholes

Table 4 summarizes the fatigue equations and parameters for the raveling and pothole field specimens from HD expressways. Compared to specimens from cracking areas, raveling areas have the smallest K value, which means the worst fatigue life. Figure 6 shows the fatigue trend lines of specimens from raveling areas. Specimens from surface layer are more sensitive to the applied loading stress compared with underneath layer. The fatigue trend lines between surface and underneath layers are overlapping with each other. These results illustrate that ten years of service has resulted in brittle pavement surface. Cracks and other failures will occur when additional traffic loading is applied.

Table 4. Fatigue equations of raveling and pothole areas in HD expressway.

Specimen No.	Fatigue Equation	Fatigue Parameters		R^2
		K	n	
H-RA-s	$N_f = 3.48 \times 10^4 \left(\frac{1}{\sigma_0}\right)^{8.1483}$	3.48×10^4	8.148	0.92
H-RA-u	$N_f = 9.10 \times 10^3 \left(\frac{1}{\sigma_0}\right)^{3.8938}$	9.10×10^3	3.894	0.89
H-PH-s	$N_f = 8.32 \times 10^4 \left(\frac{1}{\sigma_0}\right)^{3.5627}$	8.32×10^4	3.563	0.93
H-PH-u	$N_f = 8.90 \times 10^4 \left(\frac{1}{\sigma_0}\right)^{6.3209}$	8.90×10^4	6.321	0.92

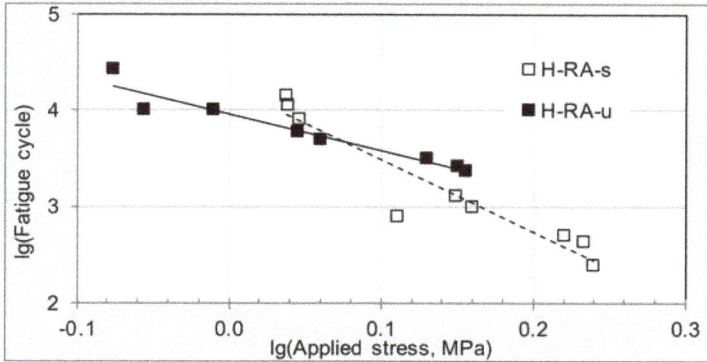

Figure 6. Fatigue trend lines of stress cycles for specimens from raveling areas.

Figure 7 compares the fatigue trend lines of specimens from potholes. Firstly, similar to results from cracking and raveling areas, surface layer from potholes is more sensitive to the applied loading stress than that of underneath layer. Secondly, specimens from pothole areas have higher fatigue life than that of specimens from raveling areas, regardless of whether surface or underneath layer is examined.

Fatigue results in this section show that surface layer has higher fatigue life under smaller stress levels, but shorter fatigue life under bigger stress levels. Asphalt binder will become hard and brittle after long-term service, leading to higher ITS values and extended fatigue life at lesser stress levels, but severely short fatigue life at larger stress levels. This signifies that after long-term service, the ability of pavement to resist low stress levels increases and to resist high stress levels drops significantly, especially when it comes to longitudinal cracking areas with the largest stress sensitivity. When a small crack appears, the crack will propagate rapidly, resulting in pavement failure.

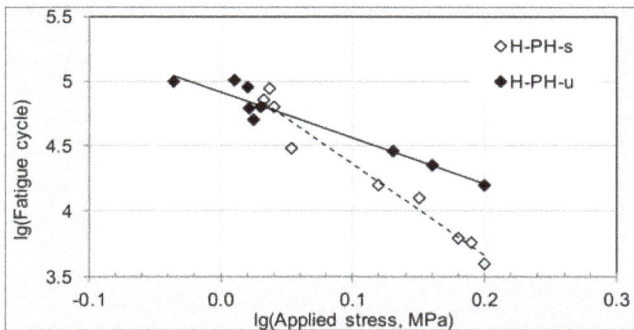

Figure 7. Fatigues trend lines of stress-cycles for specimens from pothole areas.

3.2. Fatigue Properties in Different Expressway

3.2.1. Cracking Area

Specimens from SH and TJ expressway cracking areas were tested for comparison. Table 5 presents ITS values found using fatigue tests. Like the results in Table 2, ITS values for the surface layer is higher than that of underneath layer for longitudinal cracking areas, showing that the surface layer from SH expressway has undergone severe aging and hence resulted in hard and brittle composition.

Table 5. Indirect tensile strength of asphalt samples from SH and TJ expressways.

Indirect Tensile Strength (MPa)	S-LC	S-AC	T-LC
Surface layer	3.34	3.31	2.29
Underneath layer	2.01	–	–

Figure 8 compares the fatigue trend lines of specimens from SH expressway. Fatigue trend lines of surface and underneath layers are quite close, indicating homogenous ageing index had been achieved after eight years' field service. Specimens from alligator cracking areas present the worst fatigue property, of which *n* value is more than twice that of the longitudinal cracking areas.

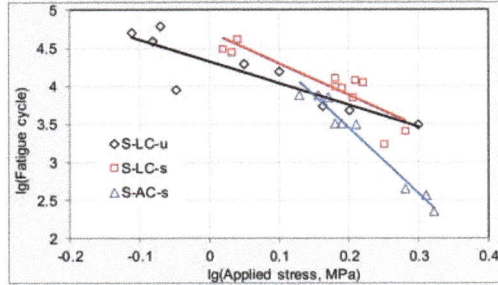

Figure 8. Fatigue trend lines for specimens from cracking areas in SH expressway.

Figure 9 compares the fatigue behaviors of specimens from longitudinal cracking areas in the HD, SH, and TJ expressways. Figure 10 compares the fatigue results of samples from alligator cracking in HD and SH expressways. The TJ expressway, which has the largest daily traffic, shows the smallest fatigue life, while HD expressway, which has a longer service time than SH and TJ, has the largest stress sensitivity to loading stress. Therefore, severe ageing can be expected on HD specimens, resulting in hard asphalt binder with significantly reduced viscoelasticity. TJ carried the busiest daily traffic with large amount of heavy trucks. Therefore, lowest fatigue life is found in Figure 9 for the TJ expressway. This relationship between traffic load and fatigue performance can also be seen in Figure 10. The traffic load of the SH expressway is higher than that of the HD expressway, which results in the samples from SH having higher stress sensitivity than HD expressway and lower fatigue life than HD expressway. Based on the previous analysis, rules can be summarized stating that heavier daily traffic results in shorter fatigue life, and longer service time results in more sensitivity to loading stress.

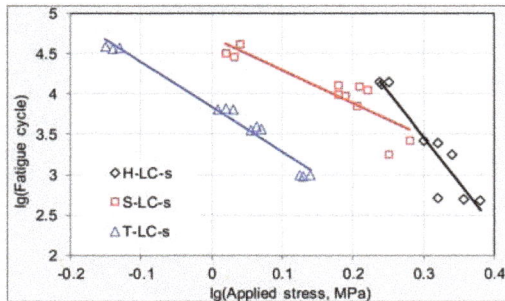

Figure 9. Comparison of fatigue trend lines between varying expressways with longitudinal cracking.

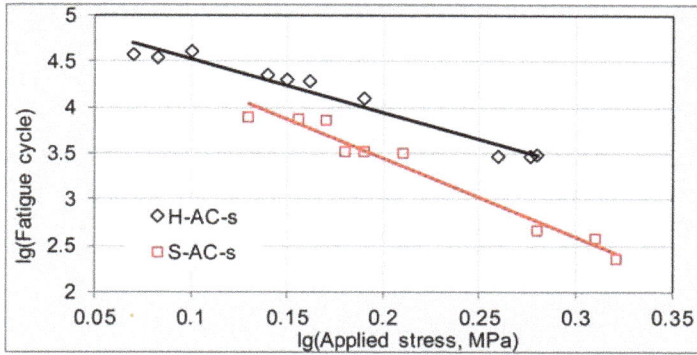

Figure 10. Comparison of fatigue trend lines between varying expressways with alligator cracking.

3.2.2. Raveling Area

Specimens from raveling areas in SH expressway were studied as well. The ITS values between the surface layer and underneath layer are 3.39 MPa and 3.42 MPa for S-RA-s and S-RA-u, respectively. Once again, the ITS values for two layers is the same, indicating that long field-aging has resulted in severe aging that went through into the second layer.

The fatigue trend lines of specimens from raveling areas in the SH and HD expressways are compared in Figure 11. In the SH expressway, specimens from the underneath and surface layers have similar n values. However, in the HD expressway, specimens from the surface layer were more sensitive to applied stresses.

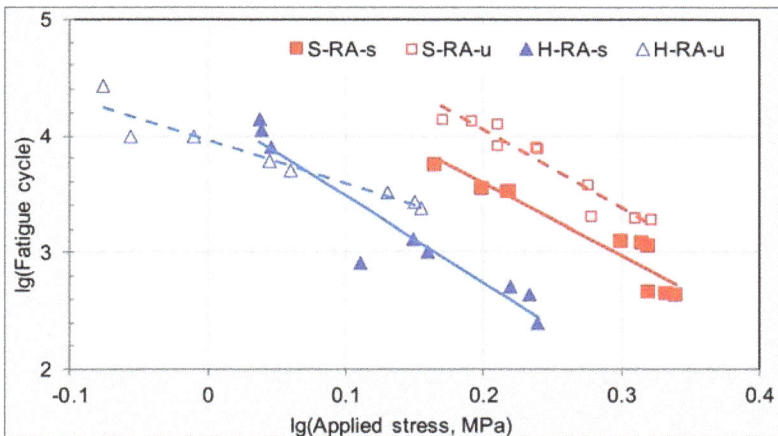

Figure 11. Comparison of fatigue trend lines between varying expressways with raveling.

3.2.3. Fatigue Characteristics

Fatigue equations and parameters of the TJ and SH expressways are shown in Table 6. According to Tables 3, 4 and 6, five groups of specimens from surface layer are found to have higher n values than the specimens from corresponding underneath layer, except groups H-LC and H-PH. This means that after long-term field service, the surface layer will become severely aged, and hence result in greater sensitivity to the traffic loading. Therefore, failure will be promoted when heavy traffic is applied onto such aged asphalt pavement. Nevertheless, other than the extremely high stress sensitivity in the longitudinal cracking areas of the HD expressway, no obvious relationship was found between the fatigue performance of field samples and the failure types.

Table 6. Fatigue equations of specimens in SH and TJ expressways.

Specimen No.	Fatigue Equation	Fatigue Parameters		R^2
		K	n	
S-LC-s	$N_f = 4.06 \times 10^4 \left(\frac{1}{\sigma_0}\right)^{3.0009}$	4×10^4	3.001	0.88
S-LC-u	$N_f = 2.13 \times 10^4 \left(\frac{1}{\sigma_0}\right)^{2.9025}$	2.13×10^4	2.903	0.78
S-AC-s	$N_f = 1.27 \times 10^5 \left(\frac{1}{\sigma_0}\right)^{8.2614}$	1.27×10^5	8.261	0.95
S-RA-s	$N_f = 2.05 \times 10^5 \left(\frac{1}{\sigma_0}\right)^{6.725}$	2.50×10^5	6.725	0.91
S-RA-u	$N_f = 6.34 \times 10^4 \left(\frac{1}{\sigma_0}\right)^{6.0854}$	6.34×10^4	6.085	0.90
T-LC-s	$N_f = 3.86 \times 10^4 \left(\frac{1}{\sigma_0}\right)^{3.3757}$	3.86×10^4	3.376	0.92

The R^2 represents the coefficient of determination for the fatigue curve. The minimum R^2 value is 0.71, which was found for H-TC-u specimen. Such high R^2 values indicate that the analytical methods for fatigue behavior used in this study are largely repeatable. Furthermore, the average R^2 value for HD expressway, which is 0.902, is higher than for the SH expressway, with a value of 0.882. SH expressway is subject to much more daily traffic than that of HD expressway, with two times as many heavy trucks using the HD expressway. Such heavy traffic loading would definitely introduce non-uniform effect on the pavement structure, hence resulting in lower repeatability for the fatigue test.

4. Conclusions

The residual fatigue properties of specimens from three long-term aged expressways were studied, according to field failure types, different traffic load, and service life. Specimens from one expressway with 10 years of service history were evaluated first, and then compared with specimens from two other expressways. Based on the research results, the following conclusions can be drawn.

(1) Residual fatigue results show that there is no clear correlation between fatigue properties and pavement failure modes, since potholes and raveling are not simply due to repeated traffic loading. Potholes and raveling are the result of moisture damage phenomenon, along with bitumen quality and adhesion between bitumen and aggregate. Furthermore, the mechanisms for longitudinal cracking and transverse cracking are quite different to what causes alligator cracking, since more than fatigue loading is involved.

(2) The minimum R^2 value for fatigue trend lines is 0.71, illustrating that the fatigue analysis with field specimens in this research has acceptable repeatability. Most specimens from surface layer perform with higher n values than the specimens from corresponding underneath layer. Surface layers have a higher fatigue life under small stress levels, but shorter fatigue life under large stress levels, indicating that the materials have been severely aged and the elastic behavior of asphalt mixture has been reduced.

(3) ITS values of surface layer samples are all higher than that of corresponding samples of their underneath layer. Specimens from transverse and longitudinal cracking areas, and their associated surface and underneath layers, have similar ITS values.

(4) In the alligator and longitudinal cracking areas of long-term field service expressways, surface and underneath layers present very close fatigue trend lines, indicating that the fatigue behavior of asphalt mixture in surface and underneath layers are aged to the same extent after ten years or eight years' field service.

(5) The fatigue performance differs for different expressways. Expressways that carried the busiest daily traffic with large amount of heavy trucks show the lowest fatigue life. While expressway that has the longest service time has the most sensitivity to loading stress. These rules can be summarized by stating that heavier daily traffic results in shorter fatigue life, and longer service time results in more sensitivity to loading stress.

Author Contributions: Yue Xiao and Zongwu Chen conceived and designed the experiments, Peide Cui performed the experiments, Peide Cui and Mingwei Yi analyzed the data, Mingliang Li, Zongwu Chen and Mingjing Fang contributed reagents/materials/analysis tools, and Yue Xiao and Peide Cui wrote the paper.

Funding: This research was funded by the Natural Science Foundation of China (No. U1733121 and 51408447), and the Fundamental Research Funds for the Central Universities (No. 173134001 and CUG170685).

Acknowledgments: The authors gratefully acknowledge Wuhan University of Technology for their materials and experimental instruments supports.

Conflicts of Interest: The authors declare no conflicts of interest.

References

1. De Almeida, A.J.; Momm, L.; Triches, G.; Shinohara, K.J. Evaluation of the influence of water and temperature on the rheological behavior and resistance to fatigue of asphalt mixtures. *Constr. Build. Mater.* **2018**, *158*, 401–409. [CrossRef]
2. Wen, H.; Wu, S.; Mohammad, L.N.; Zhang, W.; Shen, S.; Faheem, A. Long-term field rutting and moisture susceptibility performance of warm-mix asphalt pavement. *Transp. Res. Rec.* **2016**, *2575*, 103–112. [CrossRef]
3. Ma, T.; Ding, X.; Zhang, D.; Huang, X.; Chen, J. Experimental study of recycled asphalt concrete modified by high-modulus agent. *Constr. Build. Mater.* **2016**, *128*, 128–135. [CrossRef]
4. Ozer, H.; Al-Qadi, I.L.; Singhvi, P.; Bausano, J.; Carvalho, R.; Li, X.; Gibson, N. Prediction of pavement fatigue cracking at an accelerated testing section using asphalt mixture performance tests. *Int. J. Pavement Eng.* **2018**, *19*, 264–278. [CrossRef]
5. Mateos, A.; Gomez, J.A.; Hernandez, R.; Tan, Y.; Salazar, L.G.L.; Vargas-Nordcbeck, A. Application of the logit model for the analysis of asphalt fatigue tests results. *Constr. Build. Mater.* **2015**, *82*, 53–60. [CrossRef]
6. Ma, T.; Zhao, Y.; Huang, X.; Zhang, Y. Using rap material in high modulus asphalt mixture. *J. Test. Eval.* **2015**, *44*, 781–787. [CrossRef]
7. Cui, P.; Wu, S.; Xiao, Y.; Zhang, H. Study on the deteriorations of bituminous binder resulted from volatile organic compounds emissions. *Constr. Build. Mater.* **2014**, *68*, 644–649. [CrossRef]
8. Cui, P.; Wu, S.; Xiao, Y.; Wan, M.; Cui, P. Inhibiting effect of layered double hydroxides on the emissions of volatile organic compounds from bituminous materials. *J. Clean. Prod.* **2015**, *108*, 987–991. [CrossRef]
9. Zhang, H.; Xu, H.; Wang, X.; Yu, J. Microstructures and thermal aging mechanism of expanded, vermiculite modified bitumen. *Constr. Build. Mater.* **2013**, *47*, 919–926. [CrossRef]
10. Chen, Z.; Wu, S.; Xiao, Y.; Zeng, W.; Yi, M.; Wan, J. Effect of hydration and silicone resin on basic oxygen furnace slag and its asphalt mixture. *J. Clean. Prod.* **2016**, *112*, 392–400. [CrossRef]
11. Pan, P.; Wu, S.; Xiao, Y.; Liu, G. A review on hydronic asphalt pavement for energy harvesting and snow melting. *Renew. Sustain. Energy Rev.* **2015**, *48*, 624–634. [CrossRef]
12. Li, N.; Molenaar, A.A.A.; Van de ven, M.F.C.; Wu, S. Characterization of fatigue performance of asphalt mixture using a new fatigue analysis approach. *Constr. Build. Mater.* **2013**, *45*, 45–52. [CrossRef]
13. Roque, R.; Zhang, Z.; Sankar, B. Determination of crack growth rate parameters of asphalt mixtures using the superpave IDT. *J. Assoc. Asphalt Paving Technol.* **1999**, *68*, 404–433.
14. Kim, Y.R.; Lee, H.J.; Little, D.N. Fatigue characterization of asphalt concrete using viscoelasticity and continuum damage theory (with discussion). *J. Assoc. Asphalt Paving Technol.* **1997**, *66*, 520–569.
15. Hafeez, I.; Kamal, M.A.; Mirza, M.W.; Barkatullah; Bilal, S. Laboratory fatigue performance evaluation of different field laid asphalt mixtures. *Constr. Build. Mater.* **2013**, *44*, 792–797. [CrossRef]
16. Jiang, J.; Ni, F.; Dong, Q.; Wu, F.; Dai, Y. Research on the fatigue equation of asphalt mixtures based on actual stress ratio using semi-circular bending test. *Constr. Build. Mater.* **2018**, *158*, 996–1002. [CrossRef]
17. Pratico, F.G.; Casciano, A.; Tramontana, D. Pavement life-cycle cost and asphalt binder quality: Theoretical and experimental investigation. *J. Constr. Eng. Manag.* **2011**, *137*, 99–107. [CrossRef]

18. Elkashef, M.; Williams, R.C.; Cochran, E. Investigation of fatigue and thermal cracking behavior of rejuvenated reclaimed asphalt pavement binders and mixtures. *Int. J. Fatigue* **2018**, *108*, 90–95. [CrossRef]

19. AASHTO. *Standard Test Method for Determining the Resilient Modulus of Bituminous Mixtures by Indirect Tension*; AASHTO designation: TP 31; American Association of State Highway and Transportation Officials: Washington, D.C., USA, 1996.

20. Nejad, F.M.; Habibi, M.; Hosseini, P.; Jahanbakhsh, H. Investigating the mechanical and fatigue properties of sustainable cement emulsified asphalt mortar. *J. Clean. Prod.* **2017**, *156*, 717–728. [CrossRef]

21. Zaniewski, J.; Mamlouk, M. Pavement preventive maintenance: Key to quality highways. *Transp. Res. Rec. J. Transp. Res. Board* **1999**, *1680*, 26–29. [CrossRef]

22. Nejad, F.M.; Abandansari, H.F. Evaluating the effects of using recycled asphalt pavements on fatigue properties of warm mix asphalt. *Period. Polytech. Civ. Eng.* **2017**, *61*, 454–461.

23. Kim, J.; Koh, C. Development of a predictive system for estimating fatigue life of asphalt mixtures using the indirect tensile test. *J. Transp. Eng. ASCE* **2012**, *138*, 1530–1540. [CrossRef]

24. Li, N. Asphalt Mixture Fatigue Testing: Influence of Test Type and Specimen Size. Ph.D. Thesis, Delft University of Technology, Delft, The Netherlands, November 2013.

25. Al-Khateeb, G.G.; Ghuzlan, K.A. The combined effect of loading frequency, temperature, and stress level on the fatigue life of asphalt paving mixtures using the idt test configuration. *Int. J. Fatigue* **2014**, *59*, 254–261. [CrossRef]

26. Al-Khateeb, G.; Shenoy, A. A distinctive fatigue failure criterion. *J. Assoc. Asphalt Paving Technol.* **2004**, *73*, 585–622.

27. Al-Khateeb, G.; Shenoy, A. A simple quantitative method for identification of failure due to fatigue damage. *Int. J. Damage Mech.* **2011**, *20*, 3–21. [CrossRef]

28. Chen, Z.W.; Xiao, Y.; Pang, L.; Zeng, W.B.; Wu, S.P. Experimental assessment of flue gas desulfurization residues and basic oxygen furnace slag on fatigue and moisture resistance of HMA. *Fatigue Fract. Eng. Mater. Struct.* **2014**, *37*, 1242–1253. [CrossRef]

materials

MDPI

Article

Low-Temperature Performance and Damage Constitutive Model of Eco-Friendly Basalt Fiber–Diatomite-Modified Asphalt Mixture under Freeze–Thaw Cycles

Yongchun Cheng, Di Yu, Guojin Tan * and Chunfeng Zhu

College of Transportation, Jilin University, Changchun 130025, China; chengyc@jlu.edu.cn (Y.C.);
yudi16@mails.jlu.edu.cn (D.Y.); zcf-mine@163.com (C.Z.)
* Correspondence: tgj@jlu.edu.cn; Tel.: +86-0431-4186-5446

Received: 20 September 2018; Accepted: 29 October 2018; Published: 31 October 2018

Abstract: Asphalt pavement located in seasonal frozen regions usually suffers low-temperature cracking and freeze–thaw damage. For this reason, diatomite and basalt fiber were used to modify asphalt mixtures. An indirect tensile test was used to determine the low-temperature performance of the asphalt mixture. The influences of freeze–thaw (F–T) cycles on strength, tensile failure strain, stiffness modulus, and strain energy density were analyzed. The variation of the stress–strain curve under F–T cycles was analyzed. The stress–strain curve was divided into a linear zone and nonlinear zone. The linear zone stress ratio and linear zone strain ratio were proposed as indexes to evaluate the nonlinear characteristics of the stress–strain curve. The results show that the basalt fiber–diatomite-modified asphalt mixture had better low temperature crack resistance and antifreeze–thaw cycles capacity compared to the control asphalt mixture. The F–T cycles made the nonlinear characteristics of the stress–strain relationship of the asphalt mixture remarkable, and also decreased the linear zone stress ratio and linear zone strain ratio. The damage constitutive model established in this paper can describe the stress–strain relationship after F–T damage well.

Keywords: diatomite; basalt fiber; asphalt mixture; low-temperature; damage constitutive model

---

1. Introduction

The asphalt concrete pavement in seasonal frozen regions is affected by low temperature and freeze–thaw (F–T) cycles. Under low-temperature conditions, the pavement is prone to thermal cracking. Under loading and F–T cycles, cracks will develop rapidly, forming pavement distresses and damage, such as stripping, potholes, and surface deterioration, seriously effecting the service life of the road. To reduce the effect of F–T cycles on the low-temperature property of the asphalt mixture, more addition can be used to improve the low-temperature performance and frost resistance of the mixture in seasonal frozen regions.

Diatomite is a material with high porosity, high surface absorption rate, abundant resources, and low cost [1,2]. It has been widely utilized to modify asphalt, asphalt binders, and asphalt mixtures. Cheng et al. [3] demonstrated the anti-aging properties of diatomite-modified asphalt. The results indicated that diatomite is useful in improving the high-temperature stability and anti-aging properties of asphalt. Cong et al. [4,5] studied diatomite-modified asphalt binders. The results indicated that no chemical reaction occurs between asphalt and diatomite. The high-temperature properties and anti-aging properties are elevated by the addition of diatomite, while the low-temperature properties of asphalt binder are not sufficiently improved. Yang et al. [6] revealed that the addition of diatomite has a great effect on improving the high-temperature performance, fatigue performance, and moisture resistance of the asphalt mixture, but has little effect on the improvement of its low-temperature performance. However, the low-temperature performance of asphalt mixtures is very important in seasonal frozen regions, such as Jilin province in China. Therefore, it is a feasible method to deal with these distresses of diatomite-modified asphalt mixtures by adding other modifiers.

Basalt fiber (BF) is an eco-friendly mineral fiber for reinforcing the properties of asphalt mastic and asphalt mixtures that exhibits high strength, low water absorption, stable chemical characteristics, high resistance to temperature, less waste in production and no harm to the environment after abandonment [7]. Much research has investigated the effect of BF on asphalt concrete in recent years. Zheng et al. [8] evaluated the low-temperature bending property and fatigue property of basalt fiber-modified asphalt mixtures under chloride erosion and F–T cycles. The results reveal that the ultimate tensile strength and the maximum bending strain can be improved by the addition of BF. Using BF to modify asphalt mixtures can improve resistance to low temperature, fatigue, and chloride erosion and F–T action. Qin et al. [9] investigated the characterization of asphalt mastics modified by BF. The results reveal that a stable three-dimensional network would be formed in asphalt mastic and result in the improvement of the crack resistance performance by the addition of BF. Zhang et al. [10,11] studied the impact of BF distribution on the asphalt mortar through a new three-dimensional fiber distribution model. The results show that BF can effectively improve the flexural–tensile strain. Liu et al. [12] studied the variation and influence of fibers on the low-temperature performance of asphalt mastic and asphalt mixtures though the bending beam rheometer (BBR) test and low-temperature beam bending test. The results reveal that the addition of BF brings a lower creep stiffness and higher creep rate for asphalt mastic at low temperature. Gao [13] studied the performance of basalt fiber modified asphalt mixtures though low-temperature IDT test. The results show that that the addition of BF brings a higher failure strength and failure strain. Zhao [14] used the beam bending test and beam bending creep test to evaluate low-temperature performance of basalt fiber modified asphalt mixtures. The results show that the flexural strength, failure strain and creep rate were increased by adding basalt fiber, and the low-temperature cracking resistance can be improved. In summary, the low-temperature properties and crack resistance of asphalt mixtures can be improved by adding a suitable content of BF.

The effects of combination diatomite and basalt fiber on asphalt mixtures are still unclear. Cheng et al. [15] studied the properties of basalt fiber–diatomite asphalt mastic by softening point, cone penetration, viscosity, and dynamic shear rheological (DSR) tests. The results show that the high and low-temperature properties of basalt asphalt mastic were improved by the addition of basalt fiber–diatomite. Arash Davar et al. [16] studied the properties of basalt fiber–diatomite asphalt mixtures through a four-point bending beam test and indirect tensile strength test. The results show that the fatigue resistance and low-temperature properties of basalt fiber–diatomite asphalt mixtures had been improved. If the disadvantage of diatomite-modified mixtures can be solved by basalt fiber, then the basalt fiber–diatomite-modified asphalt mixture can be well-used in seasonal frozen regions.

F–T cycles have a great influence on the pavement of seasonal frozen regions. Many researchers investigated the effect of F–T cycles on asphalt mixtures through laboratory experiments, such as the indirect tensile (IDT) test, beam bending test, Marshall stability tests, fatigue tests, complex modulus test, and so on [17–21]. The engineering character of asphalt mixtures would be reduced due to the damage caused by water-freeze expansion, and the weakened bonding between mortar and aggregate under F–T cycles. The low-temperature indirect tensile test is one of the simplest tests to evaluate the low-temperature properties of asphalt mixtures [22,23], which can be adopted to analyze the modification effect of basalt fiber and diatomite on asphalt mixture under F–T cycles.

As for the effects of the F–T cycles and modifier on the change in the stress–strain curve for asphalt mixtures, few relevant studies have been published [24,25]. The statistical damage constitutive model is widely used for rock and soil under F–T and loading [26,27]. By putting forward some indexes to describe the nonlinear change of the stress–strain curve and finally establishing the damage constitutive model of basalt fiber–diatomite-modified asphalt mixtures under F–T cycles, we can describe and analysis effect of F–T cycles and modifier on the asphalt mixture at low temperature.

In this paper, a diatomite-modified asphalt mixture (DAM), basalt fiber-modified asphalt mixture (BFAM), and diatomite–basalt fiber-modified asphalt mixture (DBFAM) were studied and compared with a matrix asphalt mixture (AM). The F–T cycles were used to simulate the climatic conditions of the seasonal frozen regions. The indirect tensile test was used to determine the low-temperature performance of the asphalt mixture. The influence of the F–T cycles on IDT strength, IDT failure strain, failure stiffness modulus, and strain energy density were analyzed. The variation of the stress–strain curve under F–T cycles was analyzed. The linear zone stress ratio and linear zone strain ratio were used as indexes to evaluate the nonlinear characteristics of the stress–strain curve. A statistical damage constitutive model was established to describe the stress–strain relationship after F–T damage.

2. Materials and Methods

2.1. Raw Materials

Base asphalt AH-90, from the Panjin petrochemical industry, in Panjin City, Liaoning Province of China, was used as the binder. The physical indexes of the base asphalt are listed in Table 1.

The aggregates were basalt and the mineral filler was limestone powder. The physical properties of basalt aggregate and mineral filler are listed in Tables 2 and 3, correspondingly. The gradation was designed with 13.2 mm nominal maximum size. The selected gradation of the asphalt mixture is shown in Figure 1.

Diatomite was produced by Changchun Diatomite Products Co., Ltd., in Linjiang City, Jilin Province, China. Its physical properties are presented in Table 4.

Basalt fiber is the short fiber from Jiuxin Basalt Fiber Inc., in Jilin City, Jilin Province, China, and its properties are presented in Table 5.

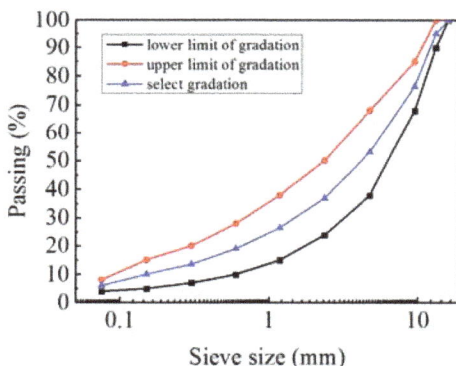

Figure 1. The Grading Curve aggregates used in this study.

Table 1. Technical parameters of tested asphalt.

Property	Value	Standard
Density (15 °C, g/cm^3)	1.018	-
Penetration (25 °C, 0.1 mm)	92.3	80~100
Softening point $T_{R\&B}$ (°C)	46.9	$\geq$42
Ductility (25 °C, cm)	>150	$\geq$100
Viscosity (135 °C, mPa·s)	306.9	-

Table 2. Properties of aggregate.

Sieve Size (mm)	13.2	9.5	4.75	2.36	1.18	0.6	0.3	0.15	0.075
Apparent density (g/cm^3)	2.811	2.805	2.815	2.817	2.808	2.805	2.778	2.777	2.768
Absorption coefficient of water (%)	0.33	0.44	0.54	0.75	-	-	-	-	-

Table 3. Physical properties of mineral powder.

Property	Hydrophilic Coefficient	Apparent Density (g/cm^3)	Gradation	
			Sieve Size (mm)	Passing (%)
Value	0.778	2.722	0.6	100
			0.15	95
			0.075	80

Table 4. Properties of diatomite.

Property	Color	Bulk Density	Specific Gravity	pH
Value	White	0.38 g/cm^3	2.1 g/cm^3	7

Table 5. Properties of basalt fiber (provided by manufacturer).

Items	Value	Standard Value
Diameter (μm)	10–13	-
Length (mm)	6	-
Water content (%)	0.030	≤0.2

2.2. Specimen Preparation

Cylindrical specimens (101.6 ± 0.2 mm diameter by 63.5 mm ± 1.3 mm high) for each mixture were prepared using a Marshall compactor. The specimens were compacted in a Marshall Compactor with 75 beats on each side. All of them were prepared with the same gradation at optimum asphalt content (OAC). The OAC of the asphalt mixture was obtained by applying the Marshall Methods [28]. Three identical samples were used to investigate low-temperature cracks under F–T cycles in asphalt mixes modified with basalt fibers and diatomite powder. According to previous research results [29], the proportions of the diatomite–basalt compound-modified asphalt mixture and optimum asphalt content are determined as follows:

- The control asphalt mixture sample (with no additives) (AM) has an OAC of 4.78%.
- The diatomite (corresponding to the volume ratio of the diatomite to entire filler is 6.5%) modified asphalt mixture (DAM) has an OAC of 5.12%.
- The basalt fiber (0.25% by weight of the asphalt mixture) modified asphalt mixture (BFAM) has an OAC of 5.09%.
- The basalt fiber (0.25%) and diatomite (6.5%) compound modified asphalt mixture (DBFAM) has an OAC of 5.22%.

The preparation procedures of DBFAM are as follows. Aggregates were mixed in the mixing oven at 160 °C. The basalt fiber was added with aggregates; they were mixed for about 60 s to disperse the basalt fiber evenly in the aggregates. Then, the asphalt was added, and they were mixed for about 90 s to make the aggregate surface uniformly coated by asphalt. Then, the mineral filler and diatomite was added and the diatomite and basalt fiber compound modified asphalt mixture was obtained after a second mix of 90 s. Except for the addition of modified materials, the preparation procedures of AM, BFAM, DAM are similar to that of DBFAM. The mixing temperature and the mixing time of each step are the same as that of DBFAM.

2.3. Process of Freeze–Thaw

Before the test, every specimen was immersed into water and under a vacuum 98.0 kPa for 15 min and soaked in atmospheric pressure for 30 min. Each cycle consisted of freezing at −18 °C for 16 h, followed by soaking in water at 60 °C for 8 h. Before freezing, the specimen was placed into a plastic bag with 10 mL water. Before thawing, the specimens were removed from the plastic bags. After 0, 3, 6, 9, 12, and 15 F–T cycles, the low-temperature indirect tensile test was carried out.

2.4. Low-Temperature Indirect Tensile Test

The low-temperature tensile property is a significant index to evaluate the crack resistance for asphalt mixture pavement, which is often investigated using the three-point bending method and indirect tensile method [30,31]. Guo et al. regarded that, while using the bending test to evaluate the crack resistance of diatomite–glass fiber-modified asphalt mixture, the effect of the modifier could not be well reflected because there were fewer fibers and less diatomite at the bottom of the middle span [1]. Therefore, the low-temperature indirect tensile test was used to evaluate the crack resistance of the DBFAM.

The specimens after the process of freeze–thaw were tested at a temperature of −10 °C with an environment box. Before the test, the specimens were placed in a chamber at −10 °C for 5 h to reach thermal equilibrium. The constant rate was 1 mm/min [28]. During the test, the vertical deformation on the top surface of the specimen and the load could be recorded by computer.

The IDT strength, R_T, the IDT failure strain ε_T, and the failure stiffness modulus S_T can be calculated by following equations [28]:

$$R_T = 0.006287 P_T / h \tag{1}$$

$$\varepsilon_T = X_T \times (0.0307 + 0.0936\mu)/(1.35 + 5\mu) \tag{2}$$

$$S_T = P_T \times (0.27 + 1.0\mu)/(h \times X_T) \tag{3}$$

$$X_T = Y_T \times (0.135 + 0.5\mu)/(1.794 - 0.0314\mu) \tag{4}$$

where R_T is the indirect tensile strength, MPa; ε_T is the tensile failure strain; S_T is the failure stiffness modulus, MPa; P_T is the indirect tensile failure load, N; Y_T is the vertical deformation, mm; X_T is the horizontal deformation, mm; μ is the Poisson ratio, which is 0.25 in this test; and h is the height of specimen, mm.

However, the conclusions obtained from R_T, ε_T, and S_T may be inconsistent sometimes. Thus, the deformation energy index was selected to evaluate the low-temperature performance as a comprehensive index [6]. The process of low-temperature cracking for the asphalt mixture is a process of dissipation. The higher is the deformation energy, the stronger is the crack resistance at low temperature.

The load–displacement curve is modified in Chinese Standard Specification (JTG E20-2011), which extends the straight line segment and takes the intersection with the abscissa as the origin of the curve, as shown in Figure 2. The deformation energy can be defined as the area of the modified stress–strain curve. The deformation energy density can be calculated by Equation (5):

$$Q_B = \frac{1}{h} \int_{Y_{T1}}^{Y_{T2}} P_T(y) dy \tag{5}$$

where Q_B is the deformation energy density, N·m; Y_{T1} the origin displacement, mm; and Y_{T2} the critical displacement, mm.

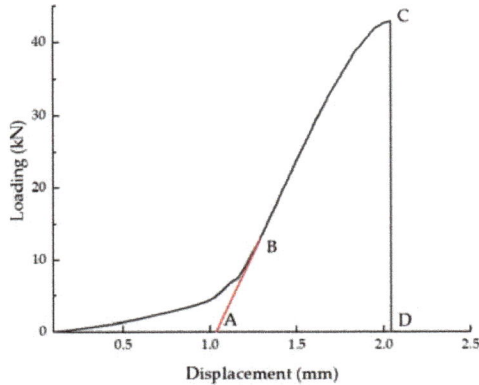

Figure 2. Load–deformation curve.

Those indexes were calculated for different asphalt mixtures under F–T cycles to analyze the change of the low-temperature performance under F–T cycles. To evaluate the F–T resistance of different asphalt mixtures, the loss ratio of each index before and after F–T cycles is given by Equation (6):

$$\Delta I_i = \left(1 - \frac{I_i}{I_0}\right) \times 100 \tag{6}$$

where ΔI_i is the loss ratio of each index after i times of F–T cycles, I_0 is without F–T cycles, and I_i is each index after i times of F–T cycles. If the index is increased under F–T cycles, then the change ratio of it can be used as $-\Delta I_i$.

2.5. The Nonlinear Evaluation Indexes of Stress–Strain Curves

The stress–strain curve was calculated from the load–displacement curve collected by the testing machine. The stress was calculated as the R_T in Equation (1). The strain was calculated as the ε_T in Equation (2). The stress–strain curves of the asphalt mixture will change after F–T cycles. In the previous study, the researchers only focused on the failure strength of the asphalt mixture under the F–T cycles [18,21], but there was no description and evaluation index of the nonlinear change of the stress–strain curve after the F–T cycles. The stress–strain curves were calculated for AM under F–T cycles, as shown in Figure 3. As can be seen in Figure 3, the stress–strain curve of the asphalt mixture exhibits nonlinear softening characteristics under the action of F–T cycles. Asphalt mixtures are no longer similar to elastomers at low temperature under F–T cycles.

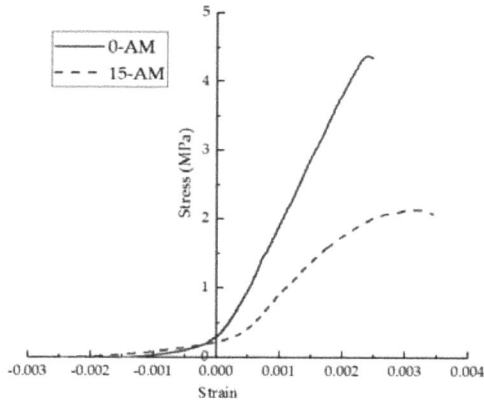

Figure 3. Stress–strain curve under freeze–thaw (F–T) cycles.

Therefore, the stress–strain curve obtained by the low-temperature indirect tensile test of the asphalt mixture after F–T cycles is divided into a linear zone and a nonlinear zone in this paper. The linear fitting is performed through the straight line segment in the stress–strain curve to ensure that R^2 is more than 0.998, to determine the linear zone slope and boundary point location. When R^2 does not meet the accuracy requirement, it iteratively reduces the boundary of the zone and solves the end point of the straight segment, as shown in Figure 4. The specific steps are as follow:

Step 1: Set the left initial border of the line segment. This paper focuses on the end point of the straight line segment, so we can take the left end of the fitted area on the straight line segment. To simplify the calculation, the left end point of the fitted area is guaranteed to be on the straight line segment. Therefore, the left end point of the fitting zone is taken as the corresponding point of the 0.3 times stress peak, set $\varepsilon = \varepsilon_0$.

Step 2: Set the right initial border of the fitted area. The left end point of the fitting zone is taken as the corresponding point of the stress peak, set $\varepsilon = \varepsilon_1$.

Step 3: Determine whether the curve in the interval $[\varepsilon_0, \varepsilon_i]$ (i = 1, 2, 3 … n) is a straight line. The curve in the interval $[\varepsilon_0, \varepsilon_i]$ is linearly fitted by the least squares method. Determine whether the R^2 is large enough. If $R^2 > 0.998$, the interval can be seen as a linear zone, and Step 5 will be executed. Otherwise, the interval is still obtained as the nonlinear zone, so Step 4 will be executed.

Step 4: Reduce the right border of the fitted area. Calculate the biggest right intersection point between fitting straight line and curve, set $\varepsilon=\varepsilon_i$. Go back to Step 3.

Step 5: Output right border of the fitted area. Set $\varepsilon_i = \varepsilon_L$. Set stress as R_L.

The linear zone stress ratio and the linear zone strain ratio are defined as indexes for describing the nonlinear problem under the F–T cycles. Therefore, the influence of different additives in asphalt mixtures on the two indexes under F–T cycles can be analyzed. The variation of the stress–strain curve under F–T cycles is analyzed. The linear zone stress ratio and linear zone strain ratio are used as indexes to evaluate the nonlinear characteristics of stress–strain curves.

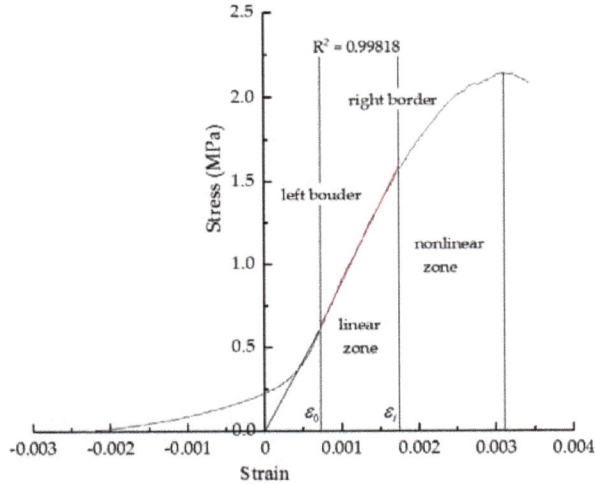

Figure 4. Linear zone and nonlinear zone in stress–strain curve.

The linear zone stress ratio (RR_L) is the ratio of the linear zone stress to R_T, which can be calculated by Equation (7). The linear zone strain ratio ($R\varepsilon_L$) is the ratio of the linear zone strain to ε_T, which can be calculated by Equation (8). These two indexes are proposed to describe the end point of the linear zone of the stress–strain curve.

$$RR_L = R_L/R_T \tag{7}$$

$$R\varepsilon_L = \varepsilon_L/\varepsilon_T \tag{8}$$

Through these two indexes, the full-stage stress–strain curve and mechanical properties of the asphalt mixture after F–T cycles can be better described.

The elastic stiffness modulus of linear zone can be obtained by using the two nonlinear indictors. The elastic stiffness modulus can exclude the influence of the nonlinear zone in the loading process; in other words, the modulus damage introduced by the loading damage and the coupling effect of loading and F–T cycles damage are eliminated. The loss ratio of elastic stiffness modulus is calculated as the damage only caused by F–T cycles.

$$S_{nL} = k \cdot E_n = k \cdot \frac{\sigma_L}{\varepsilon_L} = \frac{R\varepsilon_L}{RR_L} \cdot S_T \tag{9}$$

$$D_{nL} = 1 - S_{nL}/S_{0L} = 1 - E_n/E_0 \tag{10}$$

where S_{nL} is the elastic stiffness modulus of linear zone. E_n is the slope of the stress–strain curve. k is calculated using Equations (1)–(4) as 1.721.

2.6. Damage Constitutive Model of the Asphalt Mixture under F–T Cycles

F–T cycles not only decrease the mechanical properties of the asphalt mixture, but also cause some internal damage, which causes a change to the constitutive relation of the asphalt mixture subjected to loading. Through the damage analysis of the different kinds of asphalt mixtures, we can establish the damage constitutive model, which reflects the deformation characteristics of asphalt mixtures subject to F–T cycles.

The statistical damage constitutive model is widely used for rock and soil under F–T cycles and loading [26,27]. According to Lemaitre's stress equivalence principle (1985), an equivalent relation can describe the constitutive relation between effective stress and strain of the damaged material and undamaged one.

$$\sigma = \sigma'(1 - D) = E\varepsilon(1 - D) \tag{11}$$

The constitutive relationship of the asphalt mixture under F–T cycles at low temperatures can be described as follows:

$$\sigma = \begin{cases} E_n \cdot \varepsilon & , \quad (0 < \varepsilon \leq \varepsilon_L) \\ E_n(1 - D_t) \cdot \varepsilon & , \quad (\varepsilon_L < \varepsilon) \end{cases} \tag{12}$$

where E_n is the slope of straight line of linear zone. D_t is defined to describe the damage caused by loading.

On this basis, the following assumptions are made for asphalt mixture at low temperature:

- The asphalt mixture at low temperature accords with the generalized Hooke's law.
- The strength of micro-unit conforms to the statistical law of the three-parameter Weibull:

$$\varphi(\varepsilon) = \frac{n}{m}(\varepsilon - \gamma)^{n-1}e^{[-\frac{(\varepsilon - \gamma)^n}{m}]} \tag{13}$$

where ε is the strain of the asphalt mixture and n and m are the physical and mechanical parameters of the asphalt mixture under external load, respectively. γ is the threshold parameter. Considering that the nonlinearity is caused by damage, the threshold is taken as ε_L.

The damage of the asphalt mixture at low temperature is caused by the uneven failure of the mixture micro-units. D_t can be defined as the statistical damage variable, as the ratio of damaged micro-units to all the micro-units. Then, we can obtain the loading damage variable as follows:

$$D_t = \int_{\gamma}^{\varepsilon} \varphi(x)dx = \int_{\gamma}^{\varepsilon} \frac{n}{m}(x - \gamma)^{n-1}e^{[-\frac{(x-\gamma)^n}{m}]}dx = 1 - e^{[-\frac{(\varepsilon-\varepsilon_L)^n}{m}]} \tag{14}$$

Substituting Equations (10) and (14) with Equation (12), the constitutive equation of the asphalt mixture at low temperature after F–T cycles under loading is as follows:

$$\sigma = \begin{cases} (1 - D_{nL})E_0\varepsilon & , \quad (0 < \varepsilon \leq \varepsilon_L) \\ (1 - D_{nL})E_0 e^{[-\frac{(\varepsilon-\varepsilon_L)^n}{m}]}\varepsilon & , \quad (\varepsilon_L < \varepsilon) \end{cases} \tag{15}$$

3. Results and Discussion

3.1. IDT Strength

IDT strength can reflect the failure peak stress of the asphalt mixture at low temperature. In Figure 5, we can see the IDT strength for the AM, BFAM, DAM, and DBFAM asphalt mixtures after F–T Cycles 0, 3, 6, 9, 12, 15. The loss ratio of IDT strength is shown in Figure 6.

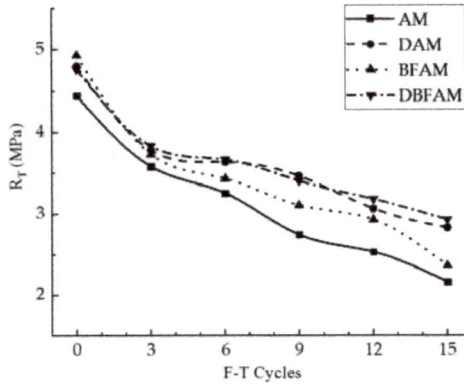

Figure 5. Indirect tensile (IDT) strength under F–T cycles. AM, Matrix asphalt mixture; DAM, diatomite-modified asphalt mixture; BFAM, basalt fiber-modified asphalt mixture; DBFAM, diatomite–basalt fiber-modified asphalt mixture.

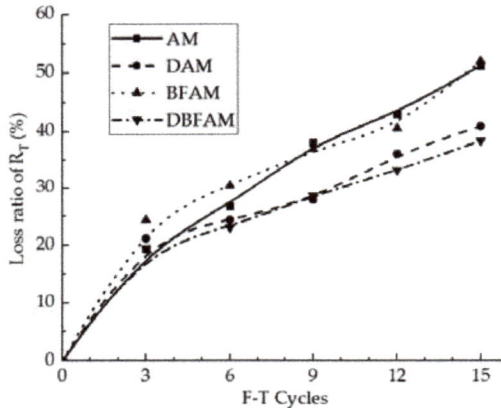

Figure 6. Loss ratio of IDT strength under F–T cycles.

As can be seen in Figures 5 and 6, the IDT strength for each kind of asphalt mixture decreases and the loss ratio of IDT strength increases with the progress of F–T cycles. As the F–T cycles increase, the strength loss increases and the rate decreases more slowly. This indicates that the F–T cycle has a significant effect on the mechanical properties of the asphalt mixture in the early stage, and tends to change slowly after six F–T cycles. The reason lies in that the F–T actions will increase the air voids and cause micro-crack formation and propagation in the asphalt mixture. This is because F–T cycles weaken the bonding between binder and stone.

Meanwhile, the strengths of DAM, BFAM, and DBFAM are all higher than that of AM under F–T cycles. The loss of strength is significantly decreasing in DAM and DBFAM, while not significantly decreasing in BFAM compared with in AM. This indicates that adding basalt fiber can improve the strength and the diatomite can improved the resistance to F–T cycles. The influence mechanism basalt fiber on strength may be due to its reinforcing, toughening and preventing cracks and redistribution of stress in asphalt mixture. The diatomite modified asphalt can also improve the strength, which may be caused by the hardening effects of adsorption between diatomite and asphalt. Diatomite is a porous material with a large specific surface area. After it is added to the asphalt mixture, the bonding ability of the asphalt mortar is improved due to the hardening effects of adsorption between diatomite and asphalt, and the water is harder to invade the bonding place between aggregate and asphalt binder. Above all, composite-adding the two modified materials can improve both low-temperature performance and resistance to F–T cycles.

3.2. IDT Failure Strain

The IDT failure strain and the change ratio of it were calculated for different specimens under F–T cycles, as shown in Figures 7 and 8.

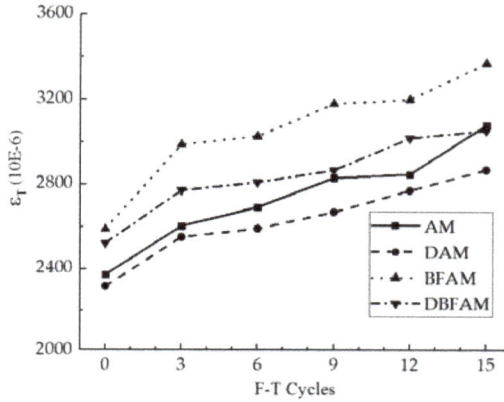

Figure 7. IDT failure strain under F–T cycles.

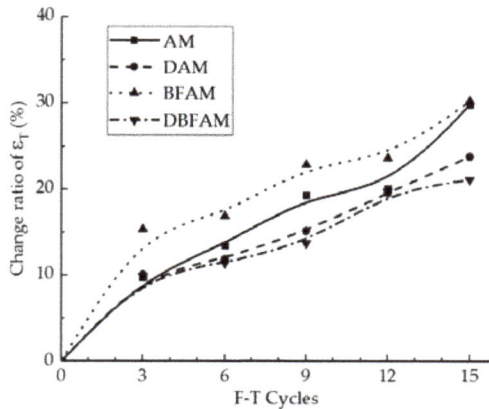

Figure 8. Change ratio of IDT failure strain under F–T cycles.

As can be seen in Figures 7 and 8, the results of IDT failure strain on each kind of asphalt mixture increase and the growth ratio also increases under F–T cycles. As the F–T cycles increase, the failure strain increases and the change rate become slow. This indicates that the F–T cycles have a significant effect on the deformation properties of the asphalt mixture in the early stage, and tend to be changed slowly after six F–T cycles.

Before the F–T cycles, the failure strains of BFAM and DBFAM are higher than that of AM, while that of DAM is less than that of AM. This indicates that adding basalt fiber can improve the failure strain, while adding diatomite is not significant. According to previous research, the failure strain would decrease through the addition of diatomite [1,16], and increase through the addition of basalt fiber [32]. The result of DBFAM reflects that adding basalt fiber can solve the problem of diatomite-modified asphalt mixture at low temperature.

Under F–T cycles, the failure strain of BFAM increases remarkably compared with that of others. The reason is that water may be easier to invade in the combination of mortar and stone through fiber in BFAM under F–T cycles. As shown in the failure strain of DAM and DBFAM, adding diatomite can solve the problem in F–T cycles by enhancing adhesion and water stability.

Above all, composite-adding the two modified materials is suitable for asphalt mixtures at low temperature under F–T cycles. It can give full play to the advantages of the two materials, and can also make up for their shortcomings.

3.3. Failure Stiffness Modulus

The failure stiffness modulus and the loss ratio of it were calculated for different specimens under F–T cycles, as shown in Figures 9 and 10.

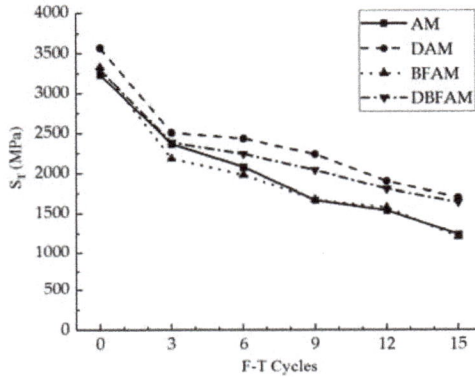

Figure 9. Failure stiffness modulus under F–T cycles.

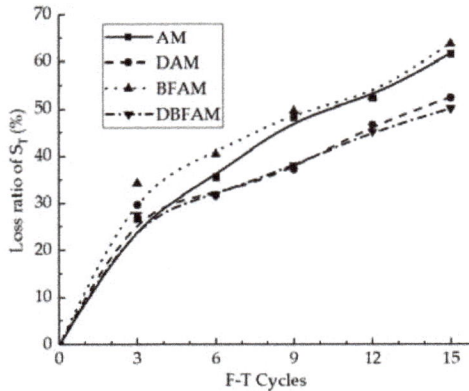

Figure 10. Loss ratio of failure stiffness modulus under F–T cycles.

As can be seen in Figures 9 and 10, the IDT stiffness modulus for each kind of asphalt mixture decreases while the loss ratio increases under F–T cycles. The stiffness modulus of each kind of sample before F–T cycles is similar. The effect of F–T cycles on the stiffness modulus is greater than the modified materials.

3.4. Deformation Energy Density

The deformation energy density and the loss ratio of it were calculated for different specimens under F–T cycles, as shown in Figures 11 and 12.

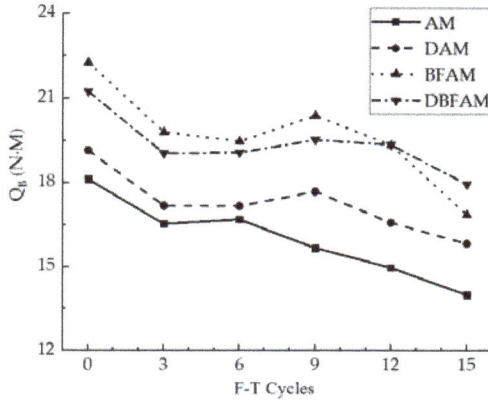

Figure 11. Deformation energy density under F–T cycles.

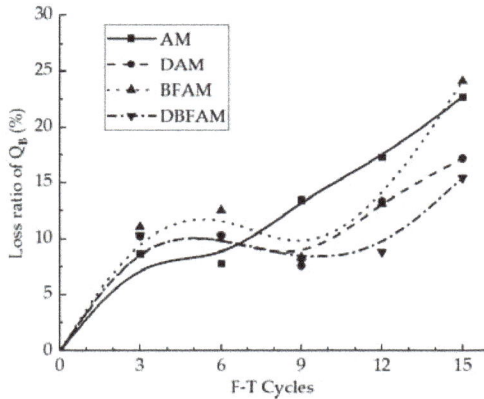

Figure 12. Loss ratio of deformation energy density under F–T cycles.

As can be seen in Figures 11 and 12, the deformation energy density for each kind of asphalt mixture decreases after the F–T cycles. The deformation energy densities of BFAM and DBFAM are higher than that of the AM group. The BFAM has the highest low-temperature crack resistance, while the DBFAM has the second highest resistance. What is more, the loss ratio of the deformation energy density of DAM and DBFAM is smaller than that of the AM group after nine F–T cycles. The DBFAM has the highest F–T cycles resistance while the DAM has the second highest resistance. The deformation energy density result reflects that the compound-modified asphalt mixture can solve the disadvantage of DAM and BFAM, so the DBFAM can be well used in seasonal frozen regions.

Above all, from the indexes changed under F–T cycles, we can see that the low temperature performance after 15 F–T cycles was significant improved by adding both diatomite and basalt fiber. The deformation energy density of DBFAM are 29% higher than AM after 15 F–T cycles, which means the DBFAM pavement can present longer service life in seasonal frozen regions. This is important for reducing the road maintenance and traffic disruption, and the cost expected for AM, BFAM, DAM,

DBFAM are 0.330, 0.364, 0.345 and 0.372 CNY/kg, respectively. It is still important in seasonal frozen regions of China although the cost of modified asphalt mixture has increased.

3.5. Stress–Strain Curves

The stress–strain curves were calculated for different specimens under 0 and 15 F–T cycles, as shown in Figure 13.

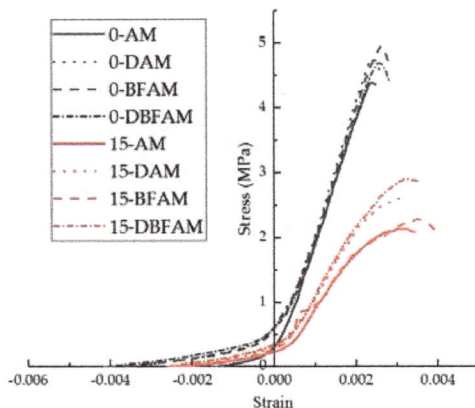

Figure 13. Stress–strain curve of IDT test before and after F–T cycles.

As can be seen in Figure 13, the stress–strain curve of the asphalt mixture exhibits softening characteristics under the action of F–T cycles. Before the F–T cycles, the IDT test of the asphalt mixture at low temperature is similar to the brittle failure. Once the specimen reaches the failure strain, the bearing capacity is quickly lost. The stress–strain curve is almost linear before the peak point.

After the F–T cycles, the brittleness of the specimens of both groups obviously decreases. The stress–strain curve is nonlinear before the peak point and the nonlinear zone is more significant under F–T cycles. After the peak, it does not completely lose its bearing capacity.

The curve of DAM and DBFAM is above that of AM and BFAM, and their nonlinear zone is less significant than AM and BFAM, which reflects that adding diatomite is effective in enhancing the resistance performance of F–T cycles for asphalt mixtures.

3.6. Linear Zone Stress Ratio and Linear Zone Strain Ratio Result

The peak stress and strain and slope of the stress–strain curve can be described by the strength, strain, and stiffness modulus for the IDT test at low temperature. It seems to treat the samples as elastomers at low temperature. However, it is difficult to reflect the softening characteristics of IDT test at low temperature after F–T cycles. The linear zone stress ratio and linear zone strain ratio are suitable indexes to inflect the nonlinear change of the stress–strain curves. The linear zone stress ratio and linear zone strain ratio were calculated for different specimens under F–T cycles, as shown in Figures 14 and 15.

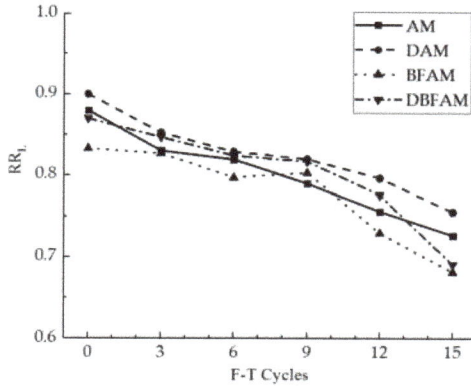

Figure 14. Linear zone stress ratio under F–T cycles.

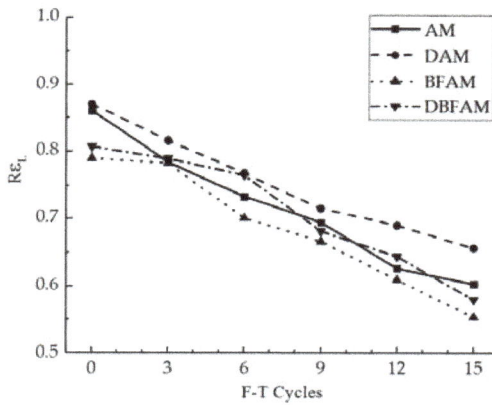

Figure 15. Linear zone strain ratio under F–T cycles.

As can be seen in Figures 14 and 15, in the progress of the F–T cycles, the linear zone stress ratio and linear zone strain ratio decrease for each kind of asphalt mixture. The decreasing trend of the linear zone strain ratio is faster than that of linear zone stress ratio. As the F–T cycle progresses, the range of the nonlinear zone gradually increases in the stress–strain curve, and the nonlinear characteristics of the stress–strain are more significant. The linear zone can reflect the elastic phase of the asphalt mixture at low temperature, while the nonlinear zone reflects the damage stage of crack propagation for the asphalt mixture at low temperature. The initial crack gradually cracks, develops, and penetrates until the specimen reaches the critical cracking point. Under the F–T cycles, the internal crack and air void will increase because of the ice volume expansion [17,33]. The bond force between the asphalt membrane and aggregates declines under the F–T cycles. Thus, it is more difficult to provide a constraint to maintain a linear relationship for the curve in high stress conditions after the F–T cycles, the relative slip happens between the aggregates, and the stress–strain curve changes into the nonlinear zone.

Before the F–T cycles, the linear zone stress ratio and linear zone strain ratio of DAM are the highest, while those of BFAM are the lowest. This indicates that adding basalt fiber can play an anchoring role in the relative slip stage in the failure strain, while adding diatomite is not significant. The IDT failure strain would decrease by adding diatomite and increase by adding basalt fiber.

The result for DBFAM reflects that adding basalt fiber can solve the problem of adding diatomite for the asphalt mixture at low temperature by playing the anchoring role in the relative slip stage.

3.7. Elastic Stiffness Modulus of Linear Zone Result

The elastic stiffness modulus of the linear zone and the loss ratio of it were calculated for different specimens under the F–T cycles, as shown in Figures 16 and 17. The loss ratio of the elastic stiffness modulus is calculated as the damage only caused by the F–T cycles.

As can be seen in Figures 16 and 17, it is similar to the reason for the IDT stiffness modulus in Figures 9 and 10. The elastic stiffness modulus of the linear zone is slightly higher than the IDT stiffness modulus and its loss ratio is also slightly lower than the loss of the IDT stiffness modulus for each kind of asphalt mixture. The loss of the elastic stiffness modulus of the linear zone, which reflects the damage only caused by the F–T cycles, is smaller and smaller than that of the IDT stiffness modulus with the increase of the F–T cycles.

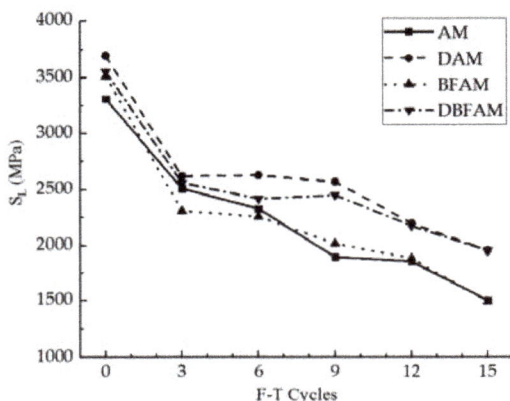

Figure 16. Elastic stiffness modulus of linear zone under F–T cycles.

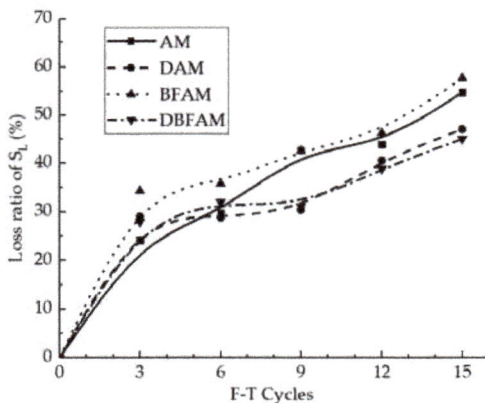

Figure 17. Loss of elastic stiffness modulus of linear zone under F–T cycles.

3.8. Experiment Verification of Constitutive Model

The D_t values in the nonlinear zone can be calculated according to Equation (12). The D_t values curves of the nonlinear zone for each group of asphalt mixture before and after 15 F–T cycles are shown in Figures 18 and 19, respectively.

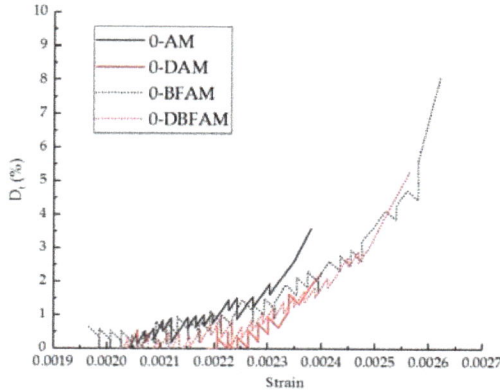

Figure 18. D_t values in the nonlinear zone before F–T cycles.

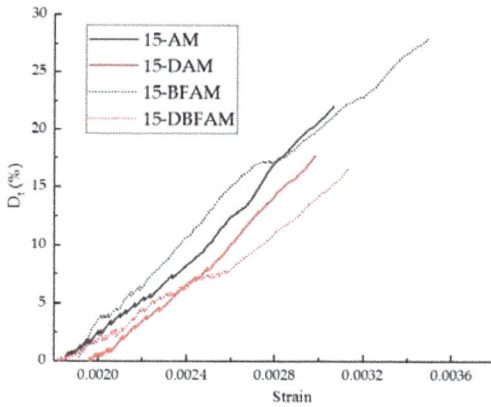

Figure 19. D_t values in the nonlinear zone after 15 F–T cycles.

According to the analysis of the damage curve, the difference of the damage degree curve is obvious before and after the 15 F–T cycles, the damage initiation stage is moved forward, and the damage degree is also obviously increased. The nonlinear zone before the peak increases significantly, which reflects the softening effect of the F–T cycle, and the nonlinearity after the F–T cycles is more significant. After the basalt fiber is added, the nonlinear zone also has obvious growth, which reflects the toughening and crack-preventing effect of the fiber. This is consistent with the effect of the two nonlinear indicators, the linear zone stress ratio and the strain ratio.

From Equation (14), the parameters of the model can be fitted, and the results are shown in Table 6. In Table 6, the effect of the F–T cycles on parameters m and n is obviously higher than that of modification. The F–T cycle makes the parameter m decrease and n increase. The R^2 indicates that the statistical damage constitutive model established by Equation (15) can better describe the stress–strain relationship after F–T damage. Comparisons of the experimental results and statistical damage constitutive model predictions of different asphalt mixture are shown in Figures 20–23.

The results show that the statistical damage constitutive model is suitable for different asphalt mixtures at low temperature.

Table 6. The parameters of the model.

F–T	Parameters	AM	BFAM	DAM	BFDAM
	m	2.47	3.54	2.48	3.36
0	n	8.13×10^{-4}	9.33×10^{-4}	5.93×10^{-4}	9.09×10^{-4}
	R^2	0.935	0.983	0.997	0.999
	m	1.33	1.2	1.27	1.28
15	n	2.73×10^{-3}	3.50×10^{-3}	3.00×10^{-3}	3.69×10^{-3}
	R^2	0.995	0.992	0.997	0.982

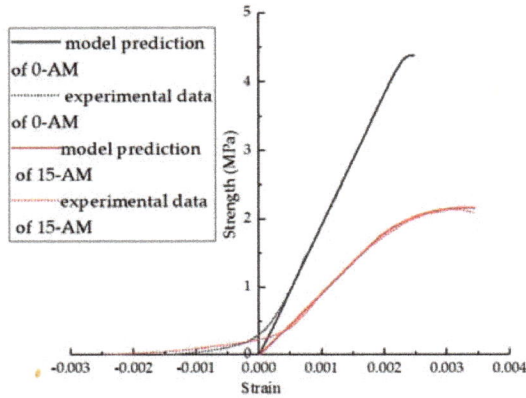

Figure 20. Comparison of experimental data and model predictions for AM under F–T cycles.

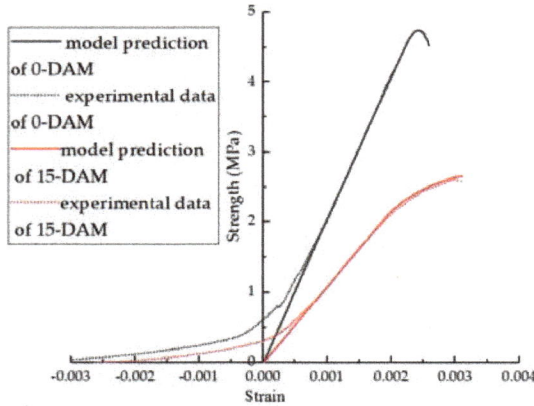

Figure 21. Comparison of experimental data and model predictions for DAM under F–T cycles.

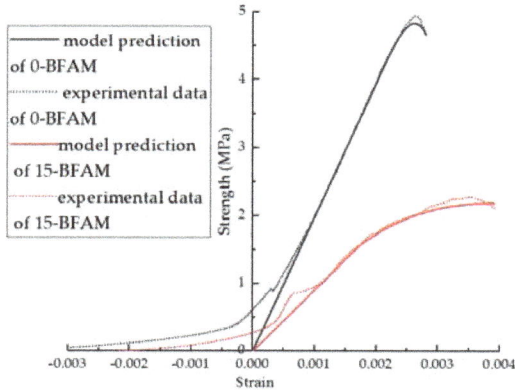

Figure 22. Comparison of experimental data and model predictions for BFAM under F–T cycles.

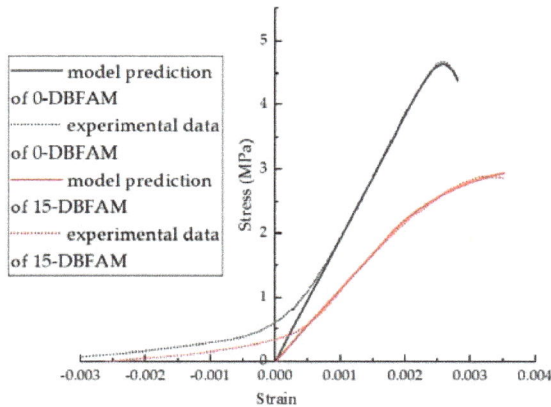

Figure 23. Comparison of experimental data and model predictions for DBFAM under F–T cycles.

4. Conclusions

In this research, the low-temperature indirect tensile test was used to investigate the variation of low-temperature performance of the asphalt mixture under F–T cycles. The influence of adding basalt fiber and diatomite on the asphalt mixture for the low-temperature properties and resistance of F–T cycles was analyzed by many indexes. We described the variation of the stress–strain curve under F–T cycles and proposed a statistical damage constitutive model under F–T cycles and loading for asphalt mixtures. From this research, the following conclusions can be drawn:

- F–T cycles will reduce the IDT strength, stiffness modulus and strain energy density of asphalt mixture, and increase the IDT failure strain. As the number of F–T cycles increases, the variation of each index decreases.
- After adding basalt fiber, the IDT strength, IDT failure strain, and strain energy density of the mixture are improved, and the low-temperature performance is improved. After adding diatomite, the loss ratio of each index decreases under F–T cycles, and the resistance of F–T cycles is improved. The low-temperature performance and resistance of F–T cycles for asphalt mixture are improved after compound modified by basalt fiber and diatomite. The eco–friendly basalt fiber–diatomite-modified asphalt mixture is suitable in seasonal frozen regions.

- The variation law and form of the stress–strain curve before and after the F–T cycle are proposed. The stress–strain curve is divided into linear zone and nonlinear zone. Under the action of the F–T cycles, the stress ratio in the linear zone is gradually reduced, the strain ratio in the nonlinear zone is gradually increased, and the nonlinear characteristics of the stress–strain are more significant. After the addition of basalt fiber, the nonlinear zone increases significantly, which reflects the reinforcement of the fiber during the cracking stage. The nonlinear point of the stress–strain curve can be described by the nonlinear index, and the mechanism of freezing and thawing and the mechanism of material modification are better reflected.
- The statistical damage constitutive model established in this paper can describe the stress–strain relationship of the asphalt mixture at low temperature after F–T cycles well.

Author Contributions: Conceptualization, Y.C. and D.Y.; Methodology, D.Y. and G.T.; Validation, Y.C.; Formal Analysis, G.T. and C.Z.; Investigation, D.Y.; Writing—Original Draft Preparation, D.Y.; Writing—Review and Editing, Y.C., G.T. and C.Z.; Project Administration, Y.C.; and Funding Acquisition, Y.C. and G.T.

Funding: This research was funded by National Natural Science Foundation of China (grant number 51678271), Science Technology Development Program of Jilin Province (grant number 20160204008SF), and Transportation Science & Technology Project of Jilin Province (grant number 2015-1-13).

Acknowledgments: The authors gratefully acknowledge Tao, J.L. in Jiangxi Traffic Science Research Institute for his technical support. In addition, Di Yu especially wish to thank Louis Cha (Jin Yong) for the novels that given him powerful spiritual support over the years.

Conflicts of Interest: The authors declare no conflict of interest.

References

1. Guo, Q.L.; Li, L.L.; Cheng, Y.C.; Jiao, Y.B.; Xu, C. Laboratory evaluation on performance of diatomite and glass fiber compound modified asphalt mixture. *Mater. Des.* **2015**, *66*, 51–59. [CrossRef]
2. Song, Y.; Che, J.; Zhang, Y. The interacting rule of diatomite and asphalt groups. *Pet. Sci. Technol.* **2011**, *29*, 254–259. [CrossRef]
3. Cheng, Y.C.; Tao, J.L.; Jiao, Y.B.; Guo, Q.L.; Li, C. Influence of Diatomite and Mineral Powder on Thermal Oxidative Ageing Properties of Asphalt. *Adv. Mater. Sci. Eng.* **2015**, *2015*, 1–10. [CrossRef]
4. Cong, P.L.; Liu, N.; Tian, Y.; Zhang, Y.H. Effects of long-term aging on the properties of asphalt binder containing diatoms. *Constr. Build. Mater.* **2016**, *123*, 534–540. [CrossRef]
5. Cong, P.L.; Chen, S.F.; Chen, H.X. Effects of diatomite on the properties of asphalt binder. *Constr. Build. Mater.* **2012**, *30*, 495–499. [CrossRef]
6. Yang, C.; Xie, J.; Zhou, X.; Liu, Q.; Pang, L. Performance Evaluation and Improving Mechanisms of Diatomite-Modified Asphalt Mixture. *Materials* **2018**, *11*, 686. [CrossRef] [PubMed]
7. Wang, W.S.; Cheng, Y.C.; Tan, G.J. Design Optimization of SBS-Modified Asphalt Mixture Reinforced with Eco-Friendly Basalt Fiber Based on Response Surface Methodology. *Materials* **2018**, *11*, 1311. [CrossRef] [PubMed]
8. Zheng, Y.X.; Cai, Y.C.; Zhang, G.H.; Fang, H.Y. Fatigue Property of Basalt Fiber-Modified Asphalt Mixture under Complicated Environment. *J. Wuhan Univ. Technol.* **2014**, *5*, 996–1004. [CrossRef]
9. Qin, X.; Shen, A.Q.; Guo, Y.C.; Li, Z.N.; Lv, Z.H. Characterization of asphalt mastics reinforced with basalt fibers. *Constr. Build. Mater.* **2018**, *159*, 508–516. [CrossRef]
10. Zhang, X.Y.; Gu, X.Y.; Lv, J.X.; Zhu, Z.K.; Zou, X.Y. Numerical analysis of the rheological behaviors of basalt fiber reinforced asphalt mortar using ABAQUS. *Constr. Build. Mater.* **2017**, *157*, 392–401. [CrossRef]
11. Zhang, X.; Gu, X.Y.; Lv, J.X.; Zou, X.Y. 3D numerical model to investigate the rheological properties of basalt fiber reinforced asphalt-like materials. *Constr. Build. Mater.* **2017**, *138*, 185–194. [CrossRef]
12. Liu, K.; Zhang, W.H.; Wang, F. Research on cryogenic properties of different fiber asphalts and mixtures. *Adv. Mater. Res.* **2011**, *146–147*, 238–242. [CrossRef]
13. Gao, C.M. Microcosmic Analysis and Performance Research of Basalt Fiber Asphalt Concrete. Ph.D. Thesis, Jilin University, Changchun, China, 2012.
14. Zhao, L.H. Study on the Influence Mechanism of Basalt Fiber on Asphalt Mixture Property. Ph.D. Thesis, Dalian University of Technology, Dalian, China, 2013.

15. Cheng, Y.C.; Zhu, C.F.; Tan, G.J.; Lv, Z.H.; Yang, J.S.; Ma, J.S. Laboratory study on properties of diatomite and basalt fiber compound modified asphalt mastic. *Adv. Mater. Sci. Eng.* **2017**, *3*, 1–10. [CrossRef]

16. Davar, A.; Tanzadeh, J.; Fadaee, O. Experimental evaluation of the basalt fibers and diatomite powder compound on enhanced fatigue life and tensile strength of hot mix asphalt at low temperatures. *Constr. Build. Mater.* **2017**, *153*, 238–246. [CrossRef]

17. Özgan, E.; Serin, S. Investigation of certain engineering characteristics of asphalt concrete exposed to freeze–thaw cycles. *Cold Reg. Sci. Technol.* **2013**, *85*, 131–136. [CrossRef]

18. Yan, K.Z.; Ge, D.D.; You, L.Y.; Wang, X.L. Laboratory investigation of the characteristics of SMA mixtures under freeze–thaw cycles. *Cold Reg. Sci. Technol.* **2015**, *119*, 68–74. [CrossRef]

19. Lachance-Tremblay, É.; Perraton, D.; Vaillancourt, M.; Benedetto, H. Degradation of asphalt mixtures with glass aggregates subjected to freeze-thaw cycles. *Cold Reg. Sci. Technol.* **2017**, *141*, 8–15. [CrossRef]

20. Gong, X.B.; Romero, P.; Dong, Z.J.; Sudbury, D. The effect of freeze–thaw cycle on the low-temperature properties of asphalt fine aggregate matrix utilizing bending beam rheometer. *Cold Reg. Sci. Technol.* **2016**, *125*, 101–107. [CrossRef]

21. Islam, M.; Asce, S.; Tarefder, R.; Asce, M. Effects of large freeze-thaw cycles on stiffness and tensile strength of asphalt concrete. *J. Cold Reg. Eng.* **2016**, *30*. [CrossRef]

22. Krcmarik, M.; Varma, S.; Kutay, M.; Jamrah, A. Development of predictive models for low-temperature indirect tensile strength of asphalt mixtures. *J. Mater. Civ. Eng.* **2016**, *28*, 04016139. [CrossRef]

23. Liu, J.; Zhao, S.; Li, L.; Li, P.; Saboundjian, S. Low temperature cracking analysis of asphalt binders and mixtures. *Cold Reg. Sci. Technol.* **2017**, *141*, 78–85. [CrossRef]

24. Yi, J.Y.; Shen, S.H.; Muhunthan, B.; Feng, D.C. Viscoelastic–plastic damage model for porous asphalt mixtures: Application to uniaxial compression and freeze–thaw damage. *Mech. Mater.* **2014**, *70*, 67–75. [CrossRef]

25. Zhang, Q.; Ze, L.; Wen, Z.; Hou, Z. Research on the damage model of asphalt mixture under synergy action of freeze-thaw and loading. *J. Xian Univ. Archit. Technol.* **2016**, *48*, 188–194.

26. Huang, S.B.; Liu, Q.S.; Cheng, A.P.; Liu, Y.Z. A statistical damage constitutive model under freeze-thaw and loading for rock and its engineering application. *Cold Reg. Sci. Technol.* **2018**, *145*, 142–150. [CrossRef]

27. Chen, S.; Qiao, C.S.; Ye, Q.; Khan, M. Comparative study on three-dimensional statistical damage constitutive modified model of rock based on power function and weibull distribution. *Environ. Earth Sci.* **2018**, *77*, 108. [CrossRef]

28. JTG E20-2011. *Standard Test Methods of Asphalt and Asphalt Mixtures for Highway Engineering*; Ministry of Transport: Beijing, China, 2011. (In Chinese)

29. Zhu, C.F. Research on Road Performance and Mechanical Properties of Diatomite-Basalt Fiber Compound Modified Asphalt Mixture. Ph.D. Thesis, Jilin University, Changchun, China, 2018.

30. Tan, Y.Q.; Sun, Z.Q.; Gong, X.B.; Xu, H.N.; Zhang, L.; Bi, Y.F. Design parameter of low-temperature performance for asphalt mixtures in cold regions. *Constr. Build. Mater.* **2017**, *155*, 1179–1187. [CrossRef]

31. Tan, Y.Q.; Zhang, L.; Xu, H.N. Evaluation of low-temperature performance of asphalt paving mixtures. *Cold Reg. Sci. Technol.* **2012**, *70*, 107–112. [CrossRef]

32. Gao, C.M.; Han, S.; Zhu, K.X.; Wang, Z.Y. Research on basalt fiber asphalt concrete's high temperature performance. *Appl. Mech. Mater.* **2014**, *505–506*, 39–42. [CrossRef]

33. Xu, H.N.; Guo, W.; Tan, Y.Q. Internal structure evolution of asphalt mixtures during freeze-thaw cycles. *Mater. Des.* **2015**, *86*, 436–446. [CrossRef]

materials

MDPI

Article

Comparative Study on the Damage Characteristics of Asphalt Mixtures Reinforced with an Eco-Friendly Basalt Fiber under Freeze-thaw Cycles

Yongchun Cheng, Wensheng Wang[iD], Yafeng Gong *, Shurong Wang, Shuting Yang and Xun Sun

College of Transportation, Jilin University, Changchun 130025, China; chengyc@jlu.edu.cn (Y.C.);
wangws17@mails.jlu.edu.cn (W.W.); wangsr@jlu.edu.cn (S.W.); yangst18@mails.jlu.edu.cn (S.Y.);
sunx18@mails.jlu.edu.cn (X.S.)
* Correspondence: gongyf@jlu.edu.cn; Tel.: +86-0431-8509-5446

Received: 15 November 2018; Accepted: 5 December 2018; Published: 7 December 2018

Abstract: The main distresses of asphalt pavements in seasonal frozen regions are due to the effects of water action, freeze-thaw cycles, traffic, and so on. Fibers are usually used to reinforce asphalt mixtures, in order to improve its mechanical properties. Basalt fiber is an eco-friendly mineral fiber with high mechanical performance, low water absorption, and an appropriate temperature range. This paper aims to address the freeze-thaw damage characteristics of asphalt mixtures (AC-13) reinforced with eco-friendly basalt fiber, with a length of 6 mm. Based on the Marshall design method and ordinary pavement performances, including rutting resistance, anti-cracking, and moisture stability, the optimum asphalt and basalt fiber contents were determined. Test results indicated that the pavement performances of asphalt mixture exhibited a trend of first increasing and then deceasing, with the basalt fiber content. Subsequently, asphalt mixtures with a basalt fiber content of 0.4% were prepared for further freeze-thaw tests. Through the comparative analysis of air voids, splitting strength, and indirect tensile stiffness modulus, it could be found that the performances of asphalt mixtures gradually declined with freeze-thaw cycles and basalt fiber had positive effects on the freeze-thaw resistance. This paper can be used as a reference for further investigation on the freeze-thaw damage model of asphalt mixtures with basalt fiber.

Keywords: asphalt mixture; basalt fiber; freeze-thaw cycle; damage characteristics

---

1. Introduction

The asphalt pavement has been widely used in flexible pavement constructions, with a rapid growing trend [1,2]. Asphalt mixture is generally considered to be a complex porous material that includes bitumen, aggregates, fillers, as well as voids [3,4]. However, due to some environmental factors, there are many distresses in asphalt pavements, such as spalling, crumble, pavement pothole, etc., especially in the seasonal frozen regions [5]. Therefore, researchers have been trying to modify asphalt mixtures and explore its freeze-thaw damage.

Fibers additives, such as cellulose fiber, polyester fiber, mineral fiber, etc., have been added into bitumen and proved to be an effective reinforcement material for asphalt mixtures [6–9]. Basalt fiber, as a novel kind of eco-friendly mineral fiber, was produced from basalt rocks with high mechanical properties, low water absorption, and its by-product can be directly degraded in the environment, without any harm [10]. Wang et al. [11,12] added basalt fiber into asphalt materials and evaluated their fatigue resistance by using direct tension, as well as fatigue tests. By means of an X-ray computed tomography technology (i.e., CT technology) and finite-element method, basalt fiber can release stress concentrations in critical areas and reduce fatigue damages. Gu et al. [13] compared and discussed basalt fiber and commonly used fibers and found that basalt fiber has a superior reinforcement effect on the high-temperature anti-rutting ability of bitumen mastic. Qin et al. [14] tested the reinforcement effects of basalt fibers, with lengths of 3, 6, and 9 mm asphalt mastics, with respect to the lignin fiber and the polyester fiber. Through leakage, penetration, strip-tensile and DSR tests, basalt fiber, especially, with a length of 6 mm, has excellent comprehensive performances, due to a steady three dimensional (3D) networking structure in bitumen mastics. Zhang et al. [15] carried out repeated and multi-stress creep tests and used Abaqus for analyzing the high-temperature performance of asphalt mastics. Then Zhang et al. [16,17] conducted the numerical simulations in Abaqus for the compressive creep and bending creep tests, for the purpose of analyzing the distribution effect and reinforcement mechanism of basalt fiber. Wang et al. [18] explored the optimization design of styrene-butadiene-styrene (SBS)-modified asphalt mixtures, with basalt fiber, with the assistance of a central composite design method. Test results indicated that asphalt mixtures, with basalt fiber, of 0.34% and a length of 6 mm, exhibited superior Marshall properties. Previous studies indicated that basalt fiber was effective in improving the mechanical properties of asphalt materials.

In recent years, experiments about the freeze-thaw cycle effects on asphalt mixtures, were also investigated by many researchers [19–21]. Xu et al. [22] employed the computed tomography (CT technology) to obtain and analyze internal images of asphalt mixtures, under different freeze-thaw cycles and investigated the influences of freeze-thaw cycles, on the evolution of internal air voids. Moreover, Xu et al. [23] studied the effects of freeze-thaw cycles on the thermodynamic characteristics of asphalt mixtures, based on the information entropy theory, CT, and digital image processing (DIP) technologies. The effects of freeze-thaw cycles on the permeability of asphalt mixtures have also been evaluated by means of a flow state, as well as water conductivity of asphalt mixtures [24]. Yan et al. [25] investigated the stone mastic asphalt (SMA) mixtures under the action of freeze-thaw cycles and evaluated the freeze-thaw resistance based on the Marshall design indicators and water stability. Badeli et al. [26] conducted the rapid freeze-thaw cycle test for asphalt mixture, using thermomechanical tests. Yi et al. [27] established the generalized Maxwell and Drucker-Prager model to evaluate the viscoelastic-plastic damage, under the condition of freeze-thaw cycles. Uniaxial compressive strength tests were carried out to investigate the mechanism of the freeze-thaw failure of asphalt mixtures. Nevertheless, efforts done for asphalt mixture with basalt fiber, under freeze-thaw cycles, are still limited in this area.

In this paper, asphalt mixtures (AC-13) reinforced with an eco-friendly basalt fiber with a length of 6 mm, were first designed by the Marshall design method, in order to determine the optimum asphalt content. Then optimum basalt fiber content could be also obtained, according to the ordinary pavement performances, such as rutting resistance, the indirect tensile stiffness modulus, and moisture stability. Subsequently, freeze-thaw cycle tests were performed for control and test groups of asphalt mixtures and the freeze-thaw damage characteristics were evaluated by a comparative analysis.

2. Materials and Methods

2.1. Raw Materials

In this paper, bitumen of AH-90 was used, which was produced by the PetroChina Liaohe Petrochemical Company (Panjin, China). The basic physical performances of the AH-90 bitumen are presented in Table 1. Andesite mineral aggregates, which came from a local quarry in the Jilin Province, were chosen. Limestone powder was selected as the mineral filler for the bitumen mixture. The physical parameters of the aggregates and the filler are given in Tables 2 and 3. Basalt fiber (shown in Figure 1) was obtained from the Jiuxin Basalt Industry Co., Ltd. (Changchun, China), the physical performances of which are listed in Table 4.

Table 1. Basic physical properties of the bitumen.

Properties		Measurement	Technical Criterion
Penetration @ 25 °C, 100 g, 5 s (0.1 mm)		88	80–100
Softening point (°C)		47	≥44
Ductility	@ 10 °C, 5 cm/min (cm)	43.5	≥30
	@ 15 °C, 5 cm/min (cm)	153	≥100
Flash point (°C)		318	≥245
Solubility (trichloroethylene, %)		99.8	≥99.5
Density @ 15 °C (g/cm^3)		1.05	–
RTFOT			
Mass loss (%)		0.22	±0.8
Penetration ratio @ 25 °C (%)		66	≥57
Ductility	@ 10 °C, 5 cm/min (cm)	28	≥8
	@ 15 °C, 5 cm/min (cm)	89.3	≥20

Table 2. Physical properties of the aggregates.

Sieve Size (mm)	13.2	9.5	4.75	2.36	1.18	0.6	0.3	0.15	0.075
Apparent density (g/cm^3)	2.803	2.781	2.774	2.760	2.713	2.720	2.699	2.647	2.700

Table 3. Physical properties of the limestone powder.

Properties	Apparent Density (g/cm^3)	Hydrophilic Coefficient	Sieving Test	
			Size (mm)	Passing (%)
Values	2.728	0.76	0.6	100
			0.15	95.3
			0.075	82.5

Table 4. Physical properties of the basalt fibers.

Properties	Color	Length	Diameter	Specific Gravity	Tensile Strength	Elastic Modulus	Elongation at Break
Units	–	mm	μm	g/cm^3	MPa	GPa	%
Value	Golden brown	6	13	2.56	3200	>40	3.2

Figure 1. The golden-brown 6 mm long basalt fibers that were used in the study.

2.2. Sample Preparation

Traditional dense-graded asphalt mixture is a frequently-used asphalt mixture and is applied widely in the asphalt pavement construction in China [28]. The standard Marshall design method was adopted to prepare the asphalt mixture specimens [29]. Figure 2 presents the gradation curve of the asphalt mixture (AC-13) used in this study, the upper and lower limits, and selected median values of AC-13 are shown in Figure 2. In this paper, basalt fibers with a length of 6 mm was added into the asphalt mixtures, at four proportions of 0.2%, 0.3%, 0.4%, and 0.5% by a mass of asphalt mixture, respectively. According to the JTG E20-2011 [30], the detailed preparation procedures are presented as follows:

(i)   The pre-heated aggregates mixed together with basalt fibers, in a mixing pot, for 90 s, in order to uniformly disperse the basalt fibers in aggregates.
(ii) The pre-heated bitumen AH-90 was weighted and poured into the mixing pot and the mixture was blended for 90 s.
(ii) The pre-weighted limestone powder was added into the mixing pot and then blended for 90 s.
(iv) Marshall specimens of AC-13, of a diameter 101.6 mm and a height of 63.5 mm, were prepared by compacting 75 blows on each side, and square slab specimens with dimensions of 300 mm × 300 mm × 50 mm, were prepared with the help of the wheel rolling [31].

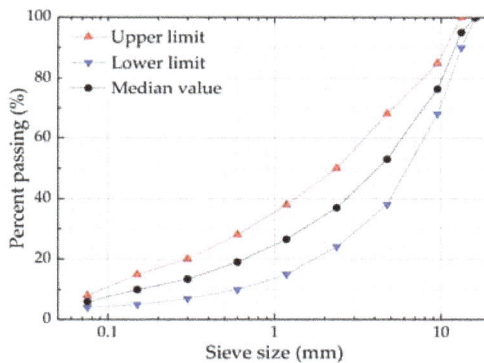

Figure 2. Gradation of the asphalt mixture (AC-13).

2.3. Testing Procedure

Figure 3 illustrates the research outline of this paper. First, raw materials of the asphalt mixtures were chosen, such as bitumen, aggregates, mineral powder, and the basalt fiber, listed in Section 2.1. Then, basalt fibers, of length 6 mm, were added into the asphalt mixtures, in different proportions of 0%, 0.2%, 0.3%, 0.4%, and 0.5% corresponding to the mass of the asphalt mixture, respectively. The optimum asphalt content for these asphalt mixtures could be determined by the Marshall design method described in Section 3.1. Subsequently, through ordinary pavement performances, including rutting resistance, anti-cracking, and moisture stability, the optimum basalt fiber content could be obtained (Section 3.2). Afterwards, the freeze-thaw cycle test was conducted for the asphalt mixtures, with the optimum basalt fiber content and without basalt fiber. Through the comparative analysis of air voids, splitting strength and indirect tensile stiffness modulus, the effects of the freeze-thaw cycles on the asphalt mixtures, could be addressed (Section 3.3).

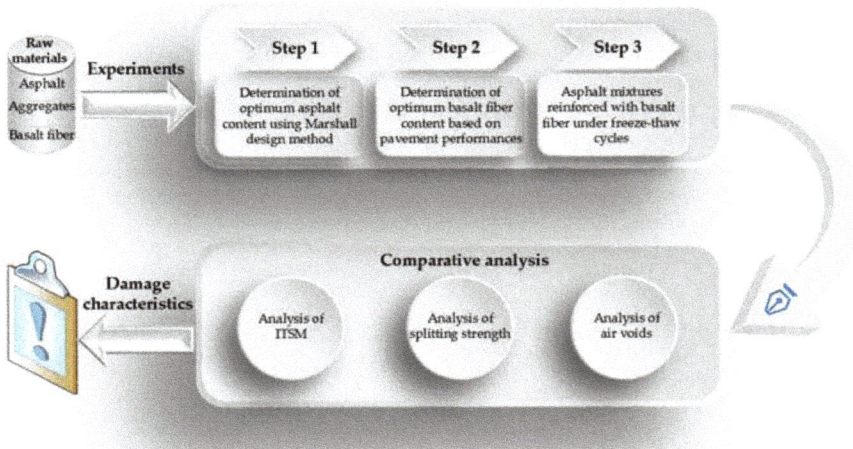

Figure 3. The research outline of this paper.

2.4. Experimental Methods

2.4.1. Marshall Design Method

Nowadays, the Marshall design method is used, extensively, in the asphalt pavement design and is also employed by the Chinese specification JTG E20-2011 [30]. The basic concepts of the Marshall mix-design method were originally developed by Bruce Marshall of the Mississippi Highway Department, around 1939, and then refined by the U.S. Army [32,33]. The Marshall design method seeks to select the optimum bitumen content at a desired density that satisfies the minimum stability and the range of flow values [29]. Compared to other design methods like the Superpave method, the Marshall design method is a proven method and requires relatively light, portable, and inexpensive equipment.

In this study, asphalt mixtures reinforced with basalt fiber were prepared on the basis of the Marshall design method and specimens were also tested, in order to obtain the Marshall design parameters, i.e., bulk density (ρ_f), air voids (VA), voids in mineral aggregates (VMA), voids filled with asphalt (VFA), Marshall stability (MS), as well as the flow value (FV). In accordance with rule T0709 JTG E20-2011, bulk density (ρ_f) can be determined through weighing the asphalt mixture specimens in air and water, following Equations (1) and (2). Afterwards, the VA, VMA, and VFA can be also calculated by using Equations (3)–(5).

$$\gamma_f = m_a / (m_f - m_w), \tag{1}$$

$$\rho_f = \gamma_f \times \rho_w, \tag{2}$$

$$VA = [1 - \gamma_f/\gamma_{TMD}] \times 100, \tag{3}$$

$$VMA = [1 - \gamma_f \times Ps/\gamma_{sb}] \times 100, \tag{4}$$

$$VFA = [(VMA - VA)/VMA] \times 100, \tag{5}$$

where ρ_w and ρ_f are the density of water and bulk density of specimens; m_a, m_w, and m_f represent the mass of the specimens in air, water, and the saturated surface dry mass, respectively; γ_f is the bulk specific gravity; γ_{TMD} is the theoretical maximum specific density which can be measured by vacuum sealing method; P_s is the aggregate content percent by weight of mixture; γ_{sb} is the bulk specific gravity of aggregates.

The Marshall test was performed to obtain the stability and flow values, according to rule T0709 in the JTG E20-2011 [30]. First, the prepared Marshall specimens (as described in Section 2.2) were conditioned in the water bath, at 60 °C, for half an hour. Afterwards, a force was applied on the side face, until the peak load. According to the indicator of the Marshall apparatus, the Marshall stability and flow could be obtained and recorded.

2.4.2. High-Temperature Rutting Test

The rutting test is usually used for the high-temperature performance of asphalt mixtures and the test (shown in Figure 4a, China Highway Engineering Instrument Institute, Beijing, China) was carried out in accordance with rule T0719 in the JTG E20-2011 [30]. The detailed experimental procedures were as follows:

(i)   Square slab specimens were placed in a dry environment of 60 ± 0.5 °C, for at least 5 h.
(ii) A rubber tire with a length of 50 mm was brought to the asphalt mixture slabs, for an hour, at a rolling speed of 42 ± 1 cycle/min, and the pressure of the loaded rubber tire was constant, i.e., 0.7 ± 0.05 MPa.
(iii) Then the rutting deflection could be measured vertically, per 20 s, by means of a linear variable differential transformer (LVDT).
(iv) The dynamic stability (DS) was defined by Equation 6 to quantitatively analyze the high-temperature rutting resistance. The rutting test was performed for three replicate specimens and the tested specimen is shown in Figure 4b.

$$DS = 15 \times N/(d_{60} - d_{45}),\tag{6}$$

where N is the rolling speed of the rubber tire and N is generally set as 42 cycle/min, d_{45} and d_{60} are the deflections at 45 min and 60 min, respectively.

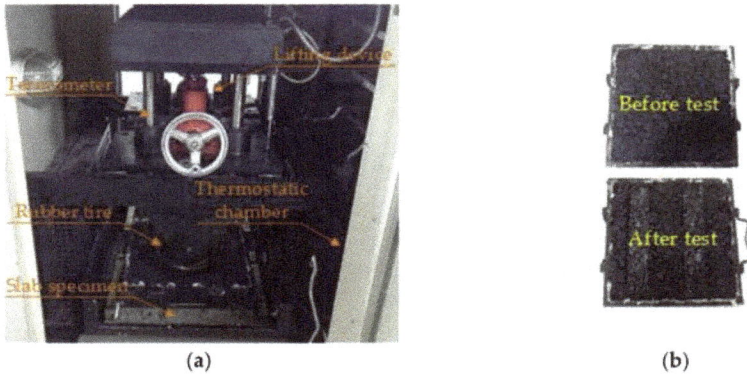

Figure 4. Rutting test in this paper: (**a**) Rutting test; and (**b**) slab specimens.

2.4.3. Low-Temperature Indirect Tensile Stiffness Modulus Test

The low-temperature tensile property is generally considered as an indicator for evaluating the anti-cracking ability and the indirect tensile stiffness modulus (ITSM) test (Cooper Research Technology Ltd., Ripley, UK) was adopted and conducted, according to the standard AASHTO TP-31, which is shown in Figure 5a [34]. A universal testing machine was used to perform the ITSM test. First, the Marshall specimens were put in an environment at 5 and 20 °C, for at least 5 h. Second, three replicate specimens were measured for ITSM and the load shown in Figure 5b was applied.

The detailed parameters of the load can be found in a previous study [34]. Then the indirect tensile stiffness modulus (S_m) could be obtained by calculation, as follows:

$$S_m = F \times (\mu + 0.27)/(h \times Z), \tag{7}$$

where F is the maximum loading (N); μ is the Poisson ratio, and $\mu = 0.25$ and 0.35, at 5 and 20 °C; h is the specimen height (mm); Z is the horizontal deformation (mm).

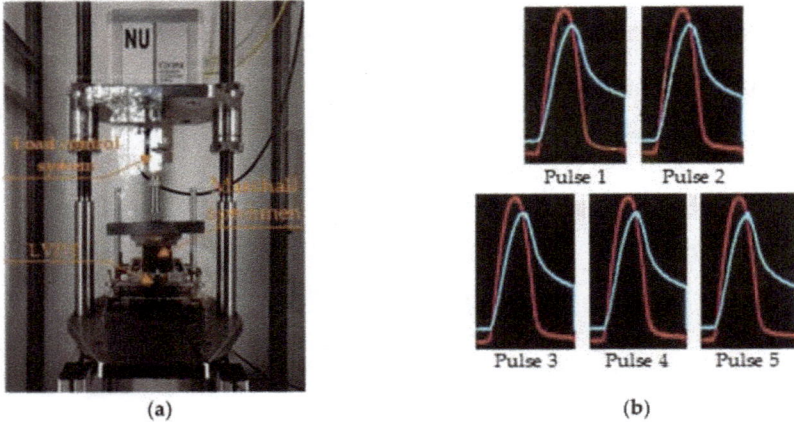

(a) (b)

Figure 5. Indirect tensile stiffness modulus test: (a) indirect tensile stiffness modulus (ITSM) test; and (b) schematic diagram of the load.

2.4.4. Moisture Stability Test

Freeze-thaw splitting test (shown in Figure 6, Nanjing Tuoxing Instrument Institute, Nanjing, China) is considered to be effective for analyzing the moisture stability of asphalt mixture and it has been widely used in many studies [22–24]. In accordance with rule T0729 in the JTG E20-2011 [30], the freeze-thaw splitting test was carried out at 25 °C, by the following steps:

(i) Marshall specimens were prepared and then divided into two groups, namely, the test group and the control group.
(ii) The test group was pretreated in a special condition, first, placed in water by vacuum saturation, after that put in the normal pressure condition.
(iii) Subsequently, the pretreated test group was conditioned at a low temperature of −18 °C for about 16 h, after that were placed in water at a temperature of 60 °C, for one day.
(iv) Both, the test and the control groups were immersed into water of 25 °C, for at least 2 h.
(v) The Marshall specimen was placed, centrally, in the Marshall apparatus and a loading force with a speed of 50 mm/min was loaded onto the specimen, until the specimen was broken.

The splitting tensile strength could be calculated by Equations (8) and (9):

$$R_{T1} = 0.006287 \times P_{T1}/h_1, \tag{8}$$

$$R_{T2} = 0.006287 \times P_{T2}/h_2, \tag{9}$$

where R_{T1} and R_{T2} are the control group and the test group, respectively; P_{T1} and P_{T2} are the maximum loads of the control and the test groups; h_1 and h_2 are the heights of the control and the test groups. Furthermore, the freeze-thaw splitting tensile strength ratio (TSR) could be obtained as follows:

$$TSR = (\overline{R}_{T2}/\overline{R}_{T1}) \times 100, \tag{10}$$

where $\overline{R}_{T1}$ and $\overline{R}_{T2}$ are the control group and the test groups, respectively.

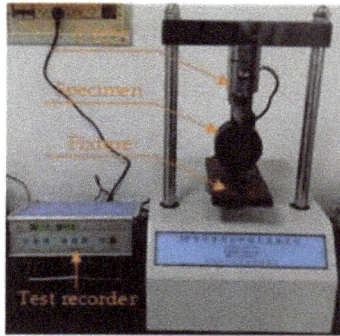

Figure 6. The freeze-thaw splitting test in this paper.

3. Results and Discussion

3.1. Determination of the Optimum Asphalt Content Using the Marshall Design Method

Asphalt mixtures with different basalt fiber proportions of 0% (control), 0.2%, 0.3%, 0.4%, and 0.5% by mass, were prepared, and were denoted by group 1 (control), 2, 3, 4, and 5, respectively. Then, the optimum asphalt content for these asphalt mixtures needed to be obtained, using the Marshall design method [29,35]. For each group, a range of the asphalt-aggregate ratios from 4.0% to 6.0% with an increment of 0.5%, was designed and tested by the Marshall design method. Figure 7 shows the Marshall design results of the asphalt mixtures (control group 1), including bulk specific gravity, *VA*, *VMA*, *VFA*, as well as *MS* and *FV*. Therefore, the optimum asphalt content (OAC) of the asphalt mixture, without the basalt fiber (control group 1) could be determined through the maximum density, maximum Marshall stability, and target air voids, and the OAC value was 5.03%. Afterwards, the OAC values of the other four groups of asphalt mixtures (i.e., group 2, 3, 4 and 5) could also be obtained and the OAC results are listed in Table 5. From Table 5, it could be seen that the OAC of different asphalt mixtures gradually increased with the basalt fiber content. This trend agrees with the results obtained in previous research, which may be attributed to the fact that basalt fiber has a larger specific surface area and the fibers can also absorb the light components in bitumen [9,36].

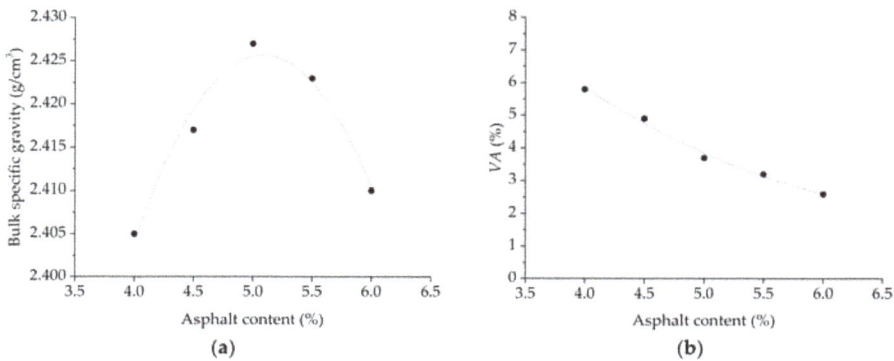

(a)

(b)

Figure 7. *Cont.*

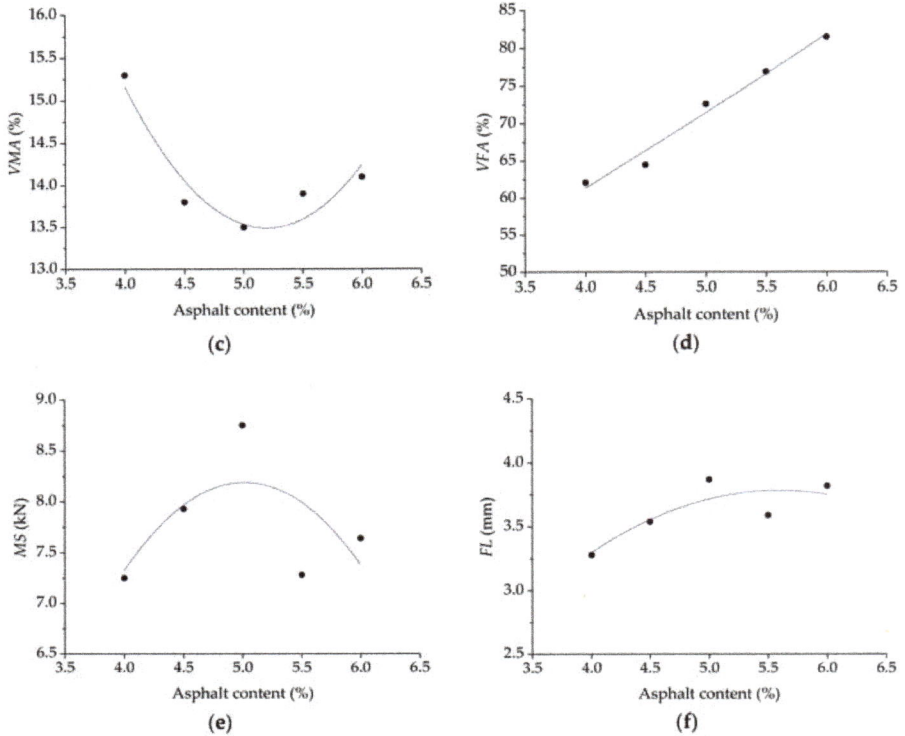

Figure 7. The Marshall design results of asphalt mixture without basalt fiber: (**a**) Bulk specific gravity; (**b**) *VA*; (**c**) *VMA*; (**d**) *VFA*; (**e**) *MS*; and (**f**) *FL*.

Table 5. Optimum asphalt content of asphalt mixtures with different basalt fiber contents.

Group	1 (Control)	2	3	4	5
Basalt Fiber Content (%)	0.0	0.2	0.3	0.4	0.5
Optimum Asphalt Content (%)	5.03	5.16	5.27	5.35	5.42

3.2. Optimum Basalt Fiber Content Based on the Pavement Performances

3.2.1. High-Temperature Rutting Resistance

The rutting test was conducted at 60 °C for the asphalt mixtures with different basalt fiber contents, at the corresponding OAC values. Figure 8 shows the high-temperature rutting test results of the five groups. It could be clearly seen that the dynamic stability results demonstrated a rising trend, first, and then came down, when the basalt fiber content was increased, gradually. Furthermore, the dynamic stability reached the largest value, at a basalt fiber content of 0.4%. Compared to the control group, the dynamic stability results of the test groups were improved by, approximately, 25.3%, 48.7%, 82.5%, and 62.1%, respectively. Ordinarily, a larger *DS* value means a preferable anti-rutting [18]. Accordingly, the basalt fiber was proved to be able to well improve the rutting resistance of the asphalt mixture. This is because the basalt fiber was uniformly dispersed in the asphalt mixture and there was a spatial networking structure. Meanwhile, the basalt fiber could absorb some light components of bitumen to improve its viscosity [13]. Thus, the stability of the asphalt mixture can be reinforced by the addition of basalt fiber. However, it should be noted that the reinforcement of the basalt fiber slightly decreased. This may be attributed to the coagulated basalt fiber or the uneven dispersion of the basalt fiber in the bitumen, leading to weak points.

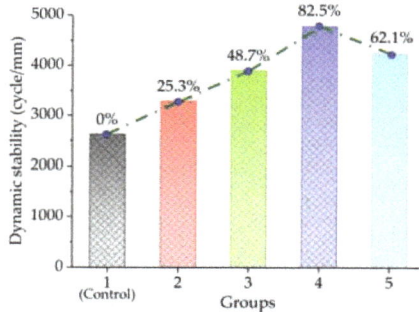

Figure 8. High-temperature rutting test results.

In addition, one-way analysis of variance (ANOVA) results using the Statistical Product and Service Solutions (SPSS) software (24.0, International Business Machines Corporation, New York, NY, USA) for high-temperature rutting test, are listed in Table 6. Tukey's HSD (honest significant difference) test was used to perform the post hoc multiple comparisons and the results are listed in Table 7. From Tables 6 and 7, the F-value was larger than $F_{0.01}(4,10) = 5.99$, indicating that the basalt fiber content had a significant influence on the high-temperature property of the asphalt mixture, which was also proved by Tukey's HSD results.

Table 6. One-way analysis of variance (ANOVA) for high-temperature rutting test.

Analysis	Sum of Squares	Degree of Freedom	Mean Square	F-Value	Significance
Between groups	8498025.6	4	2124506.4	836.8	**
Within groups	25388.0	10	2538.8		
Total	8523413.6				

Note: "**" is significant at the 0.01 level.

Table 7. Tukey's honest significant difference (HSD) test results for high-temperature rutting test.

Within Groups	1 vs. 2	1 vs. 3	1 vs. 4	1 vs. 5	2 vs. 3	2 vs. 4	2 vs. 5	3 vs. 4	3 vs. 5	4 vs. 5
Mean difference	−663	−1278	−2164	−1628	−615	−1501	−965	−886	−350	536
Significance	**	**	**	**	**	**	**	**	**	**

Note: "**" is significant at the 0.01 level.

3.2.2. Low-Temperature Indirect Tensile Stiffness Modulus

An indirect tensile stiffness modulus (ITSM) test was conducted at 5 and 20 °C, to investigate the low-temperature properties of the asphalt mixture with basalt fiber. The indirect tensile stiffness modulus could be calculated by Equation (7), based on the test data. Subsequently, the low-temperature indirect tensile stiffness modulus test results are plotted in Figure 9.

As shown in Figure 9, it could be observed that when the basalt fiber content increased continuously, the ITSM values also presented the variation trend of first increasing and then decreasing. Generally, the indirect tensile stiffness modulus is an indicator to evaluate the low- temperature anti-cracking ability, and a larger ITSM value of the asphalt pavement stands for a better anti-cracking ability. As shown in Figure 9, the low-temperature anti-cracking performance was improved with the addition of basalt fiber. With respect to the control group, the ITSM values increased by 11.2%, 18.1%, 22.5%, and 17.8% at 5 °C and 15.0%, 26.9%, 38.2%, and 30.7% at 20 °C, when adding basalt fiber of 0.2%, 0.3%, 0.4%, and 0.5% concentrations. In addition, it was evident that the ITSM had the most significant effect when the basalt fiber content was 0.4%. This variation trend might have been caused by the spatial networking structure of the basalt fiber, in the asphalt mixture. The absorption between bitumen and basalt fiber lead to a higher proportion of structural bitumen, improving the interfacial bond strength. Meanwhile, the addition of the basalt fiber could also prevent a further expansion of the cracks. The decreasing ITSM may be also attributed to the uneven dispersion of basalt fiber in bitumen.

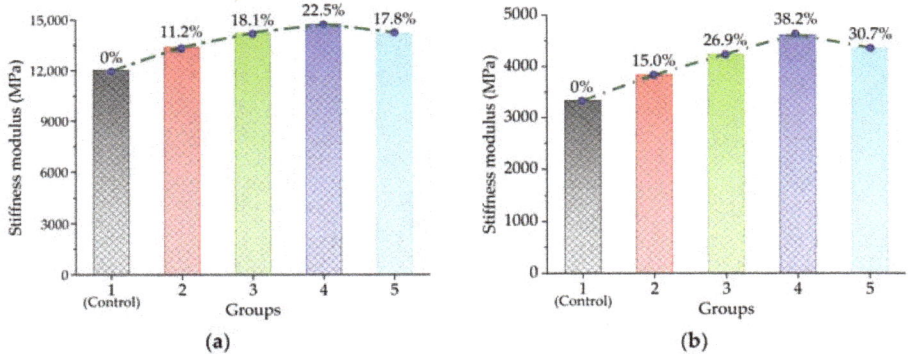

Figure 9. Low-temperature indirect tensile stiffness modulus test results: (**a**) At 5 °C; and (**b**) at 20 °C.

In addition, one-way analysis of variance (ANOVA) results, using the SPSS software for low-temperature ITSM test are listed in Table 8. Tukey's HSD (honest significant difference) test was used to perform post hoc multiple comparisons and Tukey's HSD results are listed in Table 9. From Tables 8 and 9, the F-value was larger than the $F_{0.01}(4,10) = 5.99$, indicating that the basalt fiber content had a significant influence on the low-temperature property of the asphalt mixture, which was also proved by Tukey's HSD results.

Table 8. One-way analysis of variance (ANOVA) for low-temperature ITSM.

Analysis	Sum of Squares	Degree of Freedom	Mean Square	F-Value	Significance
		5 °C			
Between groups	13339074.0	4	3334768.5	4338.8	**
Within groups	7686.0	10	768.6		
Total	13346760.0				
		20 °C			
Between groups	2992568.4	4	748142.1	613.6	**
Within groups	12192.0	10	1219.2		
Total	3004760.4				

Note: "**" is significant at the 0.01 level.

Table 9. Tukey's HSD test results for low-temperature ITSM.

Within Groups	1 vs. 2	1 vs. 3	1 vs. 4	1 vs. 5	2 vs. 3	2 vs. 4	2 vs. 5	3 vs. 4	3 vs. 5	4 vs. 5
					5 °C					
Mean difference	−1348	−2177	−2705	−2145	−829	−1357	−797	−528	32	560
Significance	**	**	**	**	**	**	**	**	**	**
					20 °C					
Mean difference	−499	−898	−1275	−1024	−399	−776	−525	−377	−126	251
Significance	**	**	**	**	**	**	**	**	**	**

Note: "**" is significant at the 0.01 level.

3.2.3. Moisture Stability Properties

The freeze-thaw splitting test was carried out at test temperature of 25 °C, so as to explore the effect of basalt fiber on the moisture stability of the asphalt mixture. The freeze-thaw splitting tensile strength ratio (*TSR*) was used as an indicator calculated by the Equation 10 and the test results are illustrated in Figure 10.

In Figure 10, the *TSR* values exhibited, approximately, similar variation trends to the *DS* and the *ITSM*. The *TSR* values of test groups were improved by 3.1%, 10.6%, 13.0%, and 10.9%, compared with the control group, and the test group 4 with a basalt fiber content of 0.4% had the highest *TSR* value. This was expected, due to the absorption effect between the bitumen and the basalt fiber, the adhesion capability between the bitumen and the aggregates were improved, significantly, so that there was a difficulty in the exfoliation of the aggregates, under the effect of water. Simultaneously, basalt fiber with a high modulus and strength formed a spatial networking structure in the asphalt mixture, playing the role of reinforcement and toughening.

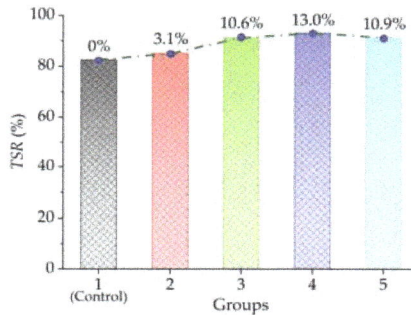

Figure 10. Moisture stability results of the freeze-thaw splitting test.

In addition, one-way analysis of variance (ANOVA) results, using the SPSS software for moisture stability tests are listed in Table 10. Tukey's HSD (honest significant difference) test was used to perform post hoc multiple comparisons and Tukey's HSD results are listed in Table 11. From Tables 10 and 11, the F-value was larger than the $F_{0.01}(4,10) = 5.99$, indicating that the basalt fiber content had a significant influence on the moisture stability of the asphalt mixture, which was also proved by Tukey's HSD results.

Table 10. One-way analysis of variance (ANOVA) for moisture stability.

Analysis	Sum of Squares	Degree of Freedom	Mean Square	F-Value	Significance
Between groups	259.1	4	64.8	1134.6	**
Within groups	0.571	10	0.057		
Total	259.7				

Note: "**" is significant at the 0.01 level.

Table 11. Tukey's HSD test results for moisture stability.

Within groups	1 vs. 2	1 vs. 3	1 vs. 4	1 vs. 5	2 vs. 3	2 vs. 4	2 vs. 5	3 vs. 4	3 vs. 5	4 vs. 5
Mean difference	−2.55	−8.77	−10.71	−8.97	−6.22	−8.16	−6.42	−1.94	−0.20	1.74
Significance	**	**	**	**	**	**	**	**	-	**

Note: "−" indicates insignificant correlation, "**" is significant at the 0.01 level.

In view of the pavement performances of the control and the test groups, when the basalt fiber increased, the rutting resistance, the anti-cracking and the moisture stability of the asphalt mixture were first improved and then slightly decreased, in which the pavement performances were improved, significantly, by adding basalt fiber of about 0.4%. Therefore, excessive basalt fiber content is not recommended and the optimal basalt fiber content was chosen as 0.4% and the corresponding optimum asphalt content was set as 5.35, in this study. The selected basalt fiber content and asphalt content were used to further investigate the water-temperature influences on asphalt mixtures.

3.3. Comparative Analysis of Damage Characteristics of the Asphalt Mixture under the Freeze-thaw Cycles

Asphalt mixtures modified by basalt fiber content of 0.4% and asphalt mixtures without basalt fiber, were prepared at the corresponding optimum asphalt content, and were divided into the test group and the control group, respectively. Before the test, the control and the test groups were immersed into water and under vacuum (98.0 kPa), for 15min, and soaked under atmospheric pressure, for 30 min. Then, the freeze-thaw cycles were carried out on both groups, in which the freezing condition was set as -18 °C, for 16 h, and the thaw condition was in water, at 60 °C for 8 h. After 0, 1, 3, 6, 9, 12, and 15 freeze-thaw cycle air voids, the splitting test, at 15 °C, and the ITSM test, at 10 °C, were carried out for further comparative analysis.

3.3.1. Analysis of Air Voids

Air voids of the control and the test groups were measured and could be calculated by CT and DIP technologies. The CT, DIP technologies, and statistical methods were adopted for the control and the test groups, before and after the freeze-thaw cycles. The process could include the following steps: (1) CT image scanning; (2) image enhancement; (3) image denoising; (4) threshold cutting and binarization of images; and (5) air voids calculation. Figure 11 shows the air voids results of both the control and the test groups, under different freeze-thaw cycles.

As shown in Figure 11, it can be clearly observed that the air voids results of asphalt mixtures gradually increased as the freeze-thaw cycles increased. Furthermore, the rising trend of air voids of the asphalt mixtures was significant but the variation presented a slow trend when the freeze-thaw cycles exceeded 9 cycles. It is worth noting that the initial air voids of the control group, without the basalt fiber, were slightly lower than that of test group with basalt fiber. This was because it was relatively difficult to compact the asphalt mixtures with basalt fiber, due to the higher elastic modulus and reinforcement effect of basalt fiber. Under the action of the freeze-thaw cycles, the internal structure of the asphalt mixtures was damaged due to the volume expansion and temperature stress. Before the nine freeze-thaw cycles, the air voids first extended and then the adjacent air voids were coalesced in the asphalt mixture, leading to the significant variation trend, however, the expansion and formation of air voids became slow after the nine freeze-thaw cycles. In addition, the air voids of the test group were significantly lower than that of the control group. It was also evident that the basalt fiber formed a spatial networking structure, playing the role of reinforcement and toughening.

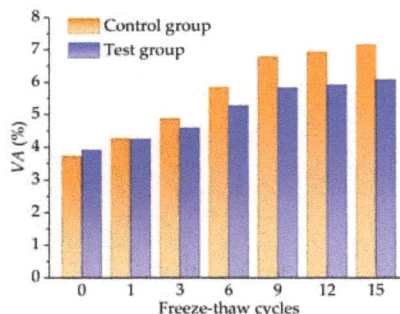

Figure 11. Comparative results of air voids of the control and the test groups, under different freeze-thaw cycles.

3.3.2. Analysis of Splitting Strength

Splitting strength of the control and the test groups could be obtained by the Equation 3, according to rule T0716 of the JTG E20-2011 [30]. Figure 12 illustrates the splitting strength results of both the control and the test groups, under the various freeze-thaw cycles.

As illustrated in Figure 12, the splitting strength values presented a decreasing trend with the freeze-thaw cycles and the splitting strength gradually decreased, slowly. Accordingly, the freeze-thaw cycle had a great effect on the mechanical properties of the asphalt mixture. This was because the adhesion capability between the bitumen and the aggregates became weaker and weaker, under the continuous action of the freeze-thaw cycles, resulting in a damaged internal structure of the asphalt mixture. Moreover, by a comparative analysis of the control and the test groups, the strength values of the test group were higher than those of the control group, under the same freeze-thaw cycles.

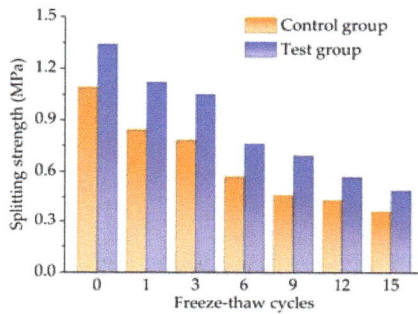

Figure 12. Comparative results of the splitting strength of the control and the test groups, under different freeze-thaw cycles.

3.3.3. Analysis of Indirect Tensile Stiffness Modulus

The indirect tensile stiffness modulus of the control and the test groups could be calculated by the Equation 7 and the experimental procedure referred to in Section 2.4.3. Figure 13 plots the indirect tensile stiffness modulus results of both the control and the test groups, under different freeze-thaw cycles.

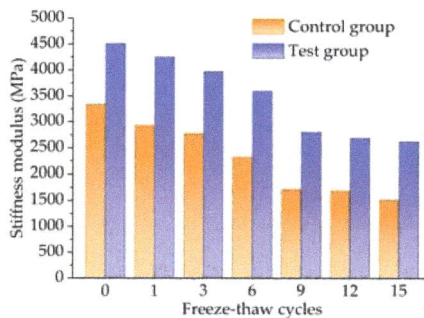

Figure 13. Comparative results of the indirect tensile stiffness modulus of the control and the test groups, under different freeze-thaw cycles.

As plotted in Figure 13, the indirect tensile stiffness modulus results presented an, approximately similar, decreasing variation trend to the splitting strength in Figure 12. It was expected that indirect tensile stiffness modulus is also considered to be an indicator of the mechanical performance of the asphalt mixture and is connected with the bearing capacity of traffic loads. The freeze-thaw cycles had a negative effect on the mechanical properties of the asphalt mixtures. Meanwhile, the addition of basalt fiber into the asphalt mixture could significantly enhance the anti-cracking and mechanical properties of the asphalt mixtures, leading to a reinforcement mechanism.

4. Conclusions

The primary objective of this paper was to study the damage characteristics of asphalt mixtures (AC-13), reinforced with an eco-friendly basalt fiber, with a length of 6 mm, under the condition of freeze-thaw cycles. The optimum asphalt content and the optimum basalt fiber content were obtained by the Marshall design method and ordinary pavement performances. Then, the freeze-thaw cycle tests were performed for the control and the test groups of asphalt mixtures. The following conclusions could be drawn:

- The optimum asphalt content gradually increased with the addition of the basalt fiber content.
- Based on ordinary pavement performances, when adding basalt fiber, the pavement performances of the asphalt mixture exhibited a trend of first, increasing and then deceasing in performance. This was due to a spatial networking structure by the basalt fiber in the asphalt mixture. However, excessive basalt fiber was not good for asphalt mixture.
- Basalt fibers with higher content would be difficult to disperse, unevenly, in the bitumen, leading to weak points. Therefore, higher basalt fiber content is not recommended and the optimum basalt fiber content could be determined as 0.4% for further freeze-thaw cycle tests, according to the pavement performances of the asphalt mixtures.
- The freeze-thaw cycles had a negative effect on the mechanical properties of the asphalt mixtures. Adding basalt fibers into asphalt mixture could significantly improve the freeze-thaw resistance and the mechanical performance of the asphalt mixture, leading to a reinforcement mechanism.

Author Contributions: Conceptualization, Y.C. and W.W.; Methodology, W.W. and Y.G.; Validation, Y.C.; Formal Analysis, Y.G. and S.W.; Investigation, W.W., S.W., S.Y., and X.S.; Writing—Original Draft Preparation, W.W.; Writing—Review & Editing, Y.C. and Y.G.; Project Administration, Y.C.; Funding Acquisition, Y.C.

Funding: This research was funded by the National Natural Science Foundation of China, Grant Number 51678271 and the Science Technology Development Program of the Jilin Province, Grant Number 20160204008SF; and was supported by the Graduate Innovation Fund of the Jilin University.

Acknowledgments: The authors would like to appreciate the anonymous reviewers for their constructive suggestions and comments to improve the quality of the paper.

Conflicts of Interest: The authors declare no conflict of interest.

References

1. Chen, M.Z.; Lin, J.T.; Wu, S.P. Potential of recycled fine aggregates powder as filler in asphalt mixture. *Constr. Build. Mater.* **2011**, *25*, 3909–3914. [CrossRef]
2. Pan, P.; Wu, S.P.; Xiao, Y.; Liu, G. A review on hydronic asphalt pavement for energy harvesting and snow melting. *Renew. Sustain. Energy Rev.* **2015**, *48*, 624–634. [CrossRef]
3. Wang, W.S.; Cheng, Y.C.; Tan, G.J.; Tao, J.L. Analysis of aggregate morphological characteristics for viscoelastic properties of asphalt mixes using simplex lattice design. *Materials* **2018**, *11*, 1908. [CrossRef] [PubMed]
4. Guo, Q.L.; Bian, Y.S.; Li, L.L.; Jiao, Y.B.; Tao, J.L.; Xiang, C.X. Stereo logical estimation of aggregate gradation using digital image of asphalt mixture. *Constr. Build. Mater.* **2015**, *94*, 458–466. [CrossRef]
5. Feng, D.C.; Yi, J.Y.; Wang, D.S.; Chen, L.L. Impact of salt and freeze-thaw cycles on performance of asphalt mixtures in coastal frozen region of China. *Cold Reg. Sci. Technol.* **2010**, *62*, 34–41. [CrossRef]
6. Ye, Q.S.; Wu, S.P.; Li, N. Investigation of the dynamic and fatigue properties of fiber-modified asphalt mixtures. *Int. J. Fatigue* **2009**, *31*, 1598–1602. [CrossRef]
7. Wu, S.P.; Ye, Q.S.; Li, N. Investigation of rheological and fatigue properties of asphalt mixtures containing polyester fibers. *Constr. Build. Mater.* **2008**, *22*, 2111–2115. [CrossRef]

8.   Tayfur, S.; Ozen, H.; Aksoy, A. Investigation of rutting performance of asphalt mixtures containing polymer modifiers. *Constr. Build. Mater.* **2007**, *21*, 328–337. [CrossRef]

9.   Xiong, R.; Fang, J.H.; Xu, A.H.; Guan, B.W.; Liu, Z.Z. Laboratory investigation on the brucite fiber reinforced asphalt binder and asphalt concrete. *Constr. Build. Mater.* **2015**, *83*, 44–52. [CrossRef]

10. Zhang, X.Y.; Gu, X.Y.; Lv, J.X. Effect of basalt fiber distribution on the flexural–tensile rheological performance of asphalt mortar. *Constr. Build. Mater.* **2018**, *179*, 307–314. [CrossRef]

11. Wang, D.; Wang, L.B.; Gu, X.Y.; Zhou, G.Q. Effect of basalt fiber on the asphalt binder and mastic at low temperature. *J. Mater. Civ. Eng.* **2013**, *25*, 355–364. [CrossRef]

12. Wang, D.; Wang, L.B.; Christian, D.; Zhou, G.Q. Fatigue properties of asphalt materials at low in-service temperatures. *J. Mater. Civ. Eng.* **2013**, *25*, 1220–1227. [CrossRef]

13. Gu, X.Y.; Xu, T.T.; Ni, F.J. Rheological behavior of basalt fiber reinforced asphalt mastic. *J. Wuhan Univ. Technol. Mater. Sci. Ed.* **2014**, *29*, 950–955. [CrossRef]

14. Qin, X.; Shen, A.Q.; Guo, Y.C.; Li, Z.N.; Lv, Z.H. Characterization of asphalt mastics reinforced with basalt fibers. *Constr. Build. Mater.* **2018**, *159*, 508–516. [CrossRef]

15. Zhang, X.Y.; Gu, X.Y.; Lv, J.X.; Zhu, Z.K.; Ni, F.J. Mechanism and behavior of fiber-reinforced asphalt mastic at high temperature. *Int. J. Pavement Eng.* **2018**, *19*, 407–415. [CrossRef]

16. Zhang, X.Y.; Gu, X.Y.; Lv, J.X.; Zhu, Z.K.; Zou, X.Y. Numerical analysis of the rheological behaviors of basalt fiber reinforced asphalt mortar using ABAQUS. *Constr. Build. Mater.* **2017**, *157*, 392–401. [CrossRef]

17. Gu, X.Y.; Zhang, X.Y.; Lv, J.X. Establishment and verification of prediction models of creep instability points of asphalt mixtures at high temperatures. *Constr. Build. Mater.* **2018**, *171*, 303–311. [CrossRef]

18. Wang, W.S.; Cheng, Y.C.; Tan, G.J. Design optimization of SBS-modified asphalt mixture reinforced with eco-friendly basalt fiber based on response surface methodology. *Materials* **2018**, *11*, 1311. [CrossRef] [PubMed]

19. Pan, P.; Wu, S.P.; Hu, X.D.; Wang, P.; Liu, Q.T. Effect of freezing-thawing and ageing on thermal characteristics and mechanical properties of conductive asphalt concrete. *Constr. Build. Mater.* **2017**, *140*, 239–247. [CrossRef]

20. Badeli, S.; Carter, A.; Dore, G.; Saliani, S. Evaluation of the durability and the performance of an asphalt mix involving Aramid Pulp Fiber (APF): Complex modulus before and after freeze-thaw cycles, fatigue, and TSRST tests. *Constr. Build. Mater.* **2018**, *174*, 60–71. [CrossRef]

21. Zheng, Y.X.; Cai, Y.C.; Zhang, G.H.; Fang, H.Y. Fatigue property of basalt fiber-modified asphalt mixture under complicated environment. *J. Wuhan. Univ. Technol. Mater. Sci. Ed.* **2014**, *29*, 996–1004. [CrossRef]

22. Xu, H.N.; Guo, W.; Tan, Y.Q. Internal structure evolution of asphalt mixtures during freeze-thaw cycles. *Mater. Des.* **2015**, *86*, 436–446. [CrossRef]

23. Xu, H.N.; Li, H.Z.; Tan, Y.Q.; Wang, L.B.; Hou, Y. A micro-scale investigation on the behaviors of asphalt mixtures under freeze-thaw cycles using entropy theory and a computerized tomography scanning technique. *Entropy* **2018**, *20*, 68. [CrossRef]

24. Xu, H.N.; Guo, W.; Tan, Y.Q. Permeability of asphalt mixtures exposed to freeze-thaw cycles. *Cold Reg. Sci. Technol.* **2016**, *123*, 99–106. [CrossRef]

25. Yan, K.Z.; Ge, D.D.; You, L.Y.; Wang, X.L. Laboratory investigation of the characteristics of SMA mixtures under freeze-thaw cycles. *Cold Reg. Sci. Technol.* **2015**, *119*, 68–74. [CrossRef]

26. Badeli, S.; Carter, A.; Dore, G. Complex modulus and fatigue analysis of asphalt mix after daily rapid freeze-thaw cycles. *J. Mater. Civ. Eng.* **2018**, *30*, 04018056. [CrossRef]

27. Yi, J.Y.; Shen, S.H.; Muhunthan, B.; Feng, D.C. Viscoelastic-plastic damage model for porous asphalt mixtures: Application to uniaxial compression and freeze-thaw damage. *Mech. Mater.* **2014**, *70*, 67–75. [CrossRef]

28. Cui, P.D.; Xiao, Y.; Fang, M.J.; Chen, Z.W.; Yi, M.W.; Li, M.L. Residual fatigue properties of asphalt pavement after long-term field service. *Materials* **2018**, *11*, 892. [CrossRef]

29. Wang, W.S.; Cheng, Y.C.; Tan, G.J.; Shi, C.L. Pavement performance evaluation of asphalt mixtures containing oil shale waste. *Road Mater. Pavement Des.* **2018**. [CrossRef]

30. *Standard Test Methods of Bitumen and Bituminous Mixtures for Highway Engineering*; JTG E20-2011; Ministry of Transport of the People's Republic of China: Beijing, China, 2011. (In Chinese)

31. Wei, H.B.; He, Q.Q.; Jiao, Y.B.; Chen, J.F.; Hu, M.X. Evaluation of anti-icing performance for crumb rubber and diatomite compound modified asphalt mixture. *Constr. Build. Mater.* **2016**, *107*, 109–116. [CrossRef]

32. Ma, T.; Ding, X.H.; Wang, H.; Zhang, W.G. Experimental study of high-performance deicing asphalt mixture for mechanical performance and anti-icing effectiveness. *J. Mater. Civ. Eng.* **2018**, *30*, 04018180. [CrossRef]

33. Ma, T.; Wang, H.; He, L.; Zhao, Y.L.; Huang, X.M.; Chen, J. Property characterization of asphalt binders and mixtures modified by different crumb rubbers. *J. Mater. Civ. Eng.* **2017**, *29*, 04017036. [CrossRef]

34. Guo, Q.L.; Li, L.L.; Cheng, Y.C.; Jiao, Y.B.; Xu, C. Laboratory evaluation on performance of diatomite and glass fiber compound modified asphalt mixture. *Mater. Des.* **2015**, *66*, 51–59. [CrossRef]

35. Cheng, Y.C.; Wang, W.S.; Tan, G.J.; Shi, C.L. Assessing high- and low-temperature properties of asphalt pavements incorporating waste oil shale as an alternative material in Jilin province, China. *Sustainability* **2018**, *10*, 2179. [CrossRef]

36. Chen, H.X.; Xu, Q.W. Experimental study of fibers in stabilizing and reinforcing asphalt binder. *Fuel* **2010**, *89*, 1616–1622. [CrossRef]

materials

MDPI

Article

Viscoelastic Mechanical Responses of HMAP under Moving Load

Yazhen Sun [1],*, Bincheng Gu [1], Lin Gao [2], Linjiang Li [1], Rui Guo [1], Qingqing Yue [1] and Jinchang Wang [3],*

[1] School of Transportation Engineering, Shenyang Jianzhu University, Shenyang 110168, China; gbc0207@stu.sjzu.edu.cn (B.G.); 1414304706@stu.sjzu.edu.cn (L.L.); gr329@stu.sjzu.edu.cn (R.G.); yqq@stu.sjzu.edu.cn (Q.Y.)
[2] College of Architecture Engineering, Chongqing University of Arts and Sciences, Chongqing 402160, China; gaolin32@163.com
[3] Institute of Transportation Engineering, Zhejiang University, Hangzhou 3100580, China
* Correspondence: syz16888@126.com (Y.S.); wjc501@zju.edu.cn (J.W.); Tel.:+86-139-4002-1867 (Y.S.); +86-135-8803-0136 (J.W.)

Received: 9 November 2018; Accepted: 6 December 2018; Published: 7 December 2018

Abstract: In order to represent the mechanical response laws of high-modulus asphalt pavement (HMAP) faithfully and objectively, the viscoelasticity of high-modulus asphalt mixture (HMAM) was considered, and the viscoelastic mechanical responses were calculated systematically based on moving load by numerical simulations. The performances of the HMAP in resistance to the deformation and the cracking at the bottom layer were compared with the ordinary asphalt pavement. Firstly, Lubao and Honeywell 7686 (H7686) were selected as the high modulus modifiers. The laboratory investigations of Asphalt mix-70 penetration, Asphalt mix-SBS (styrene-butadiene-styrene), HMAM-Lubao and HMAM-H7686 were carried out by dynamic modulus tests and wheel tracking tests. The conventional performances related to the purpose of using the HMAM were indicated. The master curves of the storage moduli were obtained and the viscoelastic parameters were fitted based on viscoelastic theories. Secondly, 3D pavement models based on moving loads for the viscoelastic structures were built using the non-linear finite element software ABAQUS. The wheel path was discretized in time and space to apply the Haversine wave load, and then the mechanical responses of four kinds of asphalt pavement were calculated. Finally, the sensitivity analysis was carried out. The results showed that the addition of the high modulus modifiers can improve the resistance to high-temperature rutting of the pavements. Except for the tensile strain and stress at the bottom of the underlayer, other responses decreased with the increases of the dynamic moduli and the change laws of the tensile strain and stress were affected by the range of the dynamic modulus. The tensile stress at the bottom of the asphalt layer would be too large if the modulus of the layer were too large, and a larger tensile strain would result. Therefore, the range of the modulus must be restricted to avoid the cracking due to excessive tension when using the HMAM. The resistance of the HMAP to deformation was better and the HMAP was less sensitive to load changes and could better withstand the adverse effects inflicted by heavy loads.

Keywords: high-modulus asphalt mixture (HMAM); dynamic tests; viscoelasticity; dynamic responses; resistance to deformations; tensile strains; tensile stresses; sensitivity analysis

---

1. Introduction

In roads with heavy traffic, the proportion of the damaged pavement is increasing [1–4]. Findings have demonstrated that increasing the modulus of the asphalt mixture was an effective way to resist damages and extend the service life of the pavements [5–7]. The-high modulus asphalt mixture

(HMAM), first developed in France, a kind of hot asphalt mixture with a dynamic modulus (15 °C, 10 Hz) greater than 14,000 MPa, is being taking seriously by researchers. In China, the material was mainly made by directly adding high-modulus modifier into the aggregates, and the modulus of asphalt mixtures could be significantly increased and the resistance to the deformations improved [8]. The asphalt pavement is actually a typical viscoelastic structure and bears moving loads. However, the pavement structure models were built based on static loads and elastic layer systems by most researchers for numerical simulations, and the inertial forces and time dependency were not taken into account, which could not reflect the actual state of the pavement [9]. How to represent the dynamic properties of the pavement, the resistances to deformations and the cracking of the asphalt layer bottom when the HMAM was used for the layer are questions that need to be answered in the promotion of the high-modulus asphalt pavement (HMAP) in China. So it is imperative to apply moving wheel loads to the pavement based on the viscoelastic properties.

Until now, researchers have carried out many studies on the mechanical properties of the HMAP. The finite element method was used to compare the mechanical responses of high modulus and ordinary asphalt pavements under different axle loads, suggesting that some mechanical responses such as the compressive strains can be effectively reduced by the HMAM [10–12]. The dynamic modulus was directly applied to analyze the shearing stress and shearing strain of the HMAP, and the conclusion was drawn that the strain level of the pavement could be significantly reduced by increasing the modulus [13–15]. A full-scale test of the asphalt pavement under repeated loads was carried out to verify this [16,17]. The creep test results were fitted to the viscoelastic parameters based on the Burgers model and the static cyclic loads were applied to simulate the rutting formation, with the results showing that the rutting deformation of the HMAP was significantly less than that of ordinary asphalt pavement when the HMAM was used as the middle surface layer [18–20]. The dynamic moduli were calculated using the LEDFAA program; it was found that the HMAP could also slow down rutting formation [21]. It can be seen that the current research on the HMAP is mainly focused on its resistance to rutting and shearing stress, and other internal responses such as the tensile strain, need to be systematically analyzed. Due to the restrictions of the experimental conditions, in most studies, the asphalt pavement was assumed to be an elastic layered system, which did not reflect the viscoelasticities of the asphalt mixtures. In a dynamic analysis of the pavement, an important factor needs to be considered: the dependency of the material properties on the loading frequency [22]. The time-domain Prony series expression of the relaxation modulus can accurately represent the long-term complex viscoelastic behaviors of the asphalt mixtures. It is important to perform the dynamic analysis of flexible pavements subjected to traffic load. The study conducted by Al-Qadi [23] reported that the maximum differences between responses obtained by quasi-static and dynamic analyses were 39%, 25% and 10% for the tensile strain at the bottom of the HMAP, the compressive stress at the top of the subgrade and the longitudinal strain, respectively. The static cyclic load could not reflect the dynamic states of the road pavements. In view of this, in this paper, the viscoelastic parameters of the HMAM need to be obtained through a dynamic modulus test, and the viscoelastic pavement model under moving load should be built to systematically study the mechanical response laws of the HMAP.

The viscoelastic mechanical responses of ordinary asphalt pavements subjected to moving loads have been studied by scholars in recent years. Such responses as the compressive strain of the subgrade at different speeds were analyzed, demonstrating that the responses increased significantly at a low vehicle speed [24,25]. The viscoelastic mechanical properties of two typical thin and thick pavements sections at different speeds were studied and the results showed that the shearing stress has a certain influence on the tensile strain at the bottom of the asphalt layer [26,27]. Based on the measured structural parameters and vehicle characteristics, a 3D viscoelastic pavement model was built using the finite element software, ABAQUS, and the mechanical responses under different loads at different speeds were analyzed. Then, a comparison was made between the measured and calculated mechanical responses, demonstrating that the responses could be reasonably simulated by the model, the shearing stress and the shearing strain tended to be concentrated in the middle surface layer, and the viscoelastic properties led to the asymmetrical mechanical response curves [28–30]. The responses under moving loads based on the Laplace transform were solved by the Boltzmann superposition principle. The results were compared with those by numerical calculations to verify the feasibility of the simulation based on viscoelasticity [31]. It can be seen that the time-domain Prony series expression of the relaxation modulus used in this study can accurately represent the long-term complex viscoelastic behaviors of the asphalt mixtures. The previous analyses were carried out using ordinary asphalt mixtures, and there was a lack of objective and systematic analysis of the HMAP, and how to represent the mechanical responses of the HMAP and the characteristics of the stresses were not discussed. In addition, the problem of the changing characteristics of the stress of the HMAP compared with other pavements urgently needs to be solved in the promotion of the HMAP in China.

Based on the previous studies, the aim of this paper was to apply real viscoelastic parameters and moving loads to the HMAP, and 3D viscoelastic pavement models based on moving loads were built. More objective and systematic studies on the mechanical responses were conducted and compared with common pavement structures and the characteristics of the HMAP were analyzed. The following studies were conducted: laboratory investigations of four kinds of asphalt mixtures were carried out by wheel tracking tests and dynamic modulus tests. The resistance to high-temperature rutting of HMAM were indicated. The viscoelastic parameters were fitted based on the viscoelastic theories. The 3D viscoelastic dynamic models were built using ABAQUS. The mechanical response laws such as the vertical deformations and the longitudinal tensile strains under standard and heavy loads were studied. The characteristics of the HMAM and the sensitivities of various pavement structures to the level of the loads were analyzed in an objective way to provide the theoretical basis for improving the HMAP.

2. Dynamic Modulus Testing

2.1. Materials

The 70-Penetration asphalt and the styrene-butadiene-styrene (SBS) asphalt (in accordance with the Chinese standard of the Technical Specifications for Construction of Highway Asphalt Pavements (JTG F40—2004)), referred to as "asphalt binder", were mixed with limestone, the aggregates to form materials used in asphalt pavements, and the properties were listed in Tables 1 and 2. The two kinds of high modulus modifiers (Lubao and H7686) were directly added into the aggregates. The modifiers were blended first with hot aggregates and then mixed with hot asphalt and mineral filler to ensure uniform dispersion of the mixture.

Table 1. Properties of 70-Penetration asphalt.

Properties	Unit	Value	Method
Penetration (25 °C, 100 g, 5 s)	0.1 mm	63.2	T0604-2011
Softening Point	°C	49.1	T0606-2011
Ductility (5 cm/min, 15 °C)	cm	>100	T0605-2011
Viscosity (177 °C)	Pa·s	3.8	T0625-2011

Table 2. Properties of styrene-butadiene-styrene (SBS) asphalt.

Properties	Unit	Value	Method
Penetration (25 °C, 100 g, 5 s)	0.1 mm	50.7	T0604-2011
Softening Point	°C	75.6	T0606-2011
Ductility (5 cm/min, 15 °C)	cm	>200	T0605-2011
Viscosity (177 °C)	Pa·s	4.1	T0625-2011

The two high modulus modifiers (Figure 1), Honeywell 7686 (H7686) and Lubao, being highly efficient, convenient, and widely used in the production process, can be directly added to the aggregates.

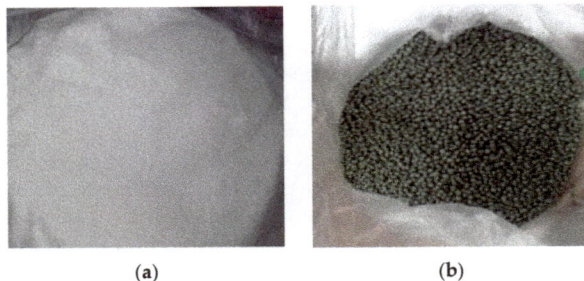

(a)   (b)

Figure 1. High modulus modifiers: (**a**) Honeywell7686; (**b**) Lubao.

Honeywell7686 (H7686), a white-powdered composite material with relatively low molecular weight, is soluble in asphalt. It is a special modifier and has the characteristics of rutting resistance, water resistance, and warm mixing construction. Lubao is a kind of high-density polyethylene material with good chemical stability and relatively high molecular quality, and is tasteless and non-toxic. The asphalt mixtures were reinforced by Lubao to improve its performance. The properties of modifiers H7686 and Lubao are shown in Tables 3 and 4.

Table 3. Properties of H7686.

Properties	Unit	Value
Droplet Point	°C	130–138
Viscosity—150 °C Brookfield	Pa·s	4.1–4.8

Table 4. Properties of Lubao.

Properties	Value	Standards
Tensile Strength/MPa	18.8	18–20
Elongation at Break/%	102.8	≥100
Density/(g·cm^{-3})	0.94	0.93–0.96
Melt Flow Rate/(g/10min)	1.2	1–4
Vicat Softening Temperature/°C	61.7	≤140
Resin Content/%	98.89	≥95
Particle Diameter/mm	3.7	3–5

2.2. Aggregate Gradation

The type of the asphalt mix is AC-20 (dense gradation asphalt concrete-20). It belongs to hot mix asphalt mixtures, in which minerals of various particle sizes were designed according to the principle of dense gradation (marking in accordance with the Chinese standard of the Technical Specifications for Construction of Highway Asphalt Pavements (JTG F40—2004)). The continuous aggregate gradation, having a nominal particle maximum size of 19 mm, is listed in Table 5. According to the Marshall volumetric mix design, the 70-penetration and the SBS asphalts were directly mixed into the aggregates and the optimum asphalt binder content were 4.4% and 4.4% by weight, respectively. The mixing amounts of H7686 and Lubao were based on the best mixing amounts recommended by the manufacturer: 0.4% and 0.5% of the total mass of the asphalt mixtures, respectively. The two kinds of high modulus modifiers (Lubao and H7686) were directly added into the aggregates and the optimal ratios of binder were 4.6% and 4.5% by weight, respectively. The mixing temperature was 160°C, and the mixing time was 90 s. The asphalt mixtures with 70-penetration asphalt binder and the asphalt mixtures with SBS asphalt binder were respectively denoted Asphalt mix-70 penetration and Asphalt mix-SBS. And the Asphalt mix-70 penetration with the Lubao modifiers and the Asphalt mix-70 penetration with the H7686 modifiers were respectively denoted HMAM-Lubao and HMAM-H7686. The properties of the four kinds of asphalt mixtures are listed in Table 6.

Table 5. Aggregate gradation.

Sieve Size/mm	Upper Limit/%	Lower Limit/%	Gradation/%
26.5	100	100	100
19	100	90	93.7
16	92	78	86.3
13.2	80	62	79.2
9.5	72	50	66.3
4.75	56	26	45.9
2.36	44	16	28.1
1.18	33	12	20.0
0.6	24	8	13.2
0.3	17	5	8.1
0.15	13	4	5.4
0.075	7	3	3.6

Table 6. The properties of asphalt mixtures.

Materials	Asphalt Contents/%	Relative Bulk Volume Density/g·cm^{-3}	Air Voids/%	Stability/kN	Flow Value/mm
Asphalt mix-70 Penetration	4.4	2.422	4.1	14.2	3.07
HMAM-Lubao	4.6	2.417	3.8	22.5	3.77
HMAM-H7686	4.5	2.420	3.9	21.32	3.69
Asphalt mix-SBS	4.4	2.411	3.8	19.38	3.46

2.3. Testing Results

A dynamic modulus test was conducted in accordance with the standard test methods of asphalt mixtures in China and the American Highway and Transportation Association standard AASHTO TP62-03. The Φ100 mm × 150 mm cylindrical test specimens were made by gyratory compaction, core drilling and cutting. Four kinds of specimens were tested for the dynamic moduli. Each specimen in this study was tested at 5, 20 and 45 °C, respectively, and the Haversine waveform was used as the loading method. At each test temperature, the load frequencies were 25, 10, 5, 1, 0.5, and 0.1 Hz, respectively, and a 60-s rest period was used between two neighboring frequencies. The dynamic test results of the four kinds of asphalt mixtures at the three temperatures and the six frequencies were automatically calculated by the microcomputer, as shown in Figure 2.

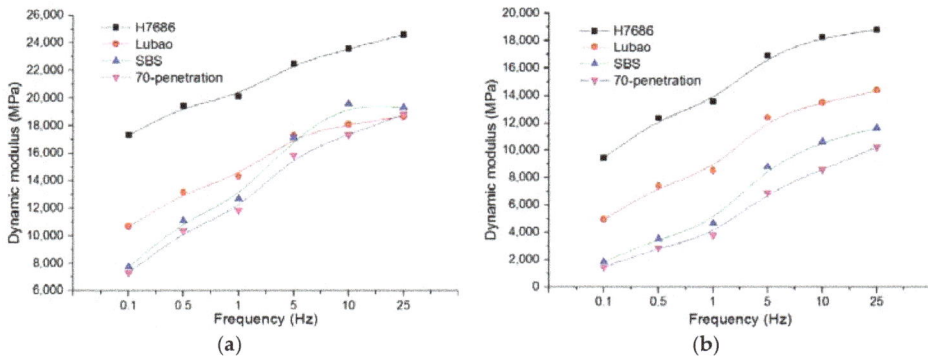

(a)   (b)

Figure 2. *Cont.*

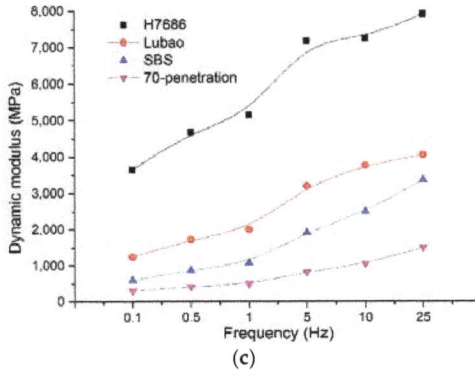

Figure 2. Dynamic modulus comparison results at different temperatures: (**a**) 5 °C; (**b**) 20 °C; (**c**) 45 °C.

As shown in Figure 2, there are similar change laws between the HMAM and other asphalt mixtures changing with temperature and frequency in viscoelasticity. At a high temperature and a low frequency, the dynamic modulus of asphalt mixtures decreases, the elasticity is weakened, and the viscosity is enhanced; at low temperature and high frequency, it is the opposite. The dynamic modulus of asphalt mixtures can be significantly increased with the adding of a high modulus modifier. The dynamic moduli of the HMAM-H7686 are 1.4 and 1.3 times more than those of the Asphalt mix-70 penetration and the Asphalt mix-SBS at 5 °C and 25 Hz, respectively, and they are 5 times and 3 times at 45 °C and 25 Hz; The reason is that the high modulus modifiers can directly provide embedded packing, reinforcement, and cementation to and enhance the stiffness of the mixtures, and so the elastic features were enhanced.

2.4. Wheel Tracking Test

Rutting damage is one of the main diseases of the road surface. The laboratory investigations of the Asphalt mix-70 penetration, the Asphalt mix-SBS, the HMAM-Lubao and the HMAM-H7686 were carried out by wheel tracking test, in order to indicate a conventional performance related to the purpose of using the HMAM. The dynamic stability (D_S), defined by Equation (1), and regarded as an indicator for directly characterizing the resistances to rutting of asphalt pavements, is positively correlated with rut resistance at high temperature. A rut board of 300 mm $\times$ 300 mm $\times$ 50 mm was made by a hydraulic sample-forming machine. The test was performed by the wheel tracking instrument at 60 °C. The resistance to deformations of the four kinds asphalt mixtures were measured based on the results of the wheel tracking test. The wheel tracking test results are listed in Table 7.

$$D_S = \frac{(t_2 - t_1) \times N}{d_2 - d_1} \tag{1}$$

where D_S is the dynamic stability, t_2 is the point at 60 min, t_1 is the point at 45 min, N is the speed, is usually 42 times·min^{-1}, d_2 is the deformation at t_2 and d_1 is the deformation at t_1.

Table 7. The wheel tracking testing results.

Materials	Rut Deformation at 45 min (mm)	Rut Deformation at 60 min (mm)	DS (times·mm^{-1})
Asphalt mix-70 Penetration	3.781	4.182	1571
Asphalt mix-SBS	2.160	2.303	4405
HMAM-Lubao	1.403	1.501	6428
HMAM-H7686	0.680	0.739	10,857

As shown in Table 7, the performance of the two HMAM at high temperature is obviously better than those of ordinary asphalt mixtures. The dynamic stabilities of the HMAM-H7686 and the HMAM-Lubao increase by 7 and 4 times, respectively, compared with the Asphalt mix-70 penetration. The high temperature performance of the HMAM-H7686 modifier is better than that of the HMAM-Lubao compared with the Asphalt mix-SBS, with increases of 2.5 and 1.5 times respectively.

3. Viscoelastic Parameters of High-Modulus Asphalt Mixture (HMAM)

Firstly, based on the results of dynamic modulus test, the master curves of the storage moduli were obtained by non-linear least-square fitting. The Wiechert mechanical model, which consists of 17 Maxwell models and a spring parallel, was then applied to describe its complex mechanical behavior. Finally, the time-domain Prony series expression of the relaxation modulus was obtained.

3.1. Master Curves of Storage Modulus

To carry out the conversion method for the Prony series, the master curves of the storage moduli were obtained, which was proposed by Park et al. [32–34]. The storage modulus was related to the dynamic modulus and the phase angle as [32]:

$$E' = |E^*| \cos \varphi \tag{2}$$

where E' is the storage modulus, $|E^*|$ is the dynamic modulus and φ is the phase angle.

The sigmoidal function (Equation (3)) was selected to describe the master curves of the storage moduli [32]:

$$\log|E'| = \delta + \frac{Max - \delta}{1 + e^{\beta + \gamma \log \omega_r}} \tag{3}$$

where ω_r is the reduce frequency, Max and δ are as lg logarithmic form for the maximum and minimum values of the dynamic modulus, respectively, β and γ are the shape parameters related to the properties of the mixtures. Parameter γ influences the steepness of the function (rate of change between the minimum and the maximum) and β is the horizontal position of the turning point.

The Arrhenius equation was used to calculate the reduced frequency ω_r at reference temperature 20° C [35] and was defined as:

$$\log \omega_r = \log \omega + \frac{\Delta E_a}{19.14714} \left(\frac{1}{T} - \frac{1}{T_r} \right) \tag{4}$$

where ω is the frequency at the reference temperature, ΔE_a is the activation energy, T is the test temperature and T_r is the reference temperature.

The data of the storage modulus were fitted by non-linear least-squares method and the parameters δ and β in Sigmoidal function were obtained by using the programming solving function in Excel, as shown in Table 8 [36]. The shift factor $\alpha(T)$ [37] was obtained based on the principle of time-temperature equivalence using Equation (5) [25], i.e.,

$$\log[\alpha(T)] = \frac{\Delta E_a}{19.14714} \left(\frac{1}{T} - \frac{1}{T_r} \right) \tag{5}$$

Table 8. Main parameters and shift factors of the master curves at reference temperature 20 °C.

Materials	δ	β	γ	ΔE_a	lg (Shift Factors)		
					5 °C	20 °C	45 °C
Asphalt mix-70 penetration	1.4229	−0.37382	−0.81492	169,829	1.6311	0	−2.3768
Asphalt mix-SBS	1.8628	−0.07669	−0.88450	165,627	1.5907	0	−2.3179
HMAM-Lubao	1.8669	−0.83931	−0.56673	208,895	2.0063	0	−2.9235
HMAM-H7686	2.5891	−1.04341	−0.79735	186,848	1.7946	0	−2.6149

Based on the parameters in Table 8, the master curves of the storage moduli of four kinds of asphalt mixtures at the reference temperature were obtained. The master curves of the storage moduli at 20 °C are shown in Figure 3. In addition, the sigmoidal function equations of other main temperature curves could be obtained and the master curves could be drawn basing on the same non-linear fittings.

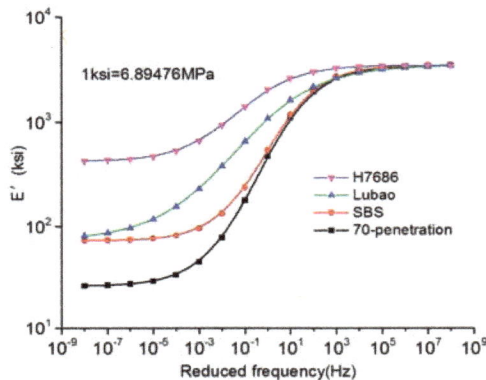

Figure 3. Master curves of the storage moduli at reference temperature 20 °C.

As shown in Figure 3, at low frequencies, the storage moduli of the HMAM were much higher than those of other asphalt mixtures. The loading frequency actually corresponded to the vehicle speed, and the speed decreased as the frequency was reduced. Therefore, the adverse effect with a low speed could effectively be resisted by the HMAP. In addition, the changing rates of the storage moduli of the HMAM were slower, which means that the HMAP was insensitive to the variation of the speeds of the driving load.

3.2. Maxwell Model Parameters

According to theoretical and experimental research, more complex and multivariate models were needed to accurately represent the long-term complex viscoelastic behaviors of asphalt mixtures [38,39]. The Maxwell model with the Wiechert mechanical model is a common mechanical analysis model, which is composed of several Maxwell models and a spring in parallel and can be used to describe more complex mechanical behaviors [40]. In this paper, the Wiechert mechanical model was used to fit the data. The correlation coefficient was greater than 0.99, which showed good agreement. The Wiechert model could also be used to obtain time-domain Prony series expression of the relaxation modulus [41]. The relaxation modulus $E(t)$ is written as [32]:

$$E(t) = E_\infty + \sum_{m=1}^{M} E_m \exp(-t/\rho_m) \qquad (6)$$

where m is the number of parallel models, E_∞ is the infinite relaxation modulus, E_m is the relaxation modulus in the m_{th} term or Prony coefficient and ρ_m is the relaxation time.

The total stress in the Wiechert model was obtained by the summation as [39]:

$$\sigma_m = \sigma_\infty + \sum_{m=1}^{M} \sigma_m \tag{7}$$

where σ is the total stress, and σ_∞ is the limit stress when angular frequency ω approaches 0 from the right side.

The stress, σ_m, in each of the Maxwell components combining a spring with a dashpot is governed by the differential equation [39]:

$$\frac{d\varepsilon}{dt} = \frac{1}{E_m} \frac{d\sigma_m}{dt} + \frac{\sigma_m}{\eta_m} \tag{8}$$

where η_m is the coefficient of viscosity, E_m is the relaxation modulus in the m_{th} term or Prony coefficient, and ε is the strain. The number of terms m used in the fitting is equal to the number of decades for which the fitting is to be done.

Due to the linearity of the material components, the total stress in the Wiechert model is obtained by Equation (9) [39]:

$$\sigma_\infty = E_\infty \varepsilon \tag{9}$$

where σ_∞ is the stress in the m_{th} term, E_∞ is the limit storage modulus when angular frequency ω approaches 0 from the right side and ε is the strain.

By using the relaxation time expression $\rho_m = \eta_m / E_m$, the time-domain could be converted into the frequency-domain. The Prony series expression of the storage modulus can be obtained from Equation (10) as [32]:

$$E'(\omega) = E_\infty + \sum_{m=1}^{M} \frac{\omega^2 \rho_m^2 E_m}{\omega^2 \rho_m^2 + 1}, \ m = 1, 2, \dots M \tag{10}$$

The mechanical parameters of the Prony series of the relaxation modulus could be fitted according to the master curves of the storage modulus after determining the relationship between the storage modulus and the Prony series of the relaxation modulus [42,43]. To solve E_m and ρ_m, the collocation method was usually used rather than solving a nonlinear system of equations with $2m$ unknowns because of the 10^{-8}–10^8 frequency range of the dynamic modulus master curves. A series of relaxation time points were set in advance, and then the parameters corresponding to these relaxation time points could be solved [29]. According to the research, if the distance between relaxation time points was too small, more points needed to be taken, and if the distance was too large, there would be a large fluctuation about the relaxation modulus curve. The relaxation modulus curve determined was stable and there was no fluctuation when the distance between relaxation time points was about 1 on the $\log(\rho_m)$ axis [33,34]. Therefore, when the distance on the $\log(\rho_m)$ axis was taken as unit 1, 17 parallel Maxwell models, namely 17 value groups of E_ms and ρ_ms, would be generated. The relaxation time points of ρ_1–ρ_{17} were determined first, and then the fitting was carried out. In this way the calculations were simplified and the accuracy of the calculated relaxation curve was ensured. The relaxation time points were usually determined in the following form [32]:

$$\rho_m = 2 \times 10^{(m-c)}, \ m = 1, 2, \dots 17 \tag{11}$$

where c is determined according to the range of test specimens and the purpose of the research, in the case of asphalt mixtures, more than 10 relaxation time points should usually be preconfigured.

The fitting could be carried out according to the master curves of the storage moduli after the parameters were determined. In this paper, the parameters of the Prony series expression at 20 °C were obtained, as shown in Table 9. The viscoelastic variation laws of different asphalt mixtures were obtained, and the mechanical response analysis of asphalt mixtures based on actual parameters could be carried out.

Table 9. Parameters in Prony series representations for relaxation moduli at 20 °C.

Relaxation Time Points	ρ_m/s	Asphalt Mix-70 Penetration	Asphalt Mix-SBS	HMAM-Lubao	HMAM-H7686
		E_m/MPa			
1	2.0×10^{-8}	124.69	79.43	338.30	32.47
2	2.0×10^{-7}	195.62	137.46	380.33	50.11
3	2.0×10^{-6}	446.27	336.56	673.41	113.59
4	2.0×10^{-5}	966.78	787.18	1111.94	247.91
5	2.0×10^{-4}	2008.84	1769.92	1776.08	535.40
6	2.0×10^{-3}	3759.50	3581.76	2637.22	1116.83
7	2.0×10^{-2}	5664.62	5742.67	3496.84	2167.92
8	2.0×10^{-1}	5629.70	5783.80	3904.68	3627.44
9	2.0×10^{0}	3124.47	3134.60	3483.50	4630.18
10	2.0×10^{1}	1050.41	1086.52	2409.40	3982.63
11	2.0×10^{2}	335.83	381.05	1407.16	2417.14
12	2.0×10^{3}	104.25	125.17	616.44	1041.63
13	2.0×10^{4}	65.22	78.35	568.91	803.59
14	2.0×10^{5}	42.22	53.33	128.24	105.46
15	2.0×10^{6}	0.00	38.56	2.33	0.76
16	2.0×10^{7}	18.11	20.62	208.97	244.21
17	2.0×10^{8}	25.56	2.13	24.38	295.42
E_∞/MPa		182.5676	502.7753	507.55	2077.00

4. Calculation of Mechanical Responses of Viscoelastic HMAP under Moving Load

4.1. 3D Viscoelastic Finite Element Model (FEM) of Pavement under Moving Load

In this paper, the models of asphalt pavements for the viscoelastic structures were built using ABAQUS, and E_m can be interconverted from the expression of Prony series of the relaxation modulus. E_m was transformed into g_i (the ratio of each elastic modulus to the sum) based on the set requirements of ABAQUS. The model had a dimension of 6 m along the direction of traffic, 6 m across the transverse direction, and 6 m in depth. A 3D model of the same size was built to minimise the edge effect and achieve one full passage of the truck on the pavement to obtain a complete longitudinal strain and stress response curve including the expected compression–tension–compression sequence [24]. There are many advantages in using a 3D FEM: first, the 3D FEM allows the consideration of complex behaviors of pavement material; second, it allows the simulation of different complex situations; third, the analysis results may substitute for the tests. However, the simulation process of moving loads and dynamic analysis require a huge amount of computation [24]. In our model, the x-axis was perpendicular to the wheel path (transverse), the y-axis was along the wheel path (longitudinal), and the z-axis was vertical. To improve the rate of convergence, eight-node brick elements with reduced integration (C3D8R) were used and the pavement model consisted of 66,650 elements and 72,072 nodes. Full interface bonding was assumed between all layers, and the bottom boundary of the model was in full constraint, the side boundary was constrained in the normal direction. The traditional loading method is static loading, which is not in accordance with the actual pavement stress, so it is important to carry out the dynamic analysis of asphalt pavement subjected to traffic load [44]. In the study, the dynamic responses caused by moving load on the HMAP were considered. Therefore, to simulate a moving load, the tyre–pavement contact area was progressively shifted along the wheel path in the direction of traffic until a single tyre pass is completed [45]. The contact area of a truck tyre is in reality closer to a rectangular than to a circular shape regardless of the types of tyre [46]. For the application of moving load, the two wheel moving paths were set symmetrically along the direction of wheel moving load and were refined, which had a dimension of 4 m along the direction of traffic and 0.186 m across the transverse direction. The center distance between the two wheel moving paths was 0.314 m, which was in accordance with the standard truck of China. The wheel path had a length of 4 m along the direction of traffic, and a driving distance of 0.2s at a speed of 72 km/h. The wheel path was discretized in time and space to apply the Haversine wave load, and the Haversine function that changes over time was applied to the wheel moving path. The Haversine function was written as:

$$Q(t) = p_{max} \sin^2\left(\frac{\pi}{2} + \frac{t}{d}\right) \tag{12}$$

where d is the duration of load that depends on the speed v and the wheel contact area radius a. It is generally believed that when the load is $6a$ away from a point, the load has no effect on the point, so we have $d = 12a/v$. When the load is far from the known point, or $t = \pm d/2$, $Q(t) = 0$. When the load directly acts on the point ($t = 0$), the load reaches the peak value, and the load pressure is p_{max}. The simulation of the driving load has been shown in reference [47].

The analyses of stresses under standard load ($p_{max} = 0.7$ MPa) and heavy load ($p_{max} = 1.0$ MPa) were conducted. In the analysis, the meshes of the loading area were refined. The shearing stresses at depths from 0.04 m to 0.10 m of the asphalt pavement were the main focuses, and the resistance to rutting deformation was mainly provided by the middle surface layer, so the HMAM was set in the middle surface layer. Four kinds of asphalt pavements (the Asphalt mix-70 penetration, the Asphalt mix-SBS, the HMAM-Lubao, the HMAM-H7686) were used for the middle surface layer to analyze the mechanical responses, and their stress characteristics and change laws were studied. The FEM model is shown in Figure 4 and the selection of pavement structure parameters were referred to the Specifications for Design of Highway Asphalt Pavement (JTG D50-2017), as shown in Table 10. The viscoelastic parameters of upper and under layer materials come from reference [48].

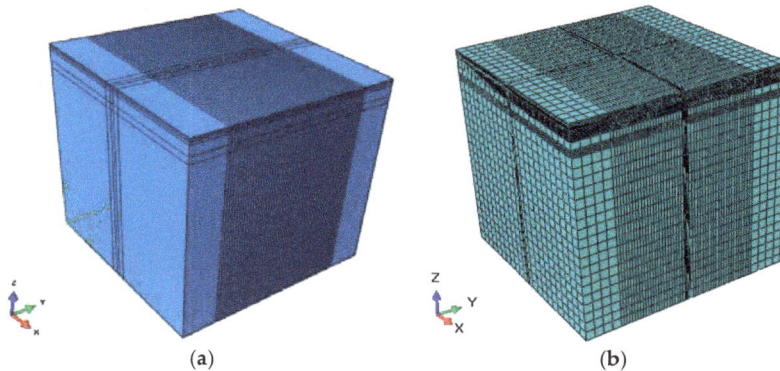

(a)   (b)

Figure 4. 3D pavement finite element model (FEM): (**a**) FEM; (**b**) mesh part.

Table 10. Parameters for pavement structure layers.

Structure Layers	Thickness/mm	Mechanical Parameters/MPa	Poisson's Ratio
Upper layer (SMA-13)	40	viscoelasticity	0.25
Middle surface layer (Four kinds of materials) (AC-20)	60	viscoelasticity	0.25
Underlayer (AC-25)	80	viscoelasticity	0.25
Base	300	7500	0.25
Subbase	200	250	0.35
Subgrade	/	100	0.4

4.2. Mechanical Responses of Pavement Structures

To verify the accuracy of the model established in this paper, the results were compared with those in reference [48]. The parameters of the pavement structures were from reference [48]. The settings of the model size, the drive speed, the drive distance and the Haversine wave load were the same as in the reference. The vertical deformations time-history curve of the road surface was obtained by using the model built in this paper, as shown in Figure 5.

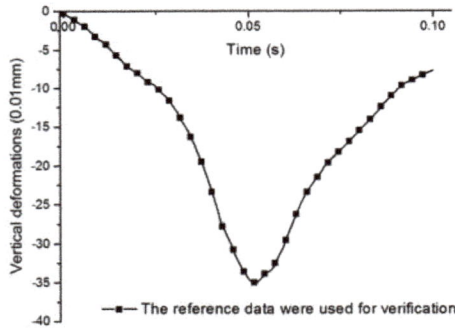

Figure 5. Curves obtained by literature data.

Comparing Figure 6 in this paper with Figure 5 in reference [48], it can be seen that the trend of two curves was same, the peak value of the deformations only differed by 0.19 mm, and the error is within the allowable range of finite element analysis, which verifies the correctness and feasibility of the numerical simulations in this paper.

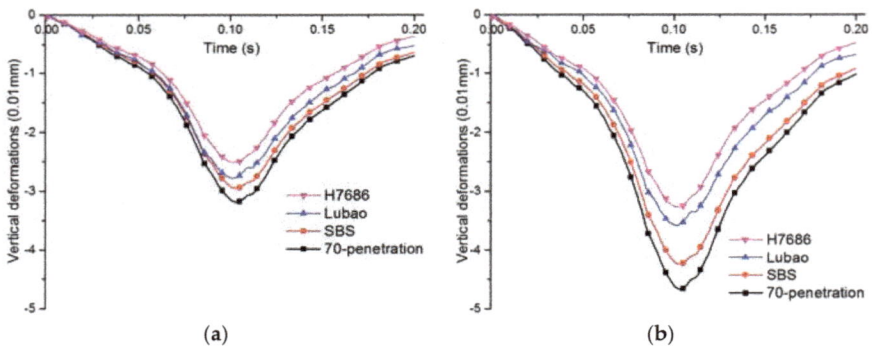

Figure 6. Vertical deformations at centers of road surface: (a) under standard load; (b) under heavy load.

4.2.1. Vertical Deformations at Road Surface

The damage of the road can be caused by the surface deformation in the process of vehicle driving, and the safety factor would reduce. In this paper, the time-varying vertical deformations at the center of different pavements surface under p_{max} = 0.7 MPa and 1.0 MPa were considered. As shown in Figure 6, there is a typical asymmetric distribution on both sides of the curve because of the viscoelasticity of asphalt mixtures. The two kinds of HMAP have more remarkable resistances to deformations under different load levels. Compared with the 70-penetration asphalt pavement, the deformation of the HMAP under 0.7 MPa reduces by about 20%–25%; the resistances to deformations of the HMAP are more prominent under heavy load, with a decline from 29% to 36%. The load levels have a great influence on the vertical deformations on the pavement surface. The vertical deformations at the pavement surface increase significantly with the load levels. The vertical deformation of the Asphalt mix-70 penetration pavement increases by about 50%, the Asphalt mix-SBS pavement increases by 45%, and the HMAP-Lubao pavement and the HMAP-H7686 pavement increase by 29 and 31%, respectively, under the heavy load. The use of the HMAM in the pavement reduces the road damages and the sensitivity to traffic loads, and at the same time increases the rutting resistance and durability.

4.2.2. Shearing Strains of Middle Surface Layers

The rutting problem of asphalt pavements was mainly caused by the shearing deformations of the asphalt layer and the compaction failure of vehicle reciprocating. The shearing strains of the asphalt pavement were mainly concentrated at depths from 0.04 m to 0.10 m, located in the middle surface layer [19] and, therefore, it is necessary to analyze the shearing strains of the middle surface layer. The time-history curves of the shearing strains are shown in Figure 7. It can be seen that the modulus of the HMAP has greater influence on the maximum shearing strain of the pavement structures. The peak values of the shearing strain of the middle surface layer decrease significantly with the increasing of the moduli. Compared with the Asphalt mix-70 penetration pavement, the shearing strain of the HMAP-H7686 pavement can be reduced by as much as 8 $\mu\varepsilon$, and the decreasing rate was 57%. The peak values of the shearing strains of the middle layer increase to different degrees with the load levels, and the similar change laws of the shearing strains under standard and heavy loads were obtained.

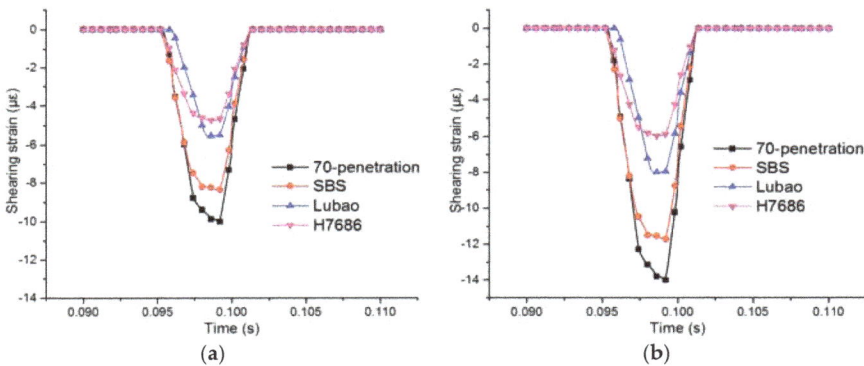

Figure 7. Shearing strains of middle surface layers: (**a**) under standard load; (**b**) under heavy load.

4.2.3. Stresses and Strains at Bottom of Underlayers

The transverse strain of the asphalt layer under the symmetrical load was 0 since the plane of the vehicle moving along the wheel path was a symmetrical plane, so the transverse strain analysis of the bottom was not performed. The time-history curves of the tensile strain at the bottom center of the underlayer under the two load levels are shown in Figure 8. As shown in Figure 8, the change laws are similar for both situations. The alternating change of pressure-pull-pressure occurs at the bottom of the asphalt layer after loading, which was consistent with the viewpoints in reference [28,29].

The fatigue damages of asphalt pavements were easily caused by extortionate tensile stress and tensile strain. The HMAM-H7686 pavement has the highest tensile strain, followed by the Asphalt mix-70 penetration pavement, the Asphalt mix-SBS pavement and the HMAM-Lubao pavement. The peak values of the tensile strains increase significantly with the loads, and the four kinds of asphalt mixtures increase by 40% to 52%. Although the dynamic modulus of the HMAM-H7686 was larger than that of the HMAM-Lubao, the tensile strains at the bottom of the underlayer cannot be effectively decreased by the HMAM-H7686 pavement, and the tensile strain of the HMAM-Lubao asphalt pavement reduces by 16% compared with that of the HMAM-H7686 pavement.

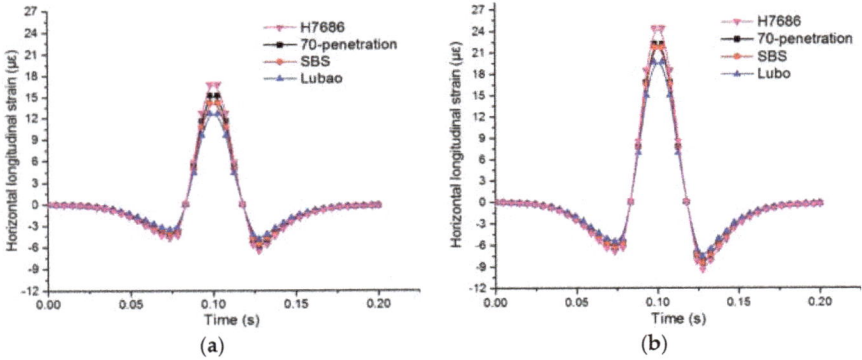

Figure 8. Horizontal longitudinal tensile strains at the bottom center of underlayers: (**a**) under standard load; (**b**) under heavy load.

The tensile stress and strain need to be considered together, for the cracking occurred at the bottom of the asphalt layer. The cracking may be associated with tensile stress at the bottom of the road surface. The tensile stresses at the bottom layer in different depths were extracted downward along the center of the wheel gap in order to explore the reason of cracking. As shown in Figure 9, there is obvious excessive tensile stress at the bottom of the underlayer of the HMAM-H7686 pavement. The tensile stresses are about 70% and 110% higher than those of the HMAM-Lubao and the Asphalt mix-70 penetration pavements, respectively. The excessive tensile stress at the bottom of the layer was produced after setting of the HMAM-H7686 layer, resulting in a larger peak value of the tensile strain. The excessive tensile stress at the bottom of the asphalt layer may be due to the too-large modulus, and the risk of cracking of the HMAM is significantly increased; therefore, it is necessary to effectively control the ranges of the moduli under the premise that the resistance to deformations of the structure should be satisfied in the selection of materials.

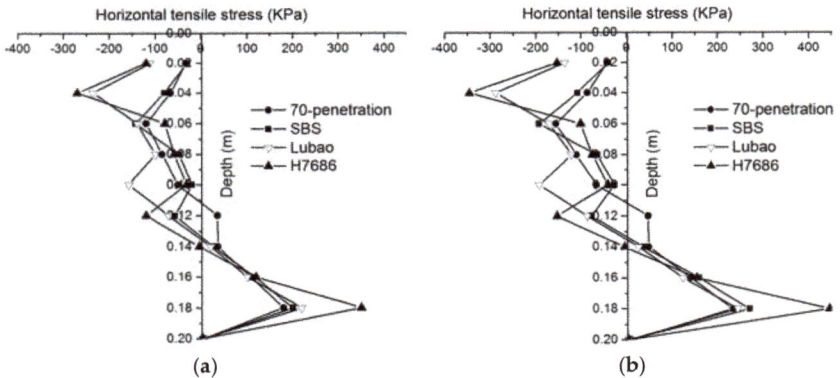

Figure 9. Horizontal longitudinal tensile stresses at the bottom center of underlayers: (**a**) under standard load; (**b**) under heavy load.

The time-history curves of the vertical strain at the bottom of the underlayers under standard and heavy loads are shown in Figure 10, and similar change laws are observed in both cases. After loading, the alternating change of pull-pressure occurs at the bottom of the asphalt layer. A typical asymmetrical distribution appears on both sides of the curve due to the viscoelasticity of asphalt mixtures. The pavements sorting by the peak values of the vertical strains in descending order are: the Asphalt mix-70 penetration pavement, the Asphalt mix-SBS pavement, the HMAM-Lubao pavement and the HMAM-H7686 pavement. HMAP show good resistances to vertical deformations. As the pavement structure with higher modulus, the vertical strains of the HMAM-H7686 asphalt pavement reduce by 6% and 14% under standard and heavy loads compared with the HMAM-Lubao pavement. It can be seen that the resistance of the asphalt pavement to vertical deformations at the bottom of the layers is significantly enhanced due to the increasing dynamic modulus.

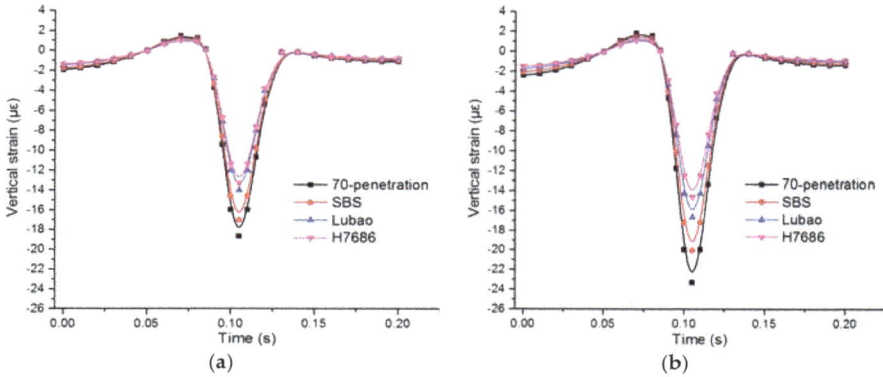

Figure 10. Vertical strains at bottom center of underlayers: (**a**) under standard load; (**b**) under heavy load.

4.2.4. Compressive Strains at Top Center of Subgrades

The resistance to overall deformations of pavements can be reflected by the compressive strain at the top center of the subgrade. The time-history curves of the compression strains at the top center of the subgrade under standard and heavy loads are shown in Figure 11. The compressive strain begin to appear and gradually increase to the peak value when the wheel is near the center, and the time-history curves are asymmetric. The pavements sorting by the peak values of the compressive strains at the top center of the subgrade in descending order are: the Asphalt mix-70 penetration pavement, the Asphalt mix-SBS pavement, the HMAM-Lubao pavement and the HMAM-H7686 pavement. The compressive strains at the top surface of the HMAM-H7686 pavement and the HMAM-Lubao pavement decrease by 40% and 42%, 35% and 26%, respectively, compared with those of the Asphalt mix-70 penetration pavement under standard and heavy loads. Above all, similar to the change laws of the vertical strain of the underlayer, there is a remarkable resistance to vertical deformations in the HMAP.

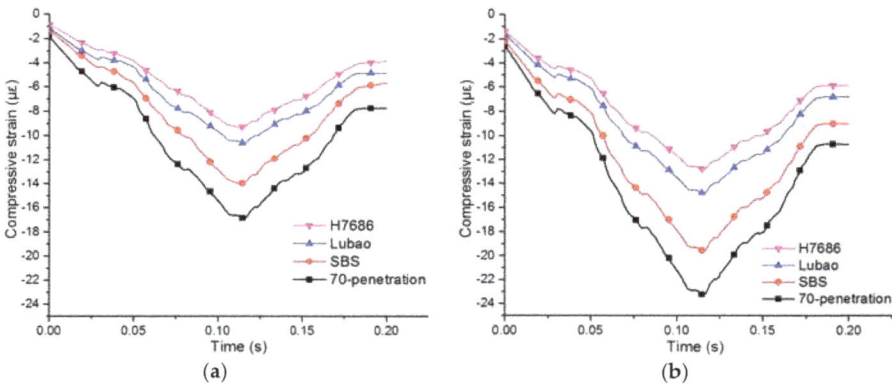

Figure 11. Compressive strains at top center of subgrades: (**a**) under standard load; (**b**) under heavy load.

4.3. Sensitivity of HMAP to Level of Loads

In order to analyze the effect of increasing load levels on the mechanical responses of different asphalt pavements, the peak values of the vertical deformations at the center of the road surface, the shearing strains of the middle surface layer, the vertical strains at the bottom center of the underlayer, and the compressive strains at the top center of the subgrade were extracted. The increasing rates of the mechanical responses are obtained with the load change, as shown in Table 11.

Table 11. Increasing of mechanical responses from standard to heavy loads.

Sample Types	Asphalt Mix-70 Penetration	Asphalt Mix-SBS	HMAM-Lubao	HMAM-H7686
	Increasing Rates of Mechanical Responses/%			
Vertical deformations of the road surface	47	44	29	30
Shearing strains of the middle surface layer	40	45	37	25
Horizontal tensile strains of asphalt layer bottom	49	52	42	45
Vertical strains of asphalt layer bottom	24	17	18	9
Compressive strains at the top of the subgrade	42	39	36	32

As shown in Table 11, the HMAP can not only improve the resistance to deformation, but also reduce sensitivity to the variation of the traffic loads, and at the same time enhance the rutting resistance and durability. However, the increasing rates of the tensile strain at the bottom of asphalt layer of the HMAP are similar to those of other pavements, reaching more than 42%, illustrating that the stress concentration at the bottom of the asphalt layer is more serious as the loads increase, and cracking at the bottom of the asphalt layer is more prone to occur. Therefore, it was necessary to control the tension at the bottom of the asphalt layer effectively under the premise that resistance to the deformations of the structure should be satisfied in practice.

5. Conclusions

A more extensive characterization of the mixtures was presented by a wheel tracking test. The dynamic moduli were tested by laboratory investigations and the viscoelastic parameters were obtained. The 3D viscoelastic FEM of pavements under moving loads were established and the mechanical responses were analyzed; the mechanical characteristics of the pavement structures after the setting of the high-modulus layer and the sensitivity of various pavement structures to the level of the loads were objectively analyzed to provide the theoretical basis for improving the structure design of the HMAP, and the following conclusions can be drawn:

(1) The wheel tracking test results indicate that the addition of the high-modulus modifiers can improve the high-temperature stability of the pavement. Dynamic modulus tests were successfully conducted to obtain the viscoelastic parameters and study the mechanical properties of the HMAM. The results will be helpful in interpreting the modifier behavior by explaining the change laws in the viscoelastic parameters with the loading temperature and frequency.

(2) The changing rates of the storage modulus curves of the HMAM were slower than those of other asphalt mixtures, which mean that the HMAP were insensitive to the variation of the speeds of the driving load.

(3) The impacts of the viscoelasticity on the mechanical responses of different pavements were identified, and the process of the mechanical responses was represented by the suggested model. The vertical deformations of the road surface, the shearing strains of the middle surface layer, the vertical strains of the underlayer, and the compressive strains of the subgrade of the two HMAP were significantly lower than those of two other pavements, and the two HMAPs perform well against the deformations. The alternating change of pull-pressure occurs at the bottom of

the asphalt layer after loading and a typical asymmetric distribution appears on both sides of the curves due to the viscoelasticity of the asphalt mixtures.

(4) The tensile strain and stress at the bottom of the underlayer of the HMAM-H7686 pavement do not decrease with the increase of the dynamic modulus, illustrating that the change laws of the tensile strain and stress are affected by the range of the dynamic modulus. The HMAM-H7686 pavements have the highest tensile strain, followed by the Asphalt mix-70 penetration pavements, the Asphalt mix-SBS pavements and the HMAM-Lubao pavement. The tensile strains at the bottom of the under layer cannot be effectively reduced by the HMAM-H7686, which has the highest dynamic modulus. There were obvious excessive tensile stress at the bottom of the underlayer for the HMAM-H7686 pavement, and the tensile stresses was about 70%, 110% higher than those of the HMAM-Lubao pavements and the Asphalt mix-70 penetration pavements, respectively. In conclusion, the range of the modulus of the materials must be controlled to avoid cracking at the bottom of the layer when the HMAM is selected.

(5) The load levels have great effects on the mechanical responses, and the degrees to which the mechanical response is affected by different load levels were discussed in detail. The HMAP is insensitive to the load changes and could better withstand the adverse effects of the heavy load. But the increasing rates of more than 42% of the tensile strain at the bottom of the underlayer of the HMAP were similar to other pavements, which means that the tensile stress at the bottom of the asphalt layer is more serious as the loads increase, and the bottom of the asphalt layer is more prone to cracking.

Author Contributions: Y.S. organized the research; B.G., L.L., R.G. and Q.Y. performed dynamic modulus tests; B.G. and L.G. carried out numerical calculation; Y.S. and B.G. wrote the manuscript; Y.S. and J.W. checked the manuscript.

Funding: The research is funded by the National Natural Science Fund (51478276), the Natural Science Foundation of Liaoning Province (20170540770).

Acknowledgments: This research was performed at the Shenyang Jianzhu University, College of Architecture Engineering of Chongqing University of Arts and Sciences and Institute of Transportation Engineering of Zhejiang University.

Conflicts of Interest: The authors declare no conflict of interest.

References

1. Tian, W.Q.; Zhou, B.; Cong, L. High-temperature performance of modified asphalt mixture and its evaluation method. *J. Build. Mater.* **2009**, *12*, 285–287.
2. Yang, C.; Xie, J.; Zhou, X.J.; Liu, Q.T.; Pang, L. Performance Evaluation and Improving Mechanisms of Diatomite-Modified Asphalt Mixture. *Materials* **2018**, *11*, 686. [CrossRef] [PubMed]
3. Pasetto, M.; Baldo, N. Resistance to Permanent Deformation of Road and Airport High Performance Asphalt Concrete Base Courses. *Adv. Mater. Res.* **2013**, *723*, 9. [CrossRef]
4. Huang, C.W.; Alrub, R.K.A.; Masad, E.A. Three-Dimensional Simulations of Asphalt Pavement Permanent Deformation Using a Nonlinear Viscoelastic and Viscoplastic Model. *J. Mater. Civ. Eng.* **2011**, *23*, 56–68. [CrossRef]
5. Sha, A.M.; Zhou, Q.H.; Yang, Q. High modulus asphalt concrete material composition design method. *J. Changan Univ. Nat. Sci. Ed.* **2009**, *3*, 1–5.
6. Moreno, F.; Sol, M.; Tomás, E. High-Modulus Asphalt Mixtures Modified with Acrylic Fibers for Their Use in Pavements under Severe Climate Conditions. *J. Cold Regions Eng.* **2016**, *30*, 04016003. [CrossRef]
7. Zhen, Y.E. Asphalt Pavement Structure Layer Selection and Evaluation Model Based on Analytic Hierarchy Process. *Res. Appl. Build. Mater.* **2014**, *2*, 14–17.
8. Wang, G.; Liu, L.P.; Sun, L.J. Study on the deformation resistance of high modulus asphalt concrete. *J. Tongji Univ. Nat Sci. Ed.* **2012**, *40*, 217–222.
9. Liu, P.; Xing, Q.; Wang, D.; Oeser, M. Application of Dynamic Analysis in Semi-Analytical Finite Element Method. *Materials* **2017**, *10*, 1010. [CrossRef]

10. Li, H.B. Performance Study of High Modulus Asphalt Mixture and Response Analysis of Pavement Structure. *Highw. Eng.* **2014**, *6*, 100–102.

11. Zhang, X.Y.; Wang, R.L. Study on mechanical response of high modulus asphalt mixture pavement. *Inner Mong. Highw. Transp.* **2016**, *3*, 1–3.

12. Zhou, C.J. Analysis of the Structural Response of Liaoning Provincial Governor under Heavy Load of Asphalt Pavement. Ph.D. Thesis, Harbin Institute of Technology, Harbin, China, June 2009.

13. Gao, M. Analysis on the influence of high modulus asphalt concrete on high temperature stress of asphalt pavement structure. *J. China Foreign Highw.* **2018**, *1*, 283–287.

14. Jurowski, K.; Grzeszczyk, S. Influence of Selected Factors on the Relationship between the Dynamic Elastic Modulus and Compressive Strength of Concrete. *Materials* **2018**, *11*, 477. [CrossRef] [PubMed]

15. Moreno-Navarro, F.; Sol-Sánchez, M.; Rubio-Gámez, M.C.; Segarra-Martínez, M. The use of additives for the improvement of the mechanical behavior of high modulus asphalt mixes. *Const. Build. Mater.* **2014**, *70*, 65–70. [CrossRef]

16. Lee, H.J.; Lee, J.H.; Park, H.M. Performance evaluation of high modulus asphalt mixtures for long life asphalt pavements. *Const. Build. Mater.* **2007**, *21*, 1079–1087. [CrossRef]

17. Motamed, A.; Bhasin, A.; Liechti, K.M. Constitutive modeling of the nonlinearly viscoelastic response of asphalt binders; incorporating three-dimensional effects. *Mech. Time-Depend. Mater.* **2013**, *17*, 83–109. [CrossRef]

18. Kumar, S.A.; Alagappan, P.; Krishnan, J.M.; Veeraragavan, A. Mechanical Response of Modified Asphalt Pavements. *Congr. Transp. Dev. Inst.* **2011**, 438–448. [CrossRef]

19. Qiu, Z.P. Numerical simulation of rutting behavior of high modulus asphalt concrete pavement. Ph.D. Thesis, Chang'an University, Xian, China, May 2009.

20. Zheng, M.; Han, L.; Wang, C.; Xu, Z.; Li, H.; Ma, Q. Simulation of Permanent Deformation in High-Modulus Asphalt Pavement with Sloped and Horizontally Curved Alignment. *Appl. Sci.* **2017**, *7*, 331. [CrossRef]

21. Espersson, M. Effect in the high modulus asphalt concrete with the temperature. *Const. Build. Mater.* **2014**, *71*, 638–643. [CrossRef]

22. Woodroofe, J.H.F.; Leblanc, P.A. Heavy vehicle axle dynamics; rig development instrumentation. In Proceedings of the International Symposium on Heavy Vehicle Weights and Dimensions, Kelowna, BC, Canada, 8–13 June 1988.

23. Al-Qadi, I.L.; Wang, H.; et al. Dynamic Analysis and In Situ Validation of Perpetual Pavement Response to Vehicular Loading. *Transp. Res. Rec.* **2008**, *2087*, 29–39. [CrossRef]

24. Sarkar, A. Numerical comparison of flexible pavement dynamic response under different axles. *Int. J. Pavement Eng.* **2016**, *17*, 377–387. [CrossRef]

25. Papagiannakis, A.T.; Amoah, N.; Taha, R. Formulation for Viscoelastic Response of Pavements under Moving Dynamic Loads. *J. Transp. Eng.* **1996**, *122*, 140–145. [CrossRef]

26. Siddharthan, R.V.; Yao, J.; Sebaaly, P.E. Pavement Strain from Moving Dynamic 3D Load Distribution. *J. Transp. Eng.* **1998**, *124*, 557–566. [CrossRef]

27. Vaitkus, A.; Paliukaitė, M. Evaluation of Time Loading Influence on Asphalt Pavement Rutting. *Procedia Eng.* **2013**, *57*, 1205–1212.

28. Zhang, H.Z.; Ren, J.D.; Ji, L. Full scale mechanical response of asphalt pavement is measured and simulated. *J. Harbin Inst. Technol.* **2016**, *48*, 41–48.

29. Huang, Z.Y.; Chen, Y.W.; Yan, K.Z.; Zhou, J. Dynamic response analysis of viscoelastic asphalt pavement under the action of moving load of heterogeneous distribution. *J. Chongqing Jiaotong Univ. Nat. Sci. Ed.* **2017**, *36*, 30–35.

30. Dong, Z.J.; Tan, Y.Q.; Ou, J.P. Dynamic response analysis of asphalt pavement under the action of moving load of three-direction non-uniform distribution. *J. civ. Eng.* **2013**, *6*, 122–130.

31. Kim, J. General Viscoelastic Solutions for Multilayered Systems Subjected to Static and Moving Loads. *J. Mater. Civ. Eng.* **2011**, *23*, 1007–1016. [CrossRef]

32. Park, S.W.; Schapery, R.A. Methods of interconversion between linear viscoelastic material functions. Part I—A numerical method based on Prony series. *Int. J. Solids Struct.* **1999**, *36*, 1653–1675. [CrossRef]

33. Mun, S.; Zi, G. Modeling the viscoelastic function of asphalt concrete using a spectrum method. *Mech. Time-Depend. Mater.* **2010**, *14*, 191–202. [CrossRef]

34. Sun, L.; Zhu, Y. A serial two-stage viscoelastic–viscoplastic constitutive model with thermodynamical consistency for characterizing time-dependent deformation behavior of asphalt concrete mixtures. *Const. Build. Mater.* **2013**, *40*, 584–595. [CrossRef]

35. Zhao, Y.Q.; Wu, J.; Wen, J. Fatigue Performance of Bridge Deck Pavement Materials. *Determ. Anal. Dyn. Modul. Princ. Curve Asph. Mix.* **2006**, *8*, 163–167.

36. Wang, H.P.; Yang, J.; Shi, X.Q. Study on dynamic modulus and principal curve of high modulus asphalt mixture. *Highw. Transp. Technol.* **2015**, *32*, 12–17.

37. Chen, H.; Luo, R.; Liu, H.Q. Dynamic modulus and phase Angle principal curves of asphalt mixture are studied based on generalized sigmoid model. *Wuhan Univ. Technol. Transp. Sci. Eng.* **2017**, *41*, 141–145.

38. Zhao, Y.Q.; Tang, J.M.; Bai, L. Study on slack modulus of asphalt mixture by complex modulus. *J. Build. Mater.* **2012**, *15*, 498–502.

39. Mun, S.; Chehab, G.; Kim, Y.R. Determination of Time-domain Viscoelastic Functions using Optimized Interconversion Techniques. *Road Mater. Pavement Des.* **2007**, *8*, 351–365. [CrossRef]

40. Ghoreishy, M.H.R. Determination of the parameters of the Prony series in hyper-viscoelastic material models using the finite element method. *Mater. Des.* **2012**, *35*, 791–797. [CrossRef]

41. Osborne, M.R.; Smyth, G.K. A Modified Prony Algorithm For Exponential Function Fitting. *SIAM J. Sci. Comp.* **1995**, *16*, 119–138. [CrossRef]

42. Park, S.W.; Kim, Y.R. Fitting Prony-Series Viscoelastic Models with Power-Law Presmoothing. *J. Mater. Civ. Eng.* **2001**, *13*, 26–32. [CrossRef]

43. Schapery, R.A. A Simple Collocation Method for Fitting Viscoelastic Models to Experimental Data. Available online: http://resolver.caltech.edu/CaltechAUTHORS:20141114-115330896 (accessed on 6 December 2018).

44. Yang, C.F.; Wang, L. Dynamic response analysis of asphalt pavement under moving non-uniform load. *Subgrade Eng.* **2012**, *5*, 93–95.

45. Yoo, P.J.; Al-Qadi, I.L.; Elseifi, M.A. Flexible pavement responses to different loading amplitudes considering layer interface condition and lateral shear forces. *Int. J. Pavement Eng.* **2006**, *7*, 14. [CrossRef]

46. Weissman, S.L. Influence of tire-pavement contact stress distribution on development of distress mechanisms in pavements. *Transp. Res. Rec.* **1999**, *1655*, 161–167. [CrossRef]

47. Zhao, Y.Q.; Zhou, C.H.; Wang, G.Z. Analysis of viscoelastic response of asphalt pavement under impulse load. *J. Dalian Univ. Technol.* **2011**, *51*, 73–77.

48. Zhao, Y.Q.; Zhong, Y. Analysis of dynamic viscoelastic response of asphalt pavement. *Vib. Impact* **2009**, *28*, 159–162.

materials

MDPI

Article

The Effect of Ultraviolet Radiation on Bitumen Aging Depth

Jinxuan Hu, Shaopeng Wu *, Quantao Liu, Maria Inmaculada García Hernández [ID], Wenbo Zeng, Shuai Nie, Jiuming Wan [ID], Dong Zhang and Yuanyuan Li

State Key Laboratory of Silicate Materials for Architectures, Wuhan University of Technology, Wuhan 430070, China; hujinxuan221@whut.edu.cn (J.H.); liuqt@whut.edu.cn (Q.L.); maria.espersson@whut.edu.cn (M.I.G.H.); zwb0212@whut.edu.cn (W.Z.); nies1993@whut.edu.cn (S.N.); wanjm@whut.edu.cn (J.W.); pytmac@whut.edu.cn (D.Z.); liyuanyuan@whut.edu.cn (Y.L.)
* Correspondence: wusp@whut.edu.cn; Tel.: +86-138-0717-6062

Received: 3 April 2018; Accepted: 3 May 2018; Published: 7 May 2018

Abstract: The aging effect of ultraviolet (UV) radiation on bitumen has gained increasing attention from researchers, resulting in the emergence of a new method to simulate the UV aging that occurs during the service life of bitumen. However, the UV aging degree is closely related to bitumen thickness and the effect of UV radiation on aging depth is not clear. The relationship between ultraviolet (UV) radiation and bitumen UV aging depth was investigated in this paper. Three groups of samples were UV aged using different aging procedures to investigate the bitumen aging mechanism of UV radiation. The results from the first group showed that UV aging depth increased along with aging time. After aging for five hours, the complex modulus of the second and third layers increased. The second group's results indicated that the aging effect of ozone was small and that the increase in aging depth was uncorrelated with ozone. The results from the third group showed that the transmittance of bitumen increased after UV aging and that the real reason why aging depth increased was permeation.

Keywords: ultraviolet radiation; bitumen; aging depth; transmittance; permeation

---

1. Introduction

Bitumen has been applied in the road construction of flexible pavement for many years. As a viscoelastic material, bitumen's properties are closely related to many aspects of road performance [1]. However, because of external environmental effects such as oxygen [2] and ultraviolet (UV) radiation [3,4], bitumenious properties do not always satisfy operating requirements [5]. During the lifetime of pavement, bitumen experiences various aging processes. After the aging process, bitumen becomes harder and more brittle, which results in the degradation of pavement properties. Bitumen aging is a very complex process but the main cause is oxidation [6,7]. Oxidation results in consistency increasing and the volatile loss of bitumen [8]. The lifetime of pavement is reduced due to hard bitumen, especially under the heavy traffic conditions [9–11].

Bitumen aging generally consists of short-term thermal oxidation aging, long-term thermal aging and UV aging [12,13]. Although UV aging only occurs on superficial layers [14], the effect of UV aging on the properties of bitumen and pavement cannot be ignored [15,16]. Short-term thermal aging is generally evaluated using aging tests, such as the Thin Film Oven Test (TFOT, ASTM D 1754) or Rolling Thin Film Oven Test (RTFOT, ASTM D 2872) [17]. Long-term thermal aging is generally assessed by the Pressure Aging Test (PAV, ASTM D 6521) [18]. Unlike thermal aging, there is still no standard UV aging method for bitumen. After UV aging, the characterization methods for bituminous properties are still not unified [19,20].

The sample thicknesses adopted by some references are shown in Table 1. Sample thickness is an important parameter of UV aging methods, which is often ignored by researchers. Bitumen, after UV aging, may show different results if samples thicknesses are different. For example, UV radiation has been shown to have a significant effect on bitumen film with a thickness of 3 μm because of photochemical reactions. If the film thickness is thicker, the aging effect will be relatively smaller [21]. Katsuyuki Yamaguchi [22] used specimens with five different thicknesses of 50, 100, 200, 500 and 1000 μm to investigate the effect of film thickness on the UV aging of bitumen. The results showed that the thicker films had a lower elastic modulus, higher viscosity and relatively less production of carbonyl groups. The aging degree increased rapidly, especially when the film thickness was below 200 μm. Shaopeng Wu applied samples with four different film thicknesses (50, 100, 200 and 500 μm) to study the thermal, chemical and rheological properties of UV aged bitumen. The results indicated that the UV aging degree of bitumen was closely related to the film thickness of the sample.

Table 1. UV aging conditions and sample thicknesses of previous research.

Author	UV Aging Conditions	Sample Thickness	References
Virginie Mouillet	Temperature: 60 °C Aging time: 60 h	Sample thickness: 10 μm	[23]
Françoise Durrieu	Temperature: 60 °C Aging time: 170 h	Sample thickness: 10 μm	[16]
Xinyu Zhao	Temperature: 60 °C Aging time: 72 h	Sample thickness: 0.5 mm	[13]
de Sá Araujo	Temperature: 60 °C Aging time: 10, 20, 30, 40, 50, 100, 150 and 200 h	Sample thickness: 0.6 mm	[24]
Henglong Zhang	Temperature: 60 °C Aging time: 12 d	Sample thickness: 1.92 mm	[25]
Zhengang Feng	Temperature: 60 °C Aging time: 6 d	Sample thickness: 2 mm	[26]
Peiliang Cong	Temperature: 60 °C Aging time: 7 d	Sample thickness: 3 mm	[27]
Song Xu	Temperature: 60 °C Aging time: 9 d	Sample thickness: 3.2 mm	[28]

From previous research, we know that UV radiation cannot influence the whole sample in a short time. Whether UV radiation influences bitumen at the bottom of samples where UV radiation cannot reach is still unknown. UV lamps release ozone during their runtime. The relationship between the aging effect on thin bitumen film and ozone is still unknown. Therefore, this paper describes a study investigating the relationship between UV radiation and UV aging depth. Three groups of samples were UV aged using different aging procedures to investigate the bitumen aging mechanism of UV radiation. Rheological properties and spectrum transmittance were tested to evaluate changes to bitumen properties.

2. Materials and Methods

2.1. Materials

The studied base bitumen was 60/80 penetration grade bitumen (B) obtained from KOCH Asphalt Co. Ltd. (Ezhou, Hubei Province, China). The physical properties of B are listed in Table 2.

Table 2. Physical properties of B.

Physical Properties	B
Softening point (°C)	49.0
Penetration (25 °C, 0.1 mm)	77.5
Ductility (5 °C, 1 cm/min)	8.9
Viscosity (60 °C, Pa s)	205
Viscosity (135 °C, Pa s)	0.46

2.2. Aging Procedure

Samples were exposed to UV radiation in a UV weathering oven (Fuzhou Meide Testing Instruments Co., Ltd., Fuzhou, China). The UV lamp was 500 W with a main wavelength of 365 nm. The aging procedures of different groups of samples were as follows.

- The first group of samples was designed to investigate whether the aging depth would increase with aging time. Bitumen were poured into an Ø90 mm glass petri dish with its lid removed. The periphery of the dish was separated from the UV radiation by an insulating layer so that the UV radiation could only penetrate from the surface. Samples were UV aged for 1 h, 5 h and 10 h at 50 °C so that the UV aging procedure would not be affected by thermal oxidation aging [3]. The average intensity of UV radiation on the samples' surface was 10 W/m^2. The peeling procedure, dissolving layer by layer, is shown in Figure 1. After UV aging, bitumen on the surface was dissolved by carbon disulfide layer by layer (a) and (b). The dissolved bitumen for one layer was almost 0.8 g and the thickness of the layer was approximately 105 μm. Then the solutions were dried for 7 d (c) and the residues (d) were tested by DSR. Every sample was dissolved three times. The first layer means that the layer was dissolved at the beginning and the third layer means that the layer was dissolved at the end. The first, second and third layers of B aged by UV for 1 h were abbreviated as B-1h-1, B-1h-2 and B-1h-3.

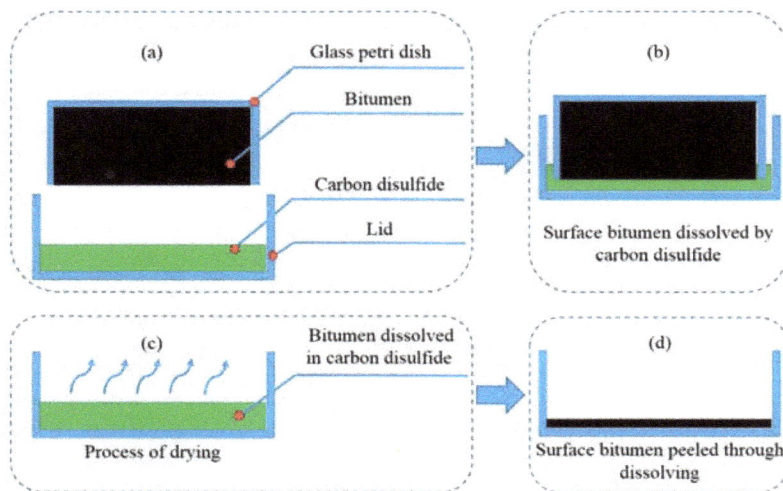

Figure 1. Peeling procedure, dissolving layer by layer.

- The second group of samples was used to explore the aging effect between samples with and without UV radiation. Bitumen was dissolved by carbon disulfide first and then the solutions were dropped onto a blank KBr slide [29,30]. After drying, the solutions turned into a thin film. The concentration of the solutions was 5 wt % and the thickness of the film after drying was approximately 10 μm. Samples were divided into two parts, as shown in Figure 2. These parts were separated by a septum, which UV radiation cannot penetrate. The region between the septum and the samples below the septum was approximately 1 cm, which guaranteed that the atmosphere could circulate. Samples were UV aged for 5 d at 50 °C. The average intensity of UV radiation on the Bu' surface was 10 W/m^2. The ozone concentration was about 0.4 ppm in the environment and about 1.2 ppm in the oven, which was tested by a gas detector (uSafe 2000/300, Sundo Shenzhen Technology Co., Ltd., Shenzhen, China). After UV aging, samples were tested by Fourier Transform Infrared (FTIR, Nexus, Thermo Nicolet Corp., Waltham, MA, USA) Spectroscopy. Samples on the septum and samples below the septum were abbreviated as Bu and Bb respectively.

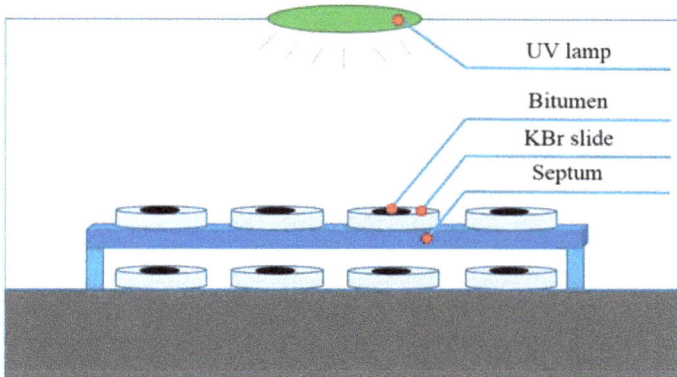

Figure 2. Samples were divided into two parts by septum.

- The third and last group of samples was used to study the UV transmittance of bitumen with aging time. Bitumen was dissolved by carbon disulfide, then solutions with different concentrations (2.5%, 5%, 10%, 15%, 20%, and 30%) were dropped into a quartz glass slice, respectively, and different thicknesses (1.1 µm, 2.3 µm, 5.6 µm, 6.3 µm, 8.1 µm, and 13.1 µm) were obtained. Then, the slice was put on the spin coater and ultrathin bitumimous film was prepared. Samples were UV aged for 0 h, 1 h, 5 h, 10 h, or 50 h. Transmittance of the blankly quartz glass slice with 1 mm thickness is shown in Figure 3. The UV radiation transmittance of the slice from 200 nm to 400 nm was larger than 85%. It was considered that the slice would not influence the UV radiation transmittance in this paper.

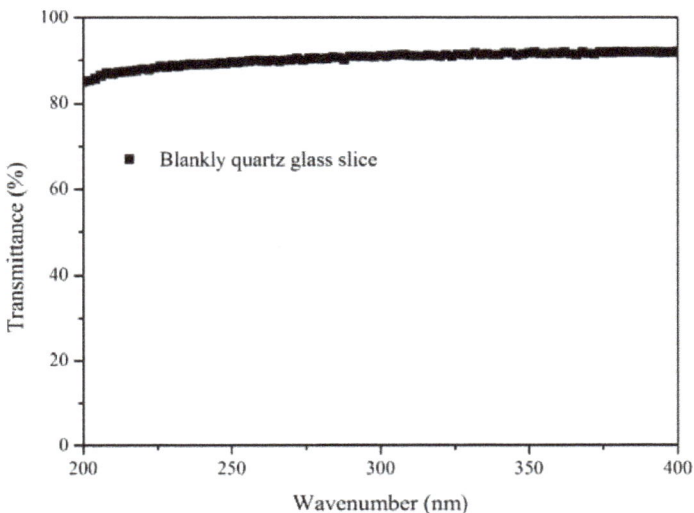

Figure 3. Transmittance of the blankly quartz glass slice with 1 mm thickness.

2.3. Characterization Method

2.3.1. Dynamic Shear Rheometer (DSR)

The rheological properties of the bitumen samples were tested by DSR (MCR101, Anton Paar Corp., Graz, Austria) under strain-controlled mode. High temperature sweep tests were adopted, with temperatures ranging from 30 to 60 °C. The constant frequency was 10 rad/s and the temperature increment was 2 °C per minute. The diameter of the plate was 25 mm and the gap between the plates was 1 mm. Essential rheological parameters such as complex modulus (G*) and phase angle (δ) can be obtained from a DSR test.

2.3.2. Fourier Transform Infrared (FTIR) Spectroscopy

FTIR test was performed under transmission mode, which provides the transmittance of the bitumen under each wavelength. The spectra of the bitumen ranging from 4000 cm^{-1} to 400 cm^{-1} can be obtained from FTIR (Nexus, Thermo Nicolet Corp., Waltham, MA, USA). The spectral resolution was 4 cm^{-1}. Chemical structures can be distinguished from the spectra, which are related to different chemical bonds. From the FTIR test, the sulfoxide group S=O (centered around 1030 cm^{-1}) and the carbonyl group C=O (centered around 1700 cm^{-1}) were monitored by studying their spectra changes. From the carbonyl group and sulfoxide group, information on the oxidation of the asphalt could be characterized. Through the area of their bonds in the spectra, the sulfoxide group S=O index ($I_{S=O}$) and carbonyl group C=O index ($I_{C=O}$) could be calculated using the following equations [31,32], which were used to characterize the aging degree:

$$I_{S=O} = \frac{\text{Area of sulphoxide band centered around} 1030 \text{ cm}^{-1}}{\sum \text{Area of spectral bands between 2000 and 600 cm}^{-1}} \quad (1)$$

$$I_{C=O} = \frac{\text{Area of carbonyl band centered around 1700 cm}^{-1}}{\sum \text{Area of spectral bands between 2000 and 600 cm}^{-1}} \quad (2)$$

2.3.3. Ultraviolet Spectrophotometry

Transmittance of the third group was tested by UV spectrophotometry (Lambda 750S, PerkinElmer, Waltham, MA, USA). The wavelength scan ranged from 200–800 nm and the average spectrum was calculated. UV radiation with different wavelengths can be absorbed by bitumen because of its chemical bonds and transmittance may be reduced. After UV aging, the chemical bonds change, which results in changes in transmittance. Consequently, the transmittance tested by UV spectrophotometry could be used to characterize the aging degree.

3. Results and Discussion

3.1. Aging Effect on Bitumen in Different Layers for Different Aging Times

3.1.1. Complex Modulus of Bitumen in Different Layers at High Temperatures

The complex modulus of B in different layers for 1 h from 30 to 60 °C is shown in Figure 4. The complex modulus of B in the first layer slightly increased after aging for one hour. This means that it was influenced by the UV radiation. Furthermore, the complex modulus of B in the second and third layers was the same as complex modulus of the original bitumen. This means that the UV radiation had no aging effect on the bitumen in these two layers.

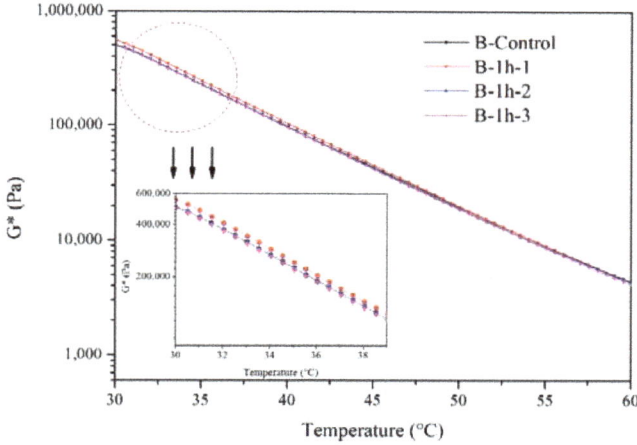

Figure 4. Complex modulus of B aged by UV radiation for 1 h.

The complex modulus of B in different layers for 5 h from 30 to 60 °C is shown in Figure 5. The complex modulus of B in the first and the second layers obviously increased after aging. The complex modulus of B in the third layer was a little higher than the complex modulus of the original bitumen. This means that after aging for 5 h, B in the second and the third layers began to be aged by the UV radiation.

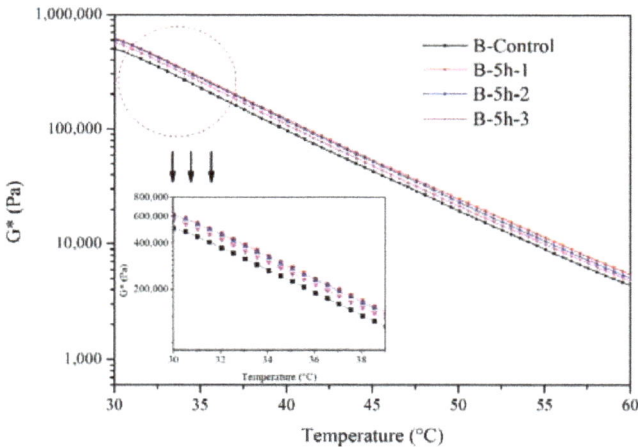

Figure 5. Complex modulus of B aged by UV radiation for 5 h.

The complex modulus of B in different layers for 10 h from 30 to 60 °C is shown in Figure 6. Furthermore, the complex modulus of bitumen at 45 °C in different layers is illustrated in Figure 7. The tendency of the UV aging of the bitumen was similar to the situation in which the bitumen was aged for 5 h. The complex modulus of B in these three layers was higher than it was in the original bitumen. This means that UV radiation influenced B in these three layers after aging for 10 h. Additionally, B in the first and the second layers showed a higher complex modulus than in the third layer, which means that the aging effect in different layers was different.

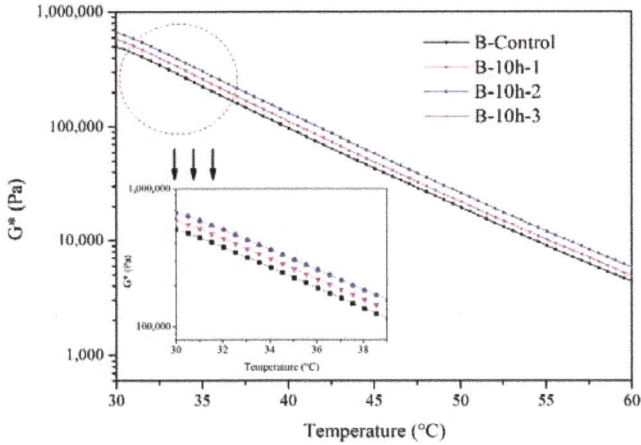

Figure 6. Complex modulus of B aged by UV radiation for 10 h.

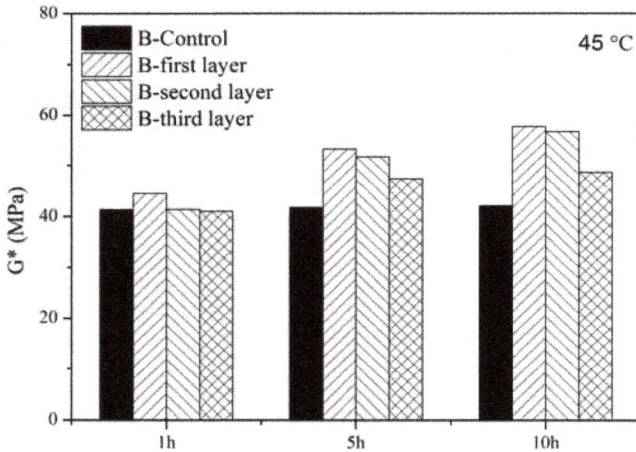

Figure 7. Complex modulus (45 °C) of B in different layers for different aging times.

3.1.2. Complex Modulus of Bitumen in the Third Layer for Different Aging Times

The complex modulus of B in the third layer, aged for different aging times from 30 to 60 °C, is shown in Figure 8. It can be seen that the complex modulus of B in the third layer aged for 1 h was the same as the original bitumen. After aging for 5 h, the complex modulus increased. The degree of aging was accelerated after aging for 10 h. This means that UV radiation only had an impact on the surface bitumen. With an increase in aging time, the aging depth due to UV radiation reached the first layer after 1 h and after UV aging for 10 h, the aging depth increased and was no longer in the first layer. In other words, aging depth accelerated due to changes in material properties.

3.2. Aging Effect on the Second Group Caused by Ozone

$I_{C=O}$ and $I_{S=O}$ of B in the second group are shown in Figures 9 and 10. After UV aging, $I_{C=O}$ and $I_{S=O}$ of Bu increased. However, $I_{C=O}$ and $I_{S=O}$ of Bb was only slightly more than original bitumen. This means that the bitumen on the septum was seriously aged by the UV radiation.

Otherwise, the bitumen below the septum was influenced by ozone and the effect was very small. Furthermore, although ozone had an effect on bitumen, the increment of aging depth was not closely related to ozone because the effect was too small.

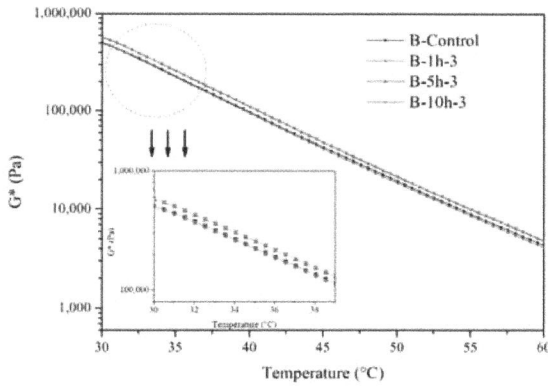

Figure 8. Complex modulus of B in the third layer's bitumen for different aging times.

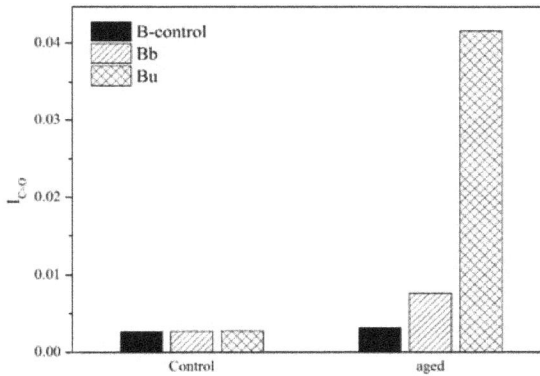

Figure 9. $I_{C=O}$ of the second group before and after UV aging.

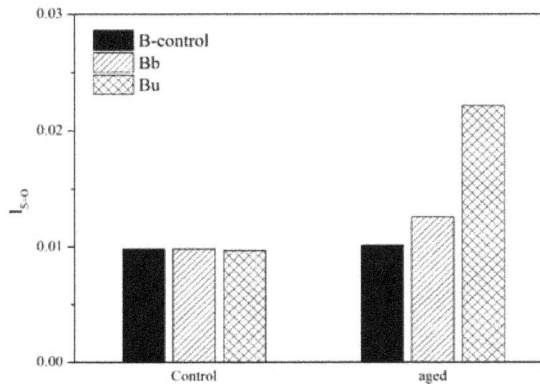

Figure 10. $I_{S=O}$ of the second group before and after UV aging.

3.3. Transmittance of Bitumen after UV Aging

3.3.1. Transmittance of Bitumen with Different Thicknesses after UV Aging

The transmittance of bitumen with different thicknesses after UV aging for 0 h is displayed in Figure 11. From Figure 12, it can be seen that the transmittance of B gradually decreased with the increase in bitumen thickness. This can be explained by the thicker bitumen film blocking light radiation more effectively. Meanwhile, transmittance gradually increased with the increase in wavelength. This means that the transmittance of the UV radiation at different wavelengths was different in this wavelength scope. Furthermore, as shown in Figure 10, the transmittance of B with 1.1 μm, 2.3 μm, 5.6 μm, 6.3 μm, 8.1 μm, and 13.1 μm thickness was 52.79%, 24.96%, 2.85%, 1.57%, 0.13%, and 0% at 400 nm, respectively. This means that the transmittance of UV radiation is very limited and that B with 13.1 μm thickness could block all UV radiation. Additionally, because of the different transmittance of different wavelength UV radiation, B with 2.3 μm, 5.6 μm, 6.3 μm and 8.1 μm thickness blocked UV radiation at the 210 nm, 322 nm, 330 nm, 372 nm wavelengths, respectively. UV radiation from 200 nm to 400 nm cannot blocked by B with 1.1 μm thickness.

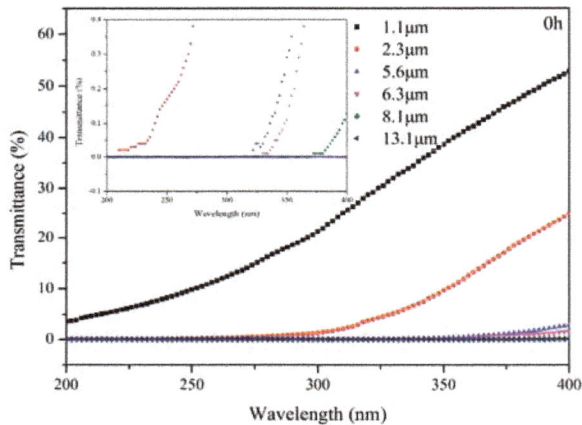

Figure 11. Transmittance of bitumen with different thicknesses after UV aging for 0 h.

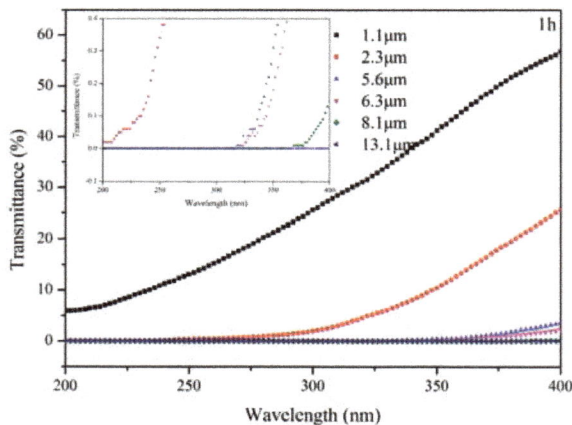

Figure 12. Transmittance of bitumen with different thickness after UV aging for 1 h.

Figure 11 shows the transmittance of bitumen with different thicknesses after UV aging for 1 h. After UV aging, the transmittance of B with different thicknesses increased. All the UV radiation from 200 nm to 400 nm could pass through B with 2.3 μm thickness. UV radiation was blocked at 318 nm, 322 nm, and 368 nm by B with 5.6 μm, 6.3 μm and 8.1 μm thickness, respectively. The transmittance of B with 13.1 μm was still 0%. This means that UV radiation still cannot pass through a bitumen film with 13.1 μm thickness, although the transmittance of B increased because of the UV radiation.

The transmittance of bitumen with different thicknesses after UV aging for 5 h is shown in Figure 13. The transmittance of B with 1.1 μm, 2.3 μm, 5.6 μm, 6.3 μm and 8.1 μm thickness at 400 nm increased to 63.74%, 25.66%, 3.92%, 2.46% and 0.45%, respectively. However, UV radiation was still blocked by B with 13.1 μm thickness.

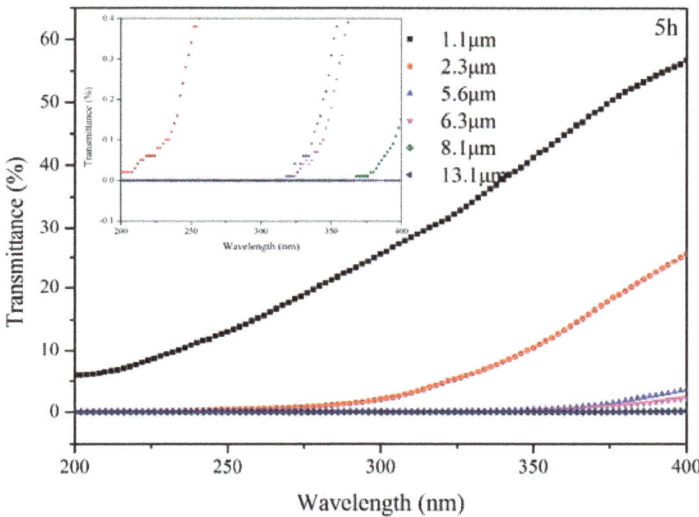

Figure 13. Transmittance of bitumen with different thickness after UV aging for 5 h.

The transmittance of bitumen with different thicknesses after UV aging for 10 h is illustrated in Figure 14. The transmittance of B further increased after UV aging. Furthermore, UV radiation from 200 nm to 400 nm wavelength could pass through B with 5.6 μm and 6.3 μm thickness after aging for 10 h. This means that after aging for 10 h, bitumen below 6.3 μm thickness of surface was aged by UV radiation at all wavelengths directly. Otherwise, B with 8.1 μm and 13.1 μm thickness blocked UV radiation at the 250 nm and 392 nm wavelengths, respectively. This means that B below 13.1 μm thickness began to be influenced by UV radiation from 392 nm to 400 nm directly and that the effect was extremely small.

Figure 15 presents the transmittance of bitumen with different thicknesses after UV aging for 50 h. After aging for 50 h, UV radiation from 200 nm to 400 nm could pass through B with different thicknesses, except for 13.1 μm thickness. This means that B under a surface of 8.1 μm thickness was aged by UV radiation directly. UV radiation was blocked at 362 nm by B with 13.1 μm thickness. B below 13.1 μm thickness could be aged by more UV radiation compared to B that was aged for 10 h.

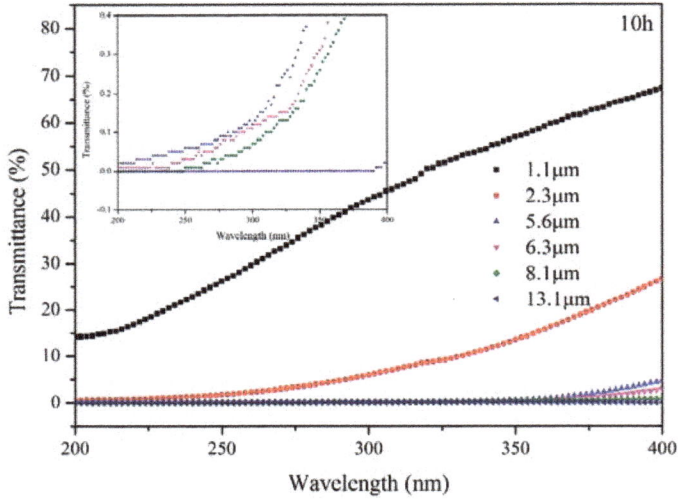

Figure 14. Transmittance of bitumen with different thickness after UV aging for 10 h.

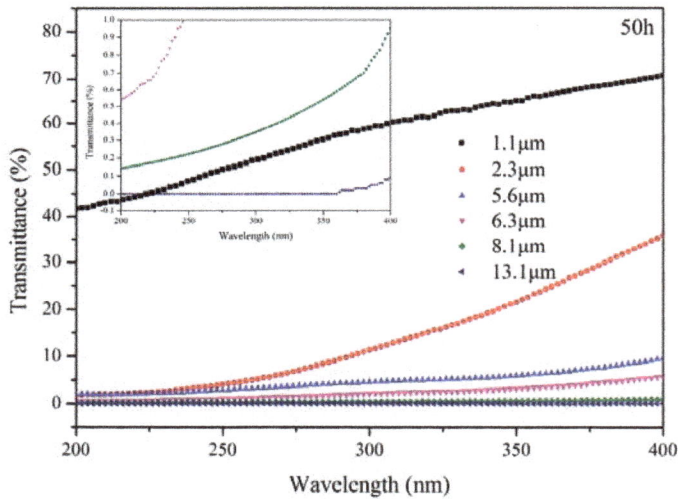

Figure 15. Transmittance of bitumen with different thickness after UV aging for 50 h.

3.3.2. UV Aging Model

The transmittance of bitumen with 13.1 μm thickness after UV aging for different aging times is illuminated in Figure 16. It was observed that UV radiation could not reach a depth greater than 13.1 μm thickness below the bitumen surface before aging for 5 h. As the aging time increased, UV radiation reached deeper. After aging for 10 h, sectional UV radiation could pass through bitumen with 13.1 μm thickness. However, from the results of the first group, it was clear that there was a UV aging effect on bitumen after aging for 5 h. This means that bitumen which had no direct contact with UV radiation was influenced by UV radiation. Furthermore, external environmental effects should be excluded according to the results of the second group. The critical point of this phenomenon is that bitumen which had no direct contact with UV radiation, had direct contact with bitumen aged by UV radiation. The real reason why bitumen which had no direct contact with UV radiation was influenced by UV radiation, is permeation. Aged bitumen permeates fresh bitumen and fresh bitumen permeates aged bitumen. The same tests were also conducted on another bitumen sample (80/100 penetration grade) and very similar results and the same conclusions were obtained. The UV aging effect is explained in Figure 17. Bitumen on the surface was subjected to various external environmental effects such as ozone and UV radiation. Bitumen at the bottom encountered the effects of permeation and ever-increasing UV radiation transmittance.

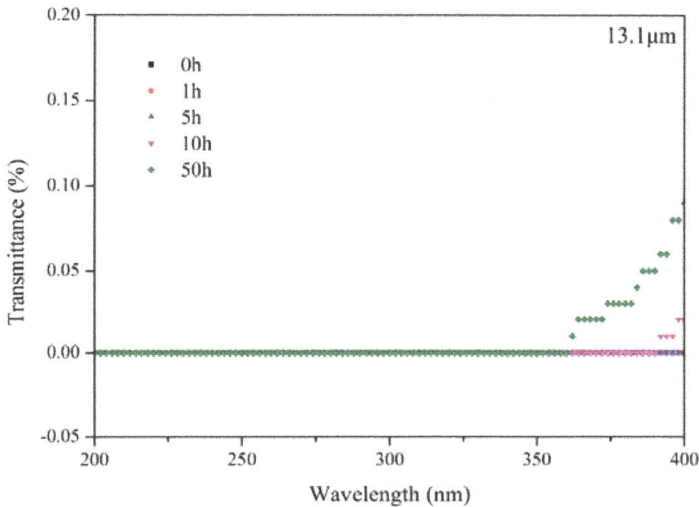

Figure 16. Transmittance of bitumen with 13.1 μm thickness after UV aging for different aging times.

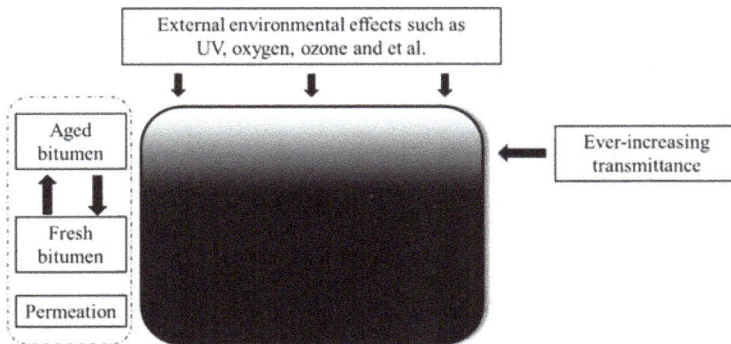

Figure 17. UV aging model.

4. Conclusions

The effect of UV radiation on bitumen aging depth was investigated. The rheological properties and FTIR results of three groups of samples were analyzed. The following conclusions can be drawn based on the results:

1. After aging for 5 h, bitumen in the second and the third layers began to be aged by UV radiation. As aging time increased, aging depth increased.
2. The aging effects of ozone were small. Bitumen is hard age if it has no contact with UV radiation.
3. The UV radiation transmittance of bitumen increased after UV aging. After UV aging for 50 h, partial UV radiation could pass through bitumen with 13.1 μm thickness.
4. After aging for 5 h, UV radiation from 200 nm to 400 nm still could not pass through bitumen with 13.1 μm thickness. The real reason why aging depth increased was permeation. Aged bitumen permeated fresh bitumen, which resulted in changes to the rheological properties of the second and third layers.
5. Permeation is closely related to the thermal stability, therefore, bitumen on road surfaces should possess high thermal stability to ensure its performance.

Author Contributions: H.J., W.S. and Q.L. conceived and designed the experiments; J.H., S.N., J.W., D.Z. and Y.L. performed the experiments; J.H. and W.Z. analyzed the data; S.W. contributed reagents/materials/analysis tools; and J.H. and M.I.G.H. wrote the paper.

Acknowledgments: This work was financially supported by National Project of Scientific and Technical Supporting Programs funded by Ministry of Science & Technology of China (No. 2011BAE28B03), the Science and Technology Plan Projects of the Ministry of Transport of China (No. 2013 318 811 250) and the National Key Scientific Apparatus Development Program from the Ministry of Science and Technology of China (No. 2013YQ160501).

Conflicts of Interest: The authors declare no conflict of interest.

References

1. Qin, Q.; Schabron, J.F.; Boysen, R.; Farrar, M.J. Field aging effect on chemistry and rheology of asphalt binders and rheological predictions for field aging. *Fuel* **2014**, *121*, 86–94. [CrossRef]
2. Calabifloody, A.; Thenoux, G. Controlling asphalt aging by inclusion of byproducts from red wine industry. *Constr. Build. Mater.* **2012**, *28*, 616–623. [CrossRef]
3. Zeng, W.; Wu, S.; Wen, J.; Chen, Z. The temperature effects in aging index of asphalt during UV aging process. *Constr. Build. Mater.* **2015**, *93*, 1125–1131. [CrossRef]
4. Xiao, F.; Amirkhanian, S.N.; Karakouzian, M.; Khalili, M. Rheology evaluations of WMA binders using ultraviolet and PAV aging procedures. *Constr. Build. Mater.* **2015**, *79*, 56–64. [CrossRef]
5. Mouillet, V.; Lamontagne, J.; Durrieu, F.; Planche, J.P.; Lapalu, L. Infrared microscopy investigation of oxidation and phase evolution in bitumen modified with polymers. *Fuel* **2008**, *87*, 1270–1280. [CrossRef]
6. Mill, T. The role of hydroaromatics in oxidative aging in asphalt. *Am. Chem. Soc. Div. Fuel Chem.* **1996**, *41*, 1245–1249, preprints of papers.
7. Petersen, J.C. A dual, sequential mechanism for the oxidation of petroleum asphalts. *Pet. Sci. Technol.* **1998**, *16*, 1023–1059. [CrossRef]
8. Apeagyei, A.K. Laboratory evaluation of antioxidants for asphalt binders. *Constr. Build. Mater.* **2011**, *25*, 47–53. [CrossRef]
9. Chen, Z.W.; Wu, S.P.; Xiao, Y.; Zeng, W.B.; Yi, M.W.; Wan, J.M. Effect of hydration and silicone resin on Basic Oxygen Furnace slag and its asphalt mixture. *J. Clean. Prod.* **2016**, *112*, 392–400. [CrossRef]
10. Pan, P.; Wu, S.P.; Xiao, Y.; Liu, G. A review on hydronic asphalt pavement for energy harvesting and snow melting. *Renew. Sus. Energy Rev.* **2015**, *48*, 624–634. [CrossRef]
11. Cui, P.Q.; Wu, S.P.; Xiao, Y.; Wan, M.; Cui, P.D. Inhibiting effect of layered double hydroxides on the emissions of volatile organic compounds from bituminous materials. *J. Clean. Prod.* **2015**, *108*, 987–991. [CrossRef]
12. Xu, S.; Li, L.; Yu, J.; Zhang, C.; Zhou, J.; Sun, Y. Investigation of the ultraviolet aging resistance of organic layered double hydroxides modified bitumen. *Constr. Build. Mater.* **2015**, *96*, 127–134. [CrossRef]
13. Zhao, X.; Wang, S.; Wang, Q.; Yao, H. Rheological and structural evolution of SBS modified asphalts under natural weathering. *Fuel* **2016**, *184*, 242–247. [CrossRef]
14. Verhasselt, A.F. A Kinetic Approach to the Aging of Bitumens. *Dev. Pet. Sci.* **2000**, *40*, 475–497. [CrossRef]
15. Lins, V.F.; Araujo, M.F.; Yoshida, M.I.; Ferraz, V.P.; Andrada, D.M.; Lameiras, F.S. Photodegradation of hot-mix asphalt. *Fuel* **2008**, *87*, 3254–3261. [CrossRef]

16. Durrieu, F.; Farcas, F.; Mouillet, V. The influence of UV aging of a Styrene/Butadiene/Styrene modified bitumen: Comparison between laboratory and on site aging. *Fuel* **2007**, *86*, 1446–1451. [CrossRef]
17. Lu, X.; Isacsson, U. Effect of ageing on bitumen chemistry and rheology. *Constr. Build. Mater.* **2002**, *16*, 15–22. [CrossRef]
18. Migliori, F.; Corte, J. Comparative Study of Rtfot and Pav Aging Simulation Laboratory Tests. *Transp. Res. Rec.* **1998**, *1638*, 56–63. [CrossRef]
19. ao, F.; Newton, D.; Putman, B.J. A long-term ultraviolet aging procedure on foamed WMA mixtures. *Mater. Struct.* **2013**, *46*, 1987–2001. [CrossRef]
20. Mouillet, V.; Farcas, F.; Chailleux, E.; Sauger, L. Evolution of bituminous mix behaviour submitted to UV rays in laboratory compared to field exposure. *Mater. Struct.* **2014**, *47*, 1287–1299. [CrossRef]
21. Traxler, R.N. *Durability of Asphalt Cements (with Discussion)*; Association of Asphalt Paving Technologists Proceedings: St. Paul, MN, USA, 1963; Volume 32, pp. 44–63.
22. Yamaguchi, K.; Sasaki, I.; Nishizaki, I.; Meiarashi, S.; Moriyoshi, A. Effects of film thickness, wavelength, and carbon black on photodegradation of asphalt. *J. Jpn. Pet. Inst.* **2005**, *48*, 150–155. [CrossRef]
23. Mouillet, V.; Farcas, F.; Besson, S. Ageing by UV radiation of an elastomer modified bitumen. *Fuel* **2008**, *87*, 2408–2419. [CrossRef]
24. Araujo, M.D.; Lins, V.D.; Pasa, V.M.; Leite, L.F. Weathering aging of modified asphalt binders. *Fuel Process. Technol.* **2013**, *115*, 19–25. [CrossRef]
25. Zhang, H.; Zhu, C.; Yu, J.; Shi, C.; Zhang, D. Influence of surface modification on physical and ultraviolet aging resistance of bitumen containing inorganic nanoparticles. *Constr. Build. Mater.* **2015**, *98*, 735–740. [CrossRef]
26. Feng, Z.; Yu, J.; Zhang, H.; Kuang, D.; Xue, L. Effect of ultraviolet aging on rheology, chemistry and morphology of ultraviolet absorber modified bitumen. *Mater. Struct.* **2013**, *46*, 1123–1132. [CrossRef]
27. Cong, P.; Wang, X.; Xu, P.; Liu, J.; He, R.; Chen, S. Investigation on properties of polymer modified asphalt containing various antiaging agents. *Polym. Degrad. Stab.* **2013**, *98*, 2627–2634. [CrossRef]
28. Xu, S.; Yu, J.; Zhang, C.; Sun, Y. Effect of ultraviolet aging on rheological properties of organic intercalated layered double hydroxides modified asphalt. *Constr. Build. Mater.* **2015**, *75*, 421–428. [CrossRef]
29. Pieri, N. Etude du vieillissement simulé et in situ des bitumes routiers par IRTF et fluorescence UV en excitation-émission synchrones: Détermination des relations structures chimiques—propriétés rhéologiques par analyse en composantes principales. **1994**, *84*, 164–165. Available online: http://www.theses.fr/1994AIX30102 (accessed on 7 May 2018).
30. Feng, Z.; Bian, H.; Li, X.; Yu, J. FTIR analysis of UV aging on bitumen and its fractions. *Mater. Struct.* **2016**, *49*, 1381–1389. [CrossRef]
31. Lamontagne, J. Vieillissement des Bitumes Modifiés Polymères à Usage Routier par Simulation et Techniques Spectroscopiques. Chimie Analytique: Aix-Marseille 3, 2002. Available online: http://www.theses.fr/2002AIX30002 (accessed on 7 May 2018).
32. Lamontagne, J.; Dumas, P.; Mouillet, V.; Kister, J. Comparison by Fourier transform infrared (FTIR) spectroscopy of different ageing techniques: Application to road bitumens. *Fuel* **2001**, *80*, 483–488. [CrossRef]

![materials logo] *materials*

MDPI

Article

Investigation of Ageing in Bitumen Using Fluorescence Spectrum

Ning Tang [1] [iD], Yu-Li Yang [1] [iD], Mei-Ling Yu [1], Wen-Li Wang [1], Shi-Yue Cao [1], Qing Wang [1,*] and Wen-Hao Pan [1,2,*]

[1]   School of Materials Science and Engineering, Shenyang Jianzhu University, Shenyang 110168, China;
     tangning@sjzu.edu.cn (N.T.); yangyuli1995@163.com (Y.-L.Y.); 15734022236@163.com (M.-L.Y.);
     13940347763@163.com (W.-L.W.); 13591599599@163.com (S.-Y.C.)
[2]   School of Materials Science and Engineering, Northeast University, Shenyang 110918, China
*   Correspondence: wangqingmxy@126.com (Q.W.); pwh@sjzu.edu.cn (W.-H.P.);
     Tel.: +86-24-2469-0309 (Q.W.); +86-24-2469-0319 (W.-H.P.)

Received: 16 June 2018; Accepted: 25 July 2018; Published: 31 July 2018

Abstract: Bitumen ageing is a very complex process and poses a threat to the performance of pavements. In the present work, a fluorescence spectrophotometer was employed to research the change rule of components and the structure of bitumen after the ageing process. The Thin Film Oven Test (TFOT) and Ultraviolet (UV) light treatment were carried out as ageing methods. The properties and components of bitumen were tested before and after aging. The 2D and 3D fluorescence spectra of bitumen were analyzed. The vector of fluorescence peak was calculated for evaluating the ageing process. The results indicated that the ideal concentration of bitumen- tetrachloromethane solution was 0.1 g/L or smaller for avoiding the fluorescence quenching. The coordinates of fluorescent peak appeared "blue-shift" after ageing due to the change of aromatics. In addition, bitumen has already occurred serious ageing when the magnitude of a vector is more than 36.

Keywords: fluorescence spectrum; bitumen; ageing; parametrization; "blue-shift"

---

1. Introduction

Bitumen is one of the primary construction materials in road engineering and waterproofing. This means bitumen is exposed to air for a long time and is directly affected by solar radiation, wind, rain and vehicle loads. As is known to many, bitumen easily ages under the actions of the natural environment of heat, light and oxygen, which can result in property degradation of bitumen and, thus greatly shortened life of asphalt pavement [1–3]. In order to prolong the life of asphalt pavement, the properties and microstructure of bitumen play a crucial role. Hence, the ageing and anti-ageing process is one of the hot topics in bitumen researches [4–8].

Bitumen ageing is primarily associated with the loss of volatile components, progressive oxidation and ultraviolet radiation during asphalt mixture construction (short-term ageing) and service (long-term ageing). Researchers try to accelerate ageing of bitumen in the laboratory for simulating the bitumen performance in the field [9]. Thin Film Oven Test (TFOT), Rolling Thin Film Oven Test (RTFOT), Pressure Ageing Vessel (PAV) and Ultraviolet light treatment (UV) are methods commonly employed for simulation in the laboratory [10–12]. The TFOT and RTFOT are adopted by American Association of State Highway and Transportation Officials (AASHTO) and American Society for Testing Materials (ASTM) as a means of evaluating the ageing of bitumen during plant mixing. The PAV was developed by the Strategic Highway Research Program (SHRP) for simulating the long-term, in-service oxidative ageing of bitumen in the field. The UV method is used to simulate the long-term ageing of bitumen under the radiation in service.

So far, a lot of evaluation methods have been established and carried out to assess the macroscale performance, such as the three conventional properties of bitumen (penetration, softening point, ductility), rheological properties and so on [13,14]. Moreover, spectroscopy has been widely used in bitumen chemistry because it can provide unique fingerprints for the bitumen at molecular (even atomic) level, for example, infrared spectroscopy (IR) to determine the change of molecular functional groups [15–19], chromatography to determine the moving regularity of the average molecular weight and molecular weight distribution curve [20–22], and nuclear magnetic resonance (NMR) to analyze the molecular structure and atomic attribution of the molecular structure by hydrogen or carbon spectrum [23,24].

The composition of bitumen is usually divided into four fractions according to similar chemical behavior, including saturates, asphaltenes, resins and aromatics (SARA) [25,26]. Hence, bitumen is a mixture of hydrocarbon compounds. The greater the C/H ratio, the more aromatic rings the bitumen usually contains. The aromatic ring has a flat conjugated ring system. The bonding between atoms is not a discrete single-double bond but is overridden by the π electron cloud [27]. The emergence and development of fluorescence spectroscopic testing techniques provide a new way for investigating the microstructure of bitumen materials. The fluorescence spectrum is based on the analysis of conjugated system compounds, which makes it possible to analyze the ageing process of bitumen by fluorescence spectroscopy [28,29].

In this study, the properties and components of bitumen were tested. The ultraviolet ageing test and thin film oven test in the laboratory were carried out to evaluate the performance of two kinds of bitumen. Furthermore, the proper concentration of the solution of bitumen-tetrachloromethane was confirmed based on the fluorescence quenching phenomenon. A fluorescence spectrophotometer was employed for obtaining the three-dimensional fluorescence spectra. Based on the coordinate of fluorescence spectra and the displacement of the vector, the aging process of bitumen was evaluated.

2. Materials and Methods

2.1. Raw Materials

PJ-70 and PJ-90 bitumen were obtained from Liaohe Oilfield, Panjin, China. The conventional properties of bitumen are as follows. For PJ-70, the softening point was 49.5 °C; penetration was 72.6 dmm (25 °C); and ductility was 39 cm (10 °C). For PJ-90, the softening point was 46.7 °C; penetration was 86.9 dmm (25 °C); and ductility was 59 cm (10 °C).

The organic solvent was carbon tetrachloride which was produced by the China Medicine Group (Beijing, China), and the purity was an analytical reagent.

2.2. Experimental Methods

● Conventional properties

The conventional properties of bitumen are softening point, penetration and ductility. These properties were evaluated for both neat bitumen and aged bitumen. Following Chinese Standard JTG E20-2011 (China) [30].

● Four components (SARA) analysis

Bitumen was separated into four components (saturates, aromatics, resins, and asphaltenes, SARA) through precipitation and chromatographic column test method. Following Chinese Standard JTG E20-2011 (China) [30].

● Thin film oven test (TFOT)

The thermal oxidation ageing of bitumen was simulated by thin film oven test (TFOT) according to the Chinese standard (JTG E20-2011) [30]. The experimental temperature was 163 °C, ageing time was 5 h.

- Ultraviolet ageing test (UV)

After the TFOT test, 15 g of the aged bitumen was moved into a stainless steel plate with 140 mm in diameter. The thickness of the bitumen film was 1 mm. Afterwards, the plate was placed into the ultraviolet ageing equipment. The UV lamp was a xenon lamp of straight pipe shape with a power of 1800 w. Experimental temperature was 60 °C. The ultraviolet radiation intensity was about 0.5 W/cm^2, the rain amounts of spray was 8 liters per day, and ageing time was 7 days.

2.3. Fluorescence Spectra

Three-dimensional fluorescence spectrum is one kind of matrix spectrum which is composed of the excitation wavelength (Y), emission wavelength (X) and fluorescence intensity (Z). The data of fluorescence intensity could be obtained when the excitation wavelength and emission wavelength changed. In the present work, a Hitachi F-7000 fluorescence spectrophotometer (made in Tokyo, Japan) was employed for obtaining fluorescence spectrum.

A 0.01 g of bitumen sample was moved into a test tube. Afterwards, the 20 mL solvent of carbon tetrachloride was added into the test tube with a tight and mild concussion. When the bitumen was completely dissolved, the solution was transferred into a volumetric flask of 50 mL. After 4 h standing the solution can be diluted into a new solution with different concentrations.

For 2D fluorescence spectrum, the emission spectrum was obtained with the fixed excitation wavelength of 260 nm and 360 nm. The scanning range was from 360 to 680 nm, the slit width was 2 nm, and the scanning speed was 2400 nm/min. The obtained data were smoothed due to the noisy influence by using the Savitzky-Golay method.

For 3D fluorescence spectrum, the scanning range of excitation wavelength (EX) and emission wavelength (EM) was from 200 to 800 nm, the slit width of excitation and emission was all 5 nm, and the scanning speed was 2400 nm/min. The original data obtained by the fluorescence spectrophotometer were converted and plotted with MATLAB (2009a, The MathWorks, Natick, MA, USA) for the fingerprint images of bitumen samples.

3. Results and Discussion

3.1. Performance of Bitumen

Figure 1 shows the surface characteristic of bitumen before and after ageing. Before ageing, the neat bitumen had a smooth surface with glossy black. After UV ageing, a lot of wrinkles appeared on the surface like a matt finish. Conversely, there was no significant change after TFOT ageing. It reveals that UV ageing can simulate long-term aging well, and TFOT aging can simulate short-term aging well.

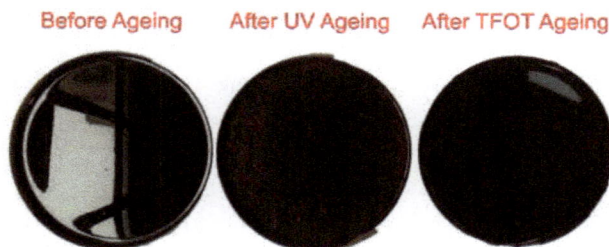

Figure 1. The plate with bitumen before and after ageing.

The performance change of bitumen after ageing was assessed by three conventional properties of bitumen, including softening point, penetration and ductility. The results of softening point, penetration and ductility of PJ-70 and PJ-90 bitumen before and after ageing were shown in Figure 2.

The softening point of PJ-70 bitumen increased after TFOT ageing and UV ageing, with increments of 4.6 °C and 10.2 °C respectively. But, the penetration and ductility decreased more significantly. The residual penetration ratio was 54.7% and 29.8%; ductility retention rate was 53.6% and 12.6%. Changes of the PJ-90 bitumen were similar to PJ-70 bitumen. The softening point increased after ageing, with increments of 6.2 °C and 12.1 °C, respectively. The residual penetration ratio was 52.5% and 28.5%, respectively. The ductility retention rate was 39.2% and 10.3%, respectively. Hence, the bitumen was easy to age through photooxidation, especially softer bitumen (PJ-90).

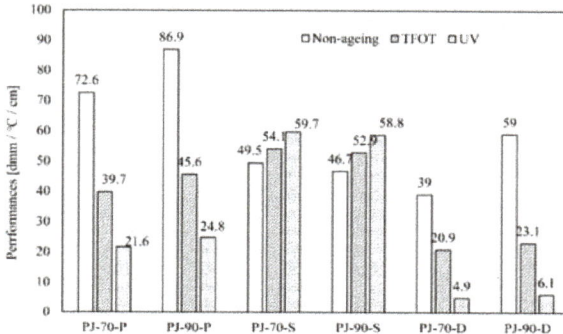

Figure 2. Conventional properties of bitumen before and after ageing.

3.2. Bitumen Fractions

Based on the widely accepted analytical approach of SARA, four components were separated from bitumen. The results of composition changes in the two bitumen samples were shown in Figure 3. In general, bitumen kept the same ageing trend as the performance changes after ageing.

The saturate fraction was nearly unchanged. It illustrated that the chains of saturates were not broken and oxidized during the ageing process. Furthermore, the aromatics are a group of small aromatic naphthenic compounds with low molecular weights. During the ageing process, the aromatics was reduced because it is easy to volatilize. It is also indicated that condensation polymerization had occurred and the chains were broken. At the same time, the oxidation of the chemical functional groups such as carbonyl and sulfoxide resulted in the increase of resins and asphaltenes fraction, and the bitumen became hard and brittle.

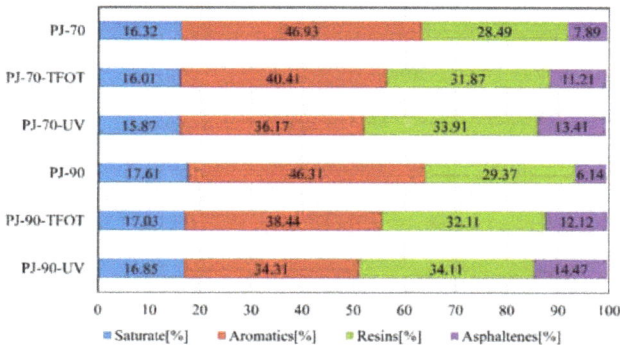

Figure 3. Compositions of bitumen before and after ageing.

3.3. Fluorescence Quenching

A variety of reasons can result in a quenching phenomenon, such as excited state reactions, energy transfer, complex-formation and collisional quenching. As a normal fluorescence fingerprint spectrum which has fluorescent features, the intensity of each data point should be a linear relationship with the concentration of fluorescent components in the bitumen.

Hence, if the fluorescence quenching can be avoided, fluorescent intensity can serve as a proof for contents change of fluorescent material, and the fingerprint picture can be used as a basis for parametrization and image recognition.

Therefore, two kinds of bitumen with different concentrations were tested, and the results are given in Table 1. There was no significant fluorescence quenching in the tests when the concentration of the bitumen diluent was 0.01 g/L, 0.1 g/L or 0.5 g/L. Comparing the different concentrations, the 0.5 g/L is a threshold value and 0.01 g/L is too low to dilute. Thus, the optimized concentration is 0.1 g/L.

Table 1. Fluorescence quenching of bitumen diluent with different concentrations.

Type	Concentration (g/L)	Peak Intensity	State
PJ-70	0.01	37	Not quenched, characteristic graphics
	0.1	110	Not quenched, characteristic graphics
	0.5	417	Not quenched, characteristic graphics
	1	689	Quenching, shape deformation,
	10	787	Quenched
PJ-90	0.01	33	Not quenched, characteristic graphics
	0.1	114	Not quenched, characteristic graphics
	0.5	458	Not quenched, characteristic graphics
	1	703	Quenching, shape deformation
	10	814	Quenched

3.4. 2D Fluorescence Spectrum of Bitumen

In order to obtain the emission spectrum of bitumen, two different excitation wavelengths were employed, 260 nm and 360 nm, respectively. A wavelength of 260 nm was commonly used for distinguishing the amounts of benzene ring in the aromatics. In addition, it was also the maximum absorption wavelength when using UV/VIS (ultraviolet and visible) spectrometer. For 360 nm, it was a commonly used excitation wavelength for petroleum. The results are shown in Figure 4.

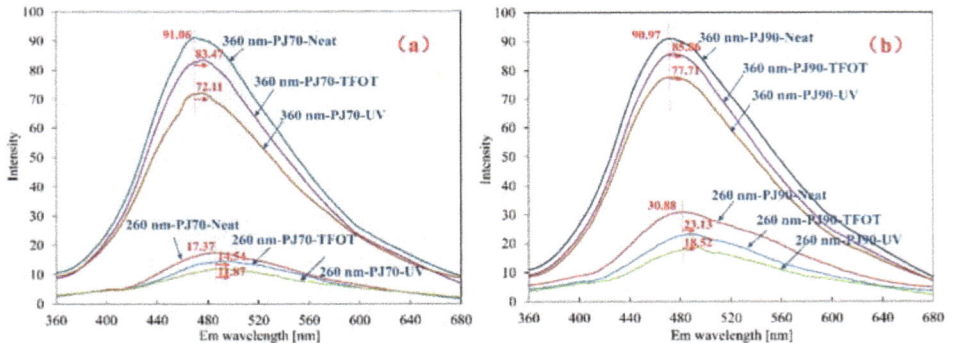

Figure 4. Emission spectrum of bitumen: (a) PJ-70 and (b) PJ-90.

For both neat bitumen samples, there was little or no difference for parabolic shape or intensity. However, a contradictory phenomenon appeared after bitumen ageing. The intensity of bitumen after TFOT ageing was higher than that of bitumen after UV ageing, both for PJ-70 and PJ-90. In fact, this was caused by the fraction changes of bitumen after ageing. As mentioned, the asphaltenes almost cannot glow fluorescent, and the content of asphaltenes fraction increased after bitumen ageing. The more the asphaltenes fraction, the lower the fluorescent intensity. According to the results of composition analysis, the content of asphaltenes fraction just had a good correspondence with the fluorescent intensity. But it was hard to know why the intensity of PJ-90 bitumen was higher than that of PJ-70 bitumen at 260 nm. Whether this variation was derived from fraction changes or due to an experimental error needs to be further elucidated. Furthermore, the peak shifts to a higher wavelength after ageing. It was suspected that the heteroatoms in the asphaltenes made bitumen polarity increase by increasing oxygen content during ageing, and this influenced the physical properties of bitumen.

In addition, the intensity also means the color of light produced by bitumen. The color of light was analyzed by OSRAM ColorCalculator (OSRAM SYLVANIA, Wilmington, MA, USA). The CIE 1931 color space was a quantitative link between wavelengths in the visible spectrum, and colors in human color vision. The results of CIE coordinates were shown in Figure 5.

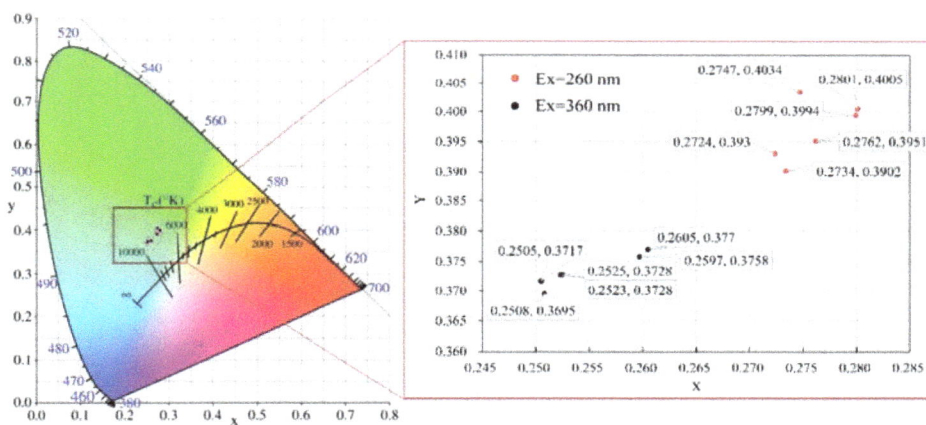

Figure 5. CIE coordinates of bitumen at 260 nm and 360 nm of excitation wavelength.

As shown, bitumen produced a greenish light at 260 nm and a bluish light at 360 nm, respectively. It revealed that the bitumen absorbed more energy at 360 nm due to the intensity of 360 nm of excitation wavelength being higher and holding a higher vibrational frequency for glowing strong light. Moreover, the CIE coordinates were irregular, but not discrete whether different excitation wavelength or ageing treatment.

3.5. 3D Fluorescence Spectrum of Bitumen

Bitumen are derived from petroleum processing, but the composition of bitumen is different due to the different geographical environment and evolution period of petroleum products. Therefore, the three-dimensional fluorescence spectrum of bitumen both has elemental similarities and peak characteristics. Meanwhile, the three-dimensional fluorescence spectrum belongs to the high-dimensional feature space. Irrelevant factors are filtered out through the debasing dimension calculation of the matrix. Thereby, the influence of bitumen on the results of the fluorescence spectrum is reduced to a minimum. The contour images were drawn based on the data of fluorescence spectra of the two bitumen samples, as shown in Figures 6 and 7.

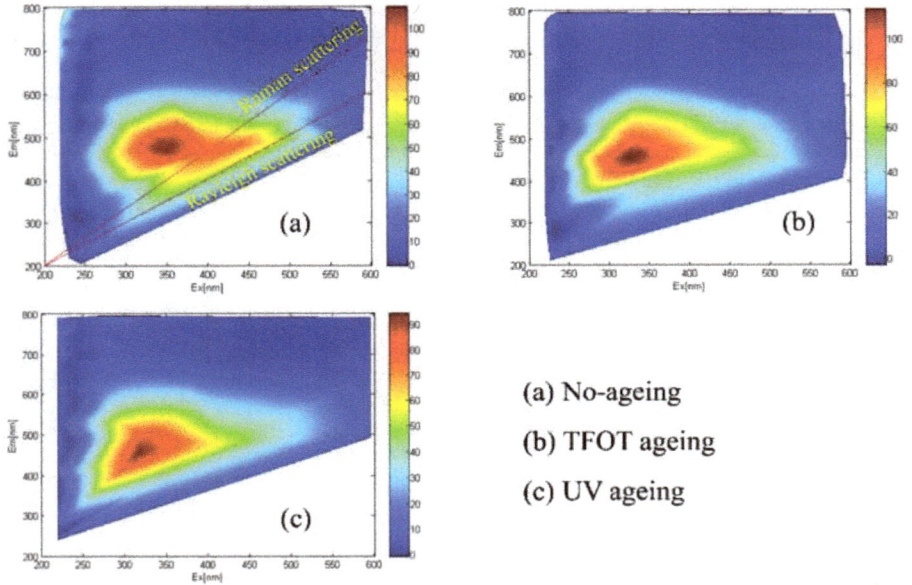

Figure 6. 3D fluorescence spectrum of PJ-70 bitumen: (**a**) No-ageing; (**b**) TFOT ageing; (**c**) UV ageing.

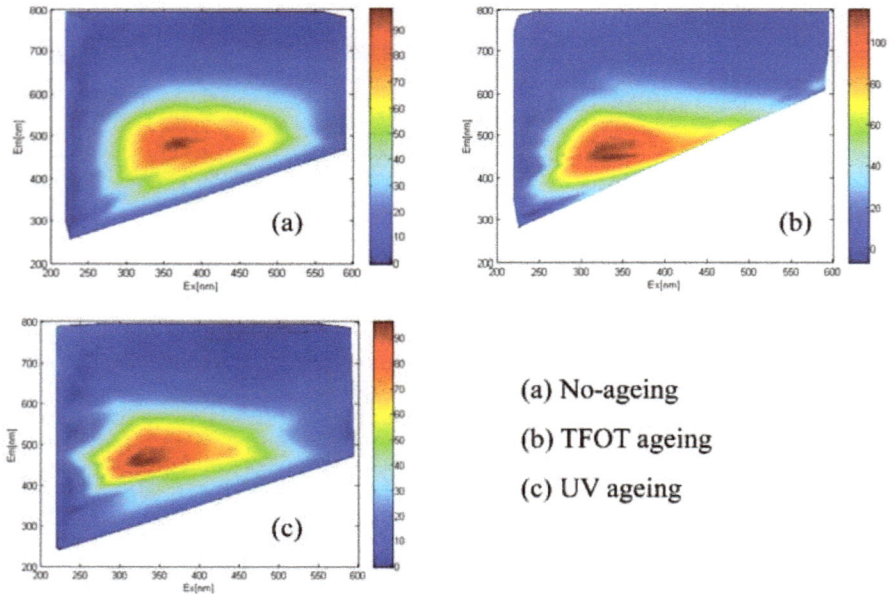

Figure 7. 3D fluorescence spectrum of PJ-90 bitumen: (**a**) No-ageing; (**b**) TFOT ageing; (**c**) UV ageing.

There were two obvious characteristic lines like a wathet blue glow, which was visible rising diagonally up to the right in the above images. For the first line, it was easy to see that the excitation wavelength (EX) of this line was equal to the emission wavelength (EM), the phenomenon is known as "Rayleigh scattering". For the other one, the EX was bigger than the EM, the phenomenon is known as "Raman scattering". Two scatterings were produced by the tetrachloride solution of bitumen. When bitumen molecules absorbed lower frequency light energy, the electron only rose to the higher vibrational level in the ground state, rather than being transited to the excited state. Afterwards, the electron returned to the original level in 10^{-15} to 10^{-12} s. If the electron reached the level safely it is Rayleigh scattering. Otherwise, it is Raman scattering.

For TFOT and UV ageing, the intensity of bitumen after UV ageing was more concentrated and more shifted. Furthermore, the intensity area of bitumen after UV ageing was smaller than that of bitumen after TFOT ageing.

Moreover, the intensity peaks of the fluorescence spectrum caused a certain displacement to short wavelength after bitumen ageing, which can be called "blue-shift". In fact, the fluorescence phenomenon mainly related to the π-conjugate system of fluorescent fraction in the organic matter [31]. If the fluorescent fraction changed after ageing, the maximum absorption wavelength was shifted to the direction of shorter or longer wavelength [27]. Hence, the greater the number of benzene rings, the greater the π electron conjugation degree and the easier the polymer can be excited to produce fluorescence.

For bitumen ageing, it was dominated by side chain breaking and substituting oxidation [1]. After ageing, the number of benzene rings and the conjugated degree of π electron increased, so the maximum absorption wavelength was supposed to shift to the longer wavelength. In fact, an opposite result was obtained. These fractured alkanes were not easy to expurgate, and continuously accumulated in the internal bitumen molecule so that products overlapped with the benzene rings of bitumen [32,33]. This phenomenon caused the illusion that the number of benzene rings in the large molecular structure of bitumen reduced and suppressed the π electron conjugate. Therefore, the intensity value of the peak dropped to the lower, and the "blue-shift" phenomenon happened. In addition, there was an intensity zone between two characteristic lines but this disappeared after bitumen ageing. It revealed that the fluorescent fraction in the bitumen changed to another non-fluorescent fraction.

3.6. Vector of Fluorescence Peak

The coordinates of fluorescence peak (EX,EM) were given in Table 2, which were obtained from Figures 6 and 7 as evaluation parameters for the bitumen ageing process. According to the (EX,EM) coordinates, the computational formula magnitude of vector |AB| was given as Equation (1).

$$|\overrightarrow{AB}| = \sqrt{(x_2 - x_1)^2 + (y_2 - y_1)^2} \tag{1}$$

where the peak coordinates of non-ageing bitumen is (x_1, y_1), and the peak coordinates of aged bitumen is (x_2, y_2). The calculated results were given in Table 2. The magnitude of the coordinate vector was all negative value after bitumen ageing due to the EX and the EM shifting to the direction of short wavelength.

Comparing the two ageing methods, the bitumen after UV ageing was more seriously aged as the magnitude of the coordinate vector was larger. This result corresponded to the changes of conventional performance of bitumen before and after ageing. It revealed that thermal oxidation was prone to ageing of bitumen for the short-term ageing in the laboratory. The aromatics fraction in the bitumen was vaporized causing the deterioration of bitumen. This was accelerated during TFOT due to the high temperature. However, the energy-rich bond of the bitumen could not absorb more energy to fracture the molecule chain during the light oxidation in 7 days [1].

Table 2. Parametrization of peak coordinate.

Ageing Method		Non-Ageing	TFOT	UV
	Coordinate	(358,472)	(332,447)	(318,456)
PJ-70	Vector	-	(−26,−25)	(−40,−16)
	Magnitude of vector	-	36.07	43.08
	Coordinate	(362,478)	(320,458)	(322,451)
PJ-90	Vector	-	(−32,−20)	(−40,−27)
	Magnitude of vector	-	37.74	48.26

From the view of different bitumen samples, PJ-90 bitumen was easy to age based on the vector calculation and the test results of performance. According to the results of bitumen compositions, the summation of resins and asphaltenes fraction of PJ-70 bitumen were more than that of PJ-90 bitumen, but the aromatics fraction had a slight difference. Furthermore, after the ageing of the thermal oxidation and photooxidation, the saturate fraction remained unchanged, but more resins and asphaltenes were produced. It revealed that the aromatics played a leading role in the ageing process. The side chain was broken and polycondensation of aromatics were more prone to ageing. In other words, the more the resins and asphaltenes fraction, the larger the magnitude of the peak vector and the more serious the bitumen ageing. In addition, the magnitude of peak vector of bitumen after UV ageing was higher than that of bitumen after TFOT ageing. This also matched the experimental results of performance after ageing.

From the view of parametrization, in combination with the conventional performance of the bitumen after ageing, when the magnitude of the vector was more than 36, the bitumen began to pose a threat to performance of the pavement. However, the correlation between the ageing process and the magnitude of the vector is still pending further study.

4. Conclusions

It is feasible to have an evaluation of bitumen ageing process by using fluorescence spectrophotometer. Based on the results obtained through the experimental investigation, the following conclusions were obtained.

The ideal concentration of bitumen-tetrachloromethane solution was 0.1 g/L or smaller for avoiding the fluorescence quenching. The bitumen had a strong energy absorption at 360 nm for producing a bluish light. The other one produced a greenish light at 260 nm.

It can be observed from the fluorescence spectra that the fluorescent intensity of bitumen had a lot to do with the composition of bitumen. Moreover, the peak coordinates shifted to the shorter wavelength because the aromatics had poor stability and were easily oxidized during the ageing process. This displacement was named "blue-shift".

The content of aromatics determined the displacement of the fluorescence spectrum peak. The vector of peak coordinates can be calculated and describes the ageing process of bitumen. When the magnitude of the vector was more than 36, it indicated that the bitumen began to pose a threat to the performance of the pavement.

Bitumen is a complicated mixture formed with fractions of different molecular weight, and it can lead to uncertainty about parametrization. Hence, the correlation between the ageing process and the magnitude of the vector is still pending further study.

Author Contributions: Conceptualization, N.T. and Q.W.; Data curation, Y.-L.Y., M.-L.Y., W.-L.W., S.-Y.C. and W.-H.P.; Funding acquisition, N.T. and Q.W.; Investigation, N.T., Y.-L.Y., M.-L.Y. and W.-L.W.; Methodology, Q.W.; Software, N.T., W.-L.W. and W.-H.P.; Writing—original draft, N.T.

Funding: This research was funded by National Natural Science Foundation of China grant number (51508344, China Postdoctoral Science Foundation grant number (2016M591458), Department of Education of Liaoning Province grant number (L2015449) and Doctoral Start-up Foundation of Liaoning Province grant number (201601212).

Acknowledgments: Thanks for Science Program and Discipline Content Education Project of Shenyang Jianzhu University. Special thanks Hong Kong Scholar Program and China Scholarship Council for sponsoring a technical visit to Politecnico di Milano.

Conflicts of Interest: The authors declare no conflict of interest.

References

1. Lu, X.; Isacsson, U. Effect of ageing on bitumen chemistry and rheology. *Constr. Build. Mater.* **2002**, *16*, 15–22. [CrossRef]

2. Lu, X.; Isacsson, U. Chemical and rheological evaluation of ageing properties of SBS polymer modified bitumens. *Fuel* **1998**, *77*, 961–972. [CrossRef]

3. Mazzoni, G.; Bocci, E.; Canestrari, F. Influence of rejuvenators on bitumen ageing in hot recycled asphalt mixtures. *J. Traffic Trans. Eng.* **2018**, *5*, 157–168. [CrossRef]

4. Liu, G.; Wu, S.; van de Ven, M.; Yu, J.; Molenaar, A. Structure and artificial ageing behavior of organo montmorillonite bitumen nanocomposites. *Appl. Clay Sci.* **2013**, *72*, 49–54. [CrossRef]

5. Wu, S.; Han, J.; Pang, L.; Yu, M.; Wang, T. Rheological properties for aged bitumen containing ultraviolet light resistant materials. *Constr. Build. Mater.* **2012**, *33*, 133–138. [CrossRef]

6. Wu, S.; Pang, L.; Liu, G.; Zhu, J. Laboratory study on ultraviolet radiation aging of bitumen. *J. Mater. Civ. Eng.* **2010**, *22*, 767–772. [CrossRef]

7. Zeng, W.; Wu, S.; Wen, J.; Chen, Z. The temperature effects in aging index of asphalt during UV aging process. *Constr. Build. Mater.* **2015**, *93*, 1125–1131. [CrossRef]

8. Pan, P.; Wu, S.P.; Xiao, Y.; Liu, G. A review on hydronic asphalt pavement for energy harvesting and snow melting. *Renew. Sustain. Energy Rev.* **2015**, *48*, 624–634. [CrossRef]

9. Airey, G.D. State of the art report on ageing test methods for bituminous pavement materials. *Int. J. Pavement Eng.* **2003**, *4*, 165–176. [CrossRef]

10. Hu, J.; Wu, S.; Liu, Q. Effect of ultraviolet radiation on bitumen by different ageing procedures. *Constr. Build. Mater.* **2018**, *163*, 73–79. [CrossRef]

11. Lesueur, D.; Teixeira, A.; Lázaro, M.M. A simple test method in order to assess the effect of mineral fillers on bitumen ageing. *Constr. Build. Mater.* **2016**, *117*, 182–189. [CrossRef]

12. Karlsson, R.; Isacsson, U. Bitumen rejuvenator diffusion as influenced by ageing. *Road Mater. Pavement Des.* **2002**, *3*, 167–182. [CrossRef]

13. Chen, M.; Leng, B.; Wu, S.; Sang, Y. Physical, chemical and rheological properties of waste edible vegetable oil rejuvenated asphalt binders. *Constr. Build. Mater.* **2014**, *66*, 286–298. [CrossRef]

14. Xiao, Y.; Wang, F.; Cui, P.D.; Lei, L.; Lin, J.T.; Yi, M.W. Evaluation of Fine Aggregate Morphology by Image Method and its Effect on Skid-resistance of Micro-surfacing. *Materials* **2018**, *11*, 920–932. [CrossRef] [PubMed]

15. Nivitha, M.R.; Prasad, E.; Krishnan, J.M. Ageing in modified bitumen using FTIR spectroscopy. *Int. J. Pavement Eng.* **2016**, *17*, 565–577. [CrossRef]

16. Zhang, F.; Yu, J.; Han, J. Effects of thermal oxidative ageing on dynamic viscosity, TG/DTG, DTA and FTIR of SBS and SBS/sulfur-modified asphalts. *Constr. Build. Mater.* **2011**, *25*, 129–137. [CrossRef]

17. Feng, Z.G.; Bian, H.J.; Li, X.J.; Yu, J.Y. FTIR analysis of UV aging on bitumen and its fractions. *Mater. Struct.* **2016**, *49*, 1381–1389. [CrossRef]

18. Feng, Z.G.; Wang, S.J.; Bian, H.J.; Guo, Q.L.; Li, X.J. FTIR and rheology analysis of aging on different ultraviolet absorber modified bitumens. *Constr. Build. Mater.* **2016**, *115*, 48–53. [CrossRef]

19. Liu, X.; Wu, S.; Liu, G.; Li, L. Effect of ultraviolet aging on rheology and chemistry of LDH-Modified bitumen. *Materials* **2015**, *8*, 5238–5249. [CrossRef] [PubMed]

20. Zargar, M.; Ahmadinia, E.; Asli, H.; Karim, M.R. Investigation of the possibility of using waste cooking oil as a rejuvenating agent for aged bitumen. *J. Hazard. Mater.* **2012**, *233*, 254–258. [CrossRef] [PubMed]

21. Asli, H.; Ahmadinia, E.; Zargar, M.; Karim, M.R. Investigation on physical properties of waste cooking oil–rejuvenated bitumen binder. *Constr. Build. Mater.* **2012**, *37*, 398–405. [CrossRef]

22. Redelius, P.; Soenen, H. Relation between bitumen chemistry and performance. *Fuel* **2015**, *140*, 34–43. [CrossRef]

23. Filippelli, L.; Gentile, L.; Rossi, C.O.; Ranieri, G.A.; Antunes, F.E. Structural change of bitumen in the recycling process by using rheology and NMR. *Ind. Eng. Chem. Res.* **2012**, *51*, 16346–16353. [CrossRef]

24. Varanda, C.; Portugal, I.; Ribeiro, J.; Silva, A.; Silva, C.M. Influence of polyphosphoric acid on the consistency and composition of formulated bitumen: Standard characterization and NMR insights. *J. Anal. Methods Chem.* **2016**, *2016*, 2915467. [CrossRef] [PubMed]
25. Eberhardsteiner, L.; Füssl, J.; Hofko, B.; Handle, F.; Hospodka, M.; Blab, R.; Grothe, H. Towards a microstructural model of bitumen ageing behaviour. *Ind. Eng. Chem. Res.* **2015**, *16*, 939–949. [CrossRef]
26. Zhu, J.; Birgisson, B.; Kringos, N. Polymer modification of bitumen: Advances and challenges. *Eur. Polym. J.* **2014**, *54*, 18–38. [CrossRef]
27. Handle, F.; Füssl, J.; Neudl, S.; Grossegger, D.; Eberhardsteiner, L.; Hofko, B.; Hospodka, M.; Blab, R.; Grothe, H. The bitumen microstructure: A fluorescent approach. *Mater. Struct.* **2016**, *49*, 167–180. [CrossRef]
28. Hofko, B.; Eberhardsteiner, L.; Füssl, J.; Grothe, H.; Handle, F.; Hospodka, M.; Grossegger, D.; Nahar, S.N.; Schmets, A.J.M.; Scarpas, A. Impact of maltene and asphaltene fraction on mechanical behavior and microstructure of bitumen. *Mater. Struct.* **2016**, *49*, 829–841. [CrossRef]
29. Grossegger, D.; Grothe, H.; Hofko, B.; Hospodka, M. Fluorescence spectroscopic investigation of bitumen aged by field exposure respectively modified rolling thin film oven test. *Road Mater. Pavement Des.* **2018**, *19*, 992–1000. [CrossRef]
30. Ministry of Transport of China. *Standard Test Methods of Bitumen and Bituminous Mixtures for Highway Engineering*; JTG E20-2011; China Communications Press: Beijing, China, 2011.
31. Hoeben, F.J.M.; Jonkheijm, P.; Meijer, E.W. About supramolecular assemblies of π-conjugated systems. *Chem. Rev.* **2005**, *105*, 1491–1546. [CrossRef] [PubMed]
32. Fujitsuka, M.; Tojo, S.; Shibahara, M. Delocalization of positive charge in π-stacked multi-benzene rings in multilayered cyclophanes. *J. Phys. Chem. A* **2010**, *115*, 741–746. [CrossRef] [PubMed]
33. Pahlavan, F.; Hung, A.M.; Zadshir, M. Alteration of π-Electron Distribution to Induce Deagglomeration in Oxidized Polar Aromatics and Asphaltenes in an Aged Asphalt Binder. *ACS Sustain. Chem. Eng.* **2018**, *6*, 6554–6569. [CrossRef]

materials

MDPI

Article

Temperature Sensitivity Characteristics of SBS/CRP-Modified Bitumen after Different Aging Processes

Rui He [1,2], Shuhua Wu [3], Xiaofeng Wang [4,*], Zhenjun Wang [1,2,*] and Huaxin Chen [1,2]

[1]   School of Materials Science and Engineering, Chang'an University, Xi'an 710061, Shaanxi, China;
      heruia@163.com (R.H.); hxchen@chd.edu.cn (H.C.)
[2]   Engineering Research Center of Transportation Materials, Ministry of Education, Chang'an University,
      Xi'an 710064, Shaanxi, China
[3]   School of Civil and Architectural Engineering, Zhengzhou University of Aeronautics,
      Zhengzhou 450046, Henan, China; wshuhua0506@126.com
[4]   Henan Provincial Communications Planning & Design Institute Co., Ltd., Zhengzhou 450052, Henan, China
*   Correspondence: wangxf0351@sina.com (X.W.); zjwang@chd.edu.cn (Z.W.)

Received: 7 October 2018; Accepted: 26 October 2018; Published: 30 October 2018

Abstract: Temperature sensitivity characteristics of bitumen can be evidently influenced by modifier types and natural aging processes. Many types of modifiers have been used to improve the temperature sensitivity performance of bitumen, but their effects are different. Therefore, different bitumen specimens as well as SBS/CRP (Styrene-butadiene-styrene polymer/crumb rubber powder)-modified bitumen were prepared and the temperature sensitivity characteristics of bitumen after different aging processes were analyzed in this study. A dynamic rheological property test and performance test at low temperature were carried out to analyze temperature sensitivity and low temperature rheological properties of bitumen. An infrared spectrum test was adopted to study the effect of functional groups under different aging process on the properties of bitumen. The relationship between macroscopic properties and microstructures of bitumen was analyzed. The results show that SBS/CRP-modified bitumen has a strong anti-aging ability in that its flexibility and structure remain in a good condition after long-term aging. The aging process has no significant effect on SBS/CRP-modified bitumen. SBS/CRP-modified bitumen has an excellent low-temperature relaxation ability and low-temperature crack resistance. In contrast to original bitumen and SBS-modified bitumen, the temperature sensitivity performance of SBS/CRP-modified bitumen is evidently enhanced. The physical blending effect is dominant in the bitumen modified process and there is no evident chemical reaction between bitumen and crumb rubber powder. SBS/CRP-modified bitumen is recommended for wide use in plateau areas with ultraviolet and cold surroundings.

Keywords: crumb rubber powder; SBS/CRP-modified bitumen; aging processes; temperature sensitivity characteristics

1. Introduction

Styrene-butadiene-styrene polymer (SBS)-modified bitumen is widely used in highway construction due to its excellent high-temperature stability, low-temperature cracking resistance, aging resistance, and fatigue resistance [1–3]. However, due to the existence of the SBS carbon-carbon double bond, its low-temperature performance and aging resistance are slightly inadequate when it is used in special areas such as plateau areas with cold and ultraviolet surroundings [4,5]. Rasool, et al. [6] analyzed the aging and rheological properties of waste rubber modified bitumen. The results showed that rubber powder could further enhance the ductility and anti-aging properties of the SBS-modified mechanism. Liang, et al. [7] studied the thermal rheology and compatibility of SBS-modified bitumen with different S and B segment ratios. The results showed that SBS-modified bitumen with a 30% S segment had excellent viscoelasticity and low-temperature sensitivity; the compatibility of SBS-modified bitumen became worse with the increase of the S segment ratio.

However, the demand for SBS modifier is increasing and leads to an increase in price with the development of highway construction. At the same time, SBS polymer is difficult to recycle, which is not conducive to sustainable development. Therefore, the use of rubber powder instead of SBS modifier has become popular. Some studies have shown that crumb rubber powder (CRP) and original bitumen can produce crack passivation at their interface and consume energy. Moreover, flexibility, elastic recovery, anti-aging, oxidation resistance, and many other advantages have been improved for original bitumen [8–12]. Crumb rubber powder modified bitumen can improve the high-temperature performance, rutting resistance, and aging resistance of bitumen pavement and can reuse waste rubber [13–15]. However, the requirement for rubberized bitumen pavement is becoming higher and higher with the development of bitumen pavement construction. Moreover, the low-temperature performance of rubber modified bitumen can not meet the requirements for bitumen pavement. On the other hand, the aging resistance of SBS-modified bitumen is not as good as that of rubber modified bitumen. Therefore, SBS/CRP-modified bitumen can not only reduce the cost of SBS modifier, but can also improve the modified effect and promote the recycling of resources.

Rubber powder is the main factor affecting the anti-aging performance of SBS/CRP-modified bitumen; and the content of rubber powder is related to the temperature sensitivity of modified bitumen. Researchers tend to focus on single performance studies and ignore the impact of performance. Therefore, it is still necessary to explore the temperature characteristics of SBS/rubber powder SBS/CRP-modified bitumen under different aging conditions. Guo, et al. [16,17] studied the best preparation technology of SBS/rubber powder SBS/CRP-modified bitumen. Li, et al. [18,19] found that SBS/CRP-modified bitumen could improve high-temperature stability. Li, et al. [20] studied the aging resistance of SBS/rubber powder SBS/CRP-modified bitumen and found that the aging resistance of SBS/rubber powder SBS/CRP-modified bitumen was inferior to that of rubber powder modified bitumen. Tan, et al. [21] found that rubber powder (or SBS) and aromatic oil could significantly improve the elastic energy storage of bitumen. Wang, et al. [22] prepared the SBS/CRP-modified bitumen with SBS and rubber powder and obtained the best preparation process. In order to improve the aging resistance of bitumen, inorganic or organic powders such as rubber powder, nano-TiO_2 and carbon black can be added. Rossi, et al. [23,24] studied natural resources such as phospholipids, ascorbic acid and organosilane, polyphosphoric acid, food grade phospholipids were homogeneously mixed to bitumen to improve aging resistance. However, these modified bitumens have a high viscosity, so it is difficult to use them in cold areas [25]. However, single modified bitumen can not meet the requirements of bitumen binder in special areas where temperature changes sharply.

The objectives of this study are to improve the temperature sensitivity and anti-aging ability of bitumen. Therefore, SBS and CRP were used to modify the bitumen taking into account material performance in this study. Softener was added to reduce the viscosity of the SBS/CRP-modified bitumen, and a variety of additives were added to improve the stability and aging resistance of the modified bitumen, which was prepared to improve the low-temperature performance and anti-aging performance of the modified bitumen. The physical properties and low-temperature rheological properties of SBS/RP-modified bitumen before and after the thin film oven test (TFOT) and pressurized aging vessel (PAV) aging were analyzed. Based on the infrared spectroscopy, the relationship between macroscopic properties and microstructure was explored; then, the anti-aging and low-temperature properties of SBS/RP-modified bitumen were evaluated.

2. Experimental Stage

2.1. Raw Materials

The properties of original bitumen are shown in Table 1. SBS was used as a modifier and the amount of SBS was 4.5% in bitumen weight. Rubber powder with a 40–60 mesh was used and its main properties are given in Table 2. Solvent enhancers were aromatic oils rich in saturated and aromatic components, which can make rubber powder and SBS swell sufficiently in bitumen and reduce the viscosity of bitumen. Its density was 1.10 g/cm^3 and the content was 7% in bitumen weight, respectively. The density of the anti-aging agent was 0.94 g/cm^3 and its content was 3% in bitumen weight. The stabilizer was sulfur powder and the amount was 0.2% in bitumen weight.

Table 1. Properties of original bitumen.

	Properties	Unit	Test Results	Specification	Test Method
	Penetration (25 °C, 100 g, 5 s)	0.1 mm	94.6	80–100	T0604
	Ductility (15 °C, 5 cm/min)	cm	>100	≮100	T0605
	Softening point (Ring-and-ball method)	°C	45.8	≮44	T0606
	Mass loss	%	+0.4	±0.8	T0610
After TFOT	Residual penetration (25 °C)	%	57.8	≮57	T0604
	Residual ductility (10 °C)	cm	12	≮8	T0605

Table 2. Properties of crumb rubber powder.

	Properties	Unit	Test Results
	Density	g/cm^3	1.15
Physical properties	Moisture	%	0.46
	Metal content	%	0.005
	Fiber content	%	0.52
	Minerals	%	6.1
Chemical properties	Acetone extract	%	7.1
	Carbon black content	%	30
	Rubber hydrocarbon	%	50.1

2.2. Preparation of SBS/CRP-Modified Bitumen

The preparation process of SBS/RP-modified bitumen is shown in Figure 1. All specimens were prepared at a temperature of 180 °C. The softeners, anti-aging agents, rubber powder, auxiliaries, and SBS were added into the original bitumen in turn; then, they were sheared using a shearing machine with 5000 rpm and the shearing time was observed during the preparation.

2.3. Preparation of Aged Bitumen

The aged bitumen specimens were obtained using a thin film oven test (TFOT), aging at 163 °C for 5 h, ASTM D 1754 and pressurized aging vessel (PAV), aging at 100 °C for 20 h under air of 2.1 MPa, ASTM D 6521, respectively. The TFOT was employed to simulate the short-term aging including the storage, transport, mixing, and paving process of bitumen in pavement construction; the PAV was used to simulate the long-term aging of the bitumen in the service period.

2.4. Physical Properties Test

The physical properties of the bitumen samples, including the softening point, penetration at 25 °C, ductility at 15 °C, and viscosity at 135 °C, were tested in accordance with the standards of ASTM D 36, ASTM D 5, ASTM D 113 and ASTM D 6925, respectively. Each specimen was tested three times and the average value was adopted as the testing result.

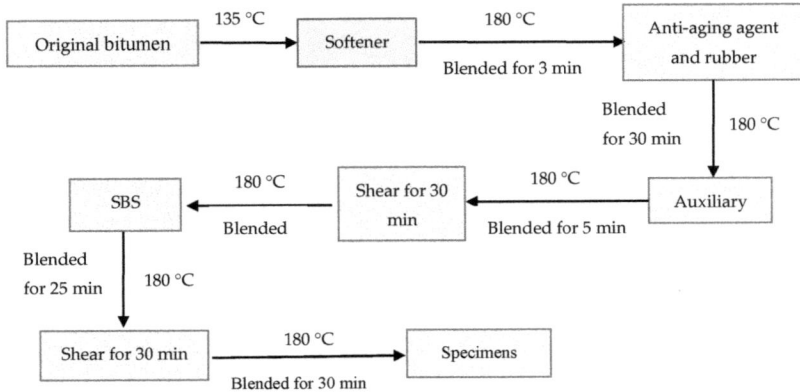

Figure 1. Preparation chart of SBS/CRP-modified bitumen.

2.5. Dynamic Rheological Property Test

The dynamic rheological properties of the bitumen were measured using a dynamic shear rheometer. The temperature sweeping test of the bitumen was performed under strain-controlled mode at a constant frequency of 10 rad/s. The temperature range was from −20 to 80 °C with a temperature increment of 2 °C per minute. The plate used was 8 mm in diameter and the gap between the parallel plates was 2 mm for each sample below 20 °C. When the temperature was above 20 °C, the plate used was 25 mm in diameter and the gap between the parallel plates was 1 mm.

2.6. Performance Test at Low Temperature

The low-temperature performance of original bitumen, SBS-modified bitumen and SBS/CRP-modified bitumen before and after aging was evaluated using the Bending Beam Rheometer (BBR) test. The test temperature was −12 °C, and the loading time and unloading time were 240 s and 10 s, respectively. The test was carried out 3 times and the average value was calculated and used as the testing result. The creep load was used to simulate the accumulated stress step by step in the performance test at low temperature. The time-deformation curve, creep rate m, and stiffness modulus at 60 s were outputted. The stiffness modulus and creep rate of bitumen at 8, 15, 30, 60, 120, and 240 s were calculated. The stiffness modulus (S) indicates the resistance to load for bitumen; and the creep rate (m) is a comprehensive reflection of the curve shape and relaxation ability of bitumen. The stiffness modulus (S) and creep rate (m) can be calculated using Equations (1) and (2), respectively.

$$S(t) = \frac{Pl^3}{4bh^3v(t)} \tag{1}$$

$$m = \frac{d \log S(t)}{d \log(t)} \tag{2}$$

where, $v(t)$ is the deformation in the middle of the beam, mm; b is the width of the beam, 12.70 mm ± 0.05 mm; P is the constant load of the beam, 980 mN ± 50 mN; h is the height of the beam, 6.35 mm ± 0.0 5 mm; and l is the span of the beam, 102 mm.

In order to further characterize the rheological properties of the modified bitumen at low temperature, Equation (3) can be used to calculate the relationship between creep compliance and time according to the original data of loading time and deformation of PAV aged bitumen, which is the reciprocal equation of Equation (1). Based on the Burgers model, the viscoelastic parameters can be obtained using Equations (4) and (5) and the comprehensive compliance parameters are used to evaluate the low-temperature relaxation characteristics of bitumen.

$$S(t) = \frac{Pl^3}{4bh^3v(t)} = \frac{1}{D(t)} \tag{3}$$

$$J(t) = \frac{1}{E_1} + \frac{1}{E_2}(1 - e^{-tE_2/\eta_2}) + \frac{t}{\eta_1} \tag{4}$$

The Burgers creep equation is expressed as:

$$J(t) = J_E + J_{De} + J_V \tag{5}$$

where $J_E = \frac{1}{E_1}$, $J_{De} = \frac{1}{E_2}(1 - e^{-tE_2/\eta_2})$, $J_V = \frac{t}{\eta_1}$; comprehensive compliance parameter $J = J_V(1 - \frac{J_E + J_{De}}{J_E + J_{De} + J_V})$ can be used to characterize the low-temperature performance of bitumen materials.

2.7. Infrared Spectrum Test

Original bitumen, SBS-modified bitumen and SBS/CRP-modified bitumen after short-term TFOT aging and the PAV pressure aging were tested using infrared spectroscopy so as to further elucidate the relationship between the macroscopic properties and microstructure of bitumen. The wavelength range of infrared spectroscopy was 400–4000 cm^{-1} and the test temperature was 20 °C.

3. Results and Discussion

3.1. Physical Properties of SBS/CRP-Modified Bitumen

Table 3 shows the properties of three kinds of bitumen after different aging processes. It shows that the addition of modifiers can greatly affect the softening point, viscosity, and penetration of original bitumen. The softening point of SBS-modified bitumen and SBS/CRP-modified bitumen increases by 58.2% and 75.8%, respectively. The penetration decreases by 19.4% and 23.1%, respectively. The rotational viscosity at 175 °C is 6.8 times and 16.9 times of that of original bitumen, respectively. The softening point of original bitumen increases obviously after modification and the penetration decreases slightly, which is mainly due to the formation of the network structure [26] in the original bitumen due to the addition of modifiers and the enhancement of the interaction between bitumen molecules. The viscosity of the modified bitumen increases greatly because the modifier absorbs the lightweight components of the original bitumen and forms the network structure [26], which increases the resistance of the rotary viscometer. In particular, the friction resistance of the rotary viscometer is further increased with the addition of rubber particles. The viscosity of the SBS/CRP-modified bitumen increases greatly with the temperature and the viscosity is greatly reduced.

Table 3. Properties of bitumen with different aging processes.

Bitumen Type		Ductility/cm 5 °C	Ductility/cm 10 °C	Softening Point/°C	Penetration/0.1 mm	Rotational Viscosity at 135 °C/mPa·s	Rotational Viscosity at 175 °C/mPa·s
Bitumen before aging	Original	-	54.6	46.7	92.1	353	56
	SBS	34.6		73.9	74.2	1400	382
	SBS/CRP	30.5		82.1	70.8	4300	945
Bitumen after TFOT	Original	-	8.0	52.6	49.5	507	101
	SBS	23.0		64.7	44.8	2440	460
	SBS/CRP	15.1		75.2	51.5	6110	1820
Bitumen after PAV	Original	-	4.2	60.7	29.8	1010	207
	SBS	1.5		65.1	21.4	3567	670
	SBS/CRP	6.0		73.4	35.0	6700	1900

After short-term aging, the softening point of original bitumen increases by 12.6%; the penetration decreases by 46.2%; and the rotational viscosity at 175 °C increase by 2 times. The softening point of SBS-modified bitumen and SBS/CRP-modified bitumen decreases by 12.4% and 9.5%; the penetration decreases by 65.6% and 37.4%, respectively; and the viscosity is 1.2 times and 1.9 times that of the original bitumen. In contrast to the original bitumen, the softening point of the modified bitumen increases; the penetration decreases; the viscosity increases; and the bitumen gradually becomes hard and brittle. The carbon-carbon double bond of the modified bitumen is broken. The network structure is damaged and the hard segment of styrene is dispersed in the modified bitumen. Therefore, the softening point decreases; the penetration decreases; and the viscosity increases.

After the long-term aging process, the ductility of three kinds of bitumen decreases obviously and the viscosity increases. The softening point of original bitumen and SBS-modified bitumen increases slightly, but the softening point of SBS/CRP-modified bitumen does not decrease significantly. This shows that the flexibility and structure of the SBS/CRP-modified bitumen are still in a good condition after long-term aging and that its anti-aging ability is still strong.

3.2. Dynamic Rheological Properties of SBS/CRP-Modified Bitumen

Figure 2 shows the composite shear modulus and phase angle of three kinds of bitumen before and after aging. With the increase in temperature, the composite shear modulus of bitumen decreases and the phase angle increases, mainly due to the softening of bitumen and the decrease in viscosity. As shown in Figure 2a,b, the composite shear modulus of bitumen increases and the phase angle decreases after the aging and long-term aging of original bitumen. The phase angles of bitumen before and after aging at the same temperature change slightly during −20–30 °C. That is, the proportion of the bitumen elastic component is larger and it tends to be elastic-plastic at the middle and low temperatures. The variation range of the composite shear modulus before and after aging at −20 and 30 °C is 299–378 MPa and 0.4–6.3 MPa, respectively. After short-term aging, the phase angle of original bitumen at 30–80 °C is smaller than that of the original bitumen, which is 70° and 87°, respectively, but the phase angle of original bitumen is large. The proportion of the viscous part is high in this temperature range. When the temperature is low, the viscosity of three kinds of bitumen after aging is very high, which is close to being elastic-plastic. Moreover, the stress hysteresis becomes weaker after loading. Therefore, the change in phase angle before and after aging is small. When the temperature is relatively high, the temperature sensitivity of the original bitumen is poor. The longer the aging time, the more volatile the light components between the bitumen molecules are, the more severe the heavy ones are, the greater the viscosity, and the weaker the stress hysteresis, which shows that the short-term aging phase angle decreases less and the long-term aging decreases more.

As shown in Figure 2c,d, the change trend of SBS-modified bitumen after short-term aging is different from that of original bitumen. After short-term aging, the composite shear modulus of bitumen decreases and the phase angle increases slightly for SBS-modified bitumen, which is more obvious at −20–18 °C. The curve of 18–80 °C almost coincides with the original curve and the change in the composite shear modulus and phase angle is not obvious, which indicates that the contribution of the composite shear modulus to the storage modulus becomes smaller and smaller; the elasticity of modified bitumen is lower and the viscosity is enhanced. The main reason for this is that the unsaturated double bond of the SBS modifier is broken and partly degraded under the action of thermal oxygen at 163 °C, which destroys its original structure and the elasticity of the modifier. The increase in the phase angle is not obvious after short-term aging. This is mainly due to the fact that the further degradation of the modifier can lead to an increase in the phase angle, while the hardening degree of original bitumen can lead to a decrease in the phase angle. With long-term aging, the elasticity reduction caused by the degradation of the modifier is reduced, and the hardening of the original bitumen is dominant. After long-term aging, the phase angle decreases and the composite shear modulus of bitumen increases.

As shown in Figure 2e,f, the change trend of SBS/CRP-modified bitumen is the same as that of SBS. Short-term aging has little effect on the composite shear modulus or phase angle of SBS/CRP-modified bitumen. The curves of unaged and short-term aged bitumen almost coincide at 40 °C. In contrast to SBS-modified bitumen at the same temperature, SBS/CRP-modified bitumen possesses a smaller composite shear modulus, which indicates that SBS/CRP-modified bitumen has lower strength and greater flexibility at a low temperature. The composite shear modulus and phase change with the change in temperature and the temperature sensitivity are good. The main reason for this is that a three-dimensional network structure is formed in the compound modification system of SBS and rubber powder [27]. When the strain of SBS-modified bitumen reaches the limit state, the fracture stress can quickly concentrate on the surface of crumb rubber powder particles. Then, the crumb rubber powder particles absorb and consume a large amount of energy, which prevents the crack from forming and expanding. At the same time, crumb rubber powder particles play a reinforcing role in bitumen, meaning that the elastic recovery ability of the two composite modifications can be significantly improved, which shows a strong low-temperature deformation capacity.

3.3. Performance of SBS/CRP-Modified Bitumen at Low Temperature

3.3.1. Effects of Aging on Stiffness Modulus and Creep Rate

Figures 3 and 4 show the change in the stiffness modulus and creep rate of bitumen. As shown in Figure 3, in contrast to the original bitumen after short-term aging, the stiffness modulus of original bitumen increases, while those of SBS-modified bitumen and SBS/CRP-modified bitumen decrease, which is consistent with the results of the composite shear modulus. The stiffness modulus of the three bitumen increases significantly after long-term aging. The main reason for this is that the original bitumen tends to be in densification during thermal-oxygen aging and pressure aging and the original bitumen shows brittle and rigid characteristics, which result in poor low-temperature relaxation ability of original bitumen. As shown in Figure 4, after short-term aging, the m value of original bitumen increases slightly, while SBS-modified bitumen and SBS/CRP-modified bitumen decrease, which is inconsistent with the result of stiffness modulus. Moreover, the creep rate of all kinds of bitumen decreases after long-term aging. However, there are some limitations in evaluating the low-temperature performance of bitumen by S and m values at a certain test temperature. The creep properties of bitumen based on the Burgers model can be used to evaluate the low-temperature performance of bitumen.

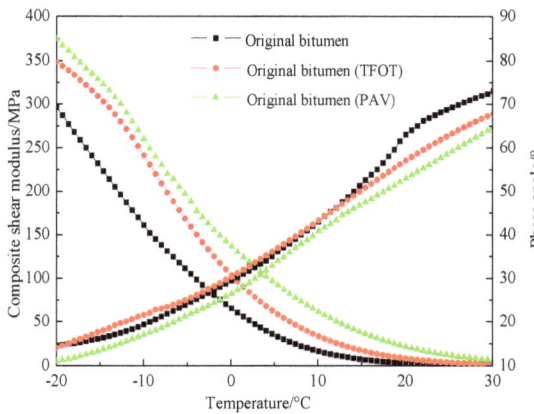

(a) Original bitumen at a temperature of −20–30 °C.

Figure 2. *Cont.*

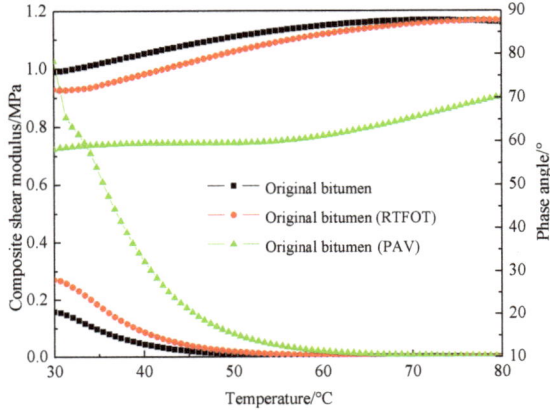

(**b**) Original bitumen at a temperature of 30–80 °C.

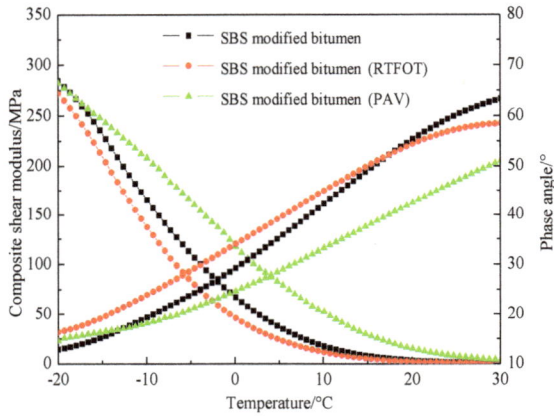

(**c**) SBS-modified bitumen at a temperature of −20–30 °C.

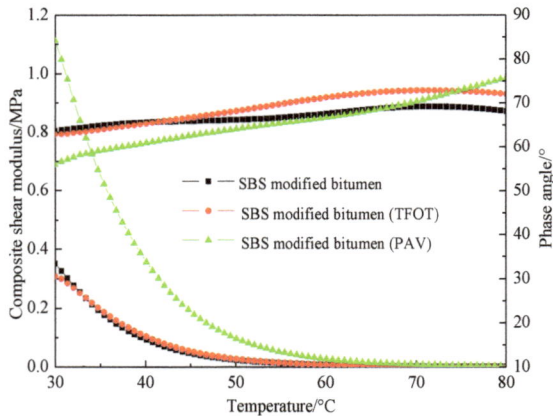

(**d**) SBS-modified bitumen at a temperature of 30–80 °C.

Figure 2. *Cont.*

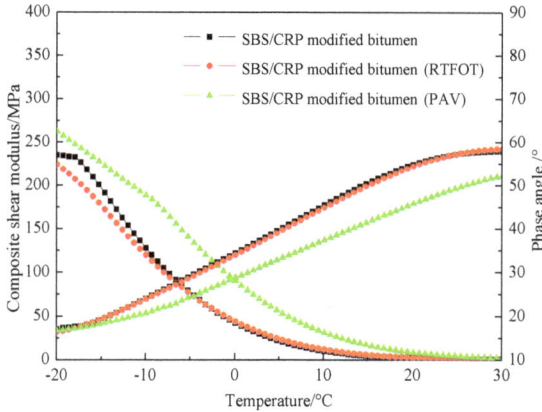

(e) SBS/CRP-modified bitumen at a temperature of −20–30 °C.

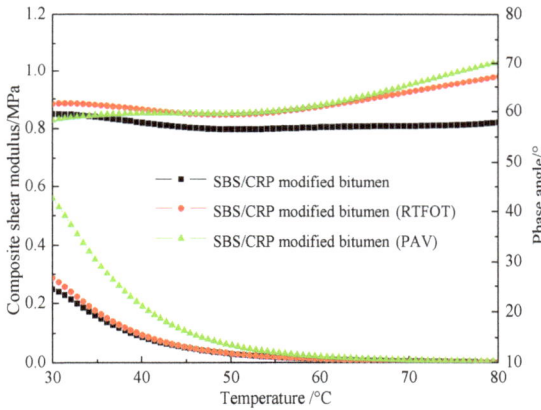

(f) SBS/CRP-modified bitumen at a temperature of 30–80 °C.

Figure 2. Dynamic rheological properties of bitumen at different temperatures.

3.3.2. Stiffness Modulus Curve of Bitumen

The stiffness modulus curves of three kinds of bitumen after PAV are shown in Figure 5, which can be calculated using the original data of performance at low temperature test. As shown in Figure 5, the logarithm of the creep stiffness modulus decreases gradually with a time delay in the process of constant force loading, and the higher the slope of creep stiffness is, the stronger the creep ability is. Three kinds of creep stiffness modulus slope in the following order: original bitumen < SBS-modified bitumen < SBS/CRP-modified bitumen. At the same temperature, the stiffness modulus of SBS-modified bitumen is close to that of original bitumen; however, the difference between SBS-modified bitumen and SBS/CRP-modified bitumen is higher than that of original bitumen. And the difference between SBS-modified bitumen and SBS/CRP-modified bitumen is more obvious with the decrease in temperature, which further indicates that SBS/CRP-modified bitumen has a stronger low-temperature relaxation ability and better low-temperature crack resistance.

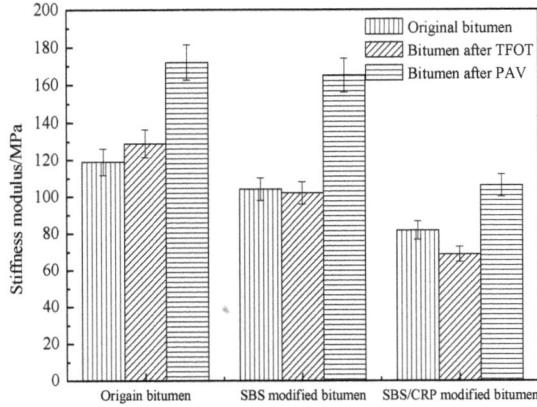

Figure 3. The effects of aging on the stiffness modulus of bitumen.

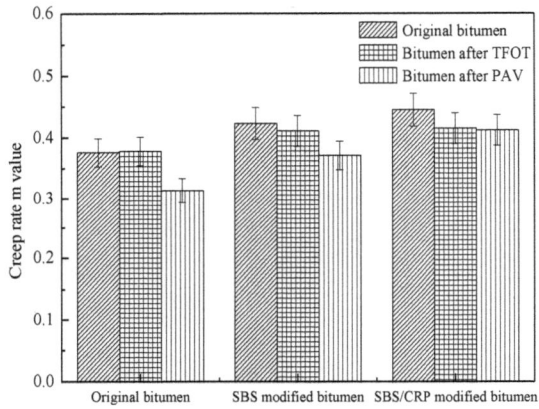

Figure 4. The effects of aging on the creep rate of bitumen.

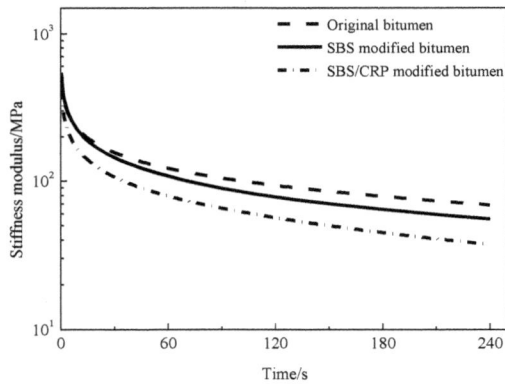

Figure 5. Creep stiffness modulus versus time curve of bitumen.

3.3.3. Low-Temperature Creep Properties of Bitumen Based on the Burgers Model

Table 4 shows the Burgers viscoelastic parameters of different kinds of bitumen. At the same temperature, as the aging degree deepens, the stiffness modulus S of bitumen increases and the creep rate m decreases. Based on the data of Table 4, the comprehensive compliance parameters of different kinds of bitumen are obtained using Equation (5), as shown in Table 5. At a low temperature, the proportion of elastic deformation is bigger and the proportion of viscous deformation is smaller. If the proportion of viscous flow is high, the compressive tensile stress of bitumen pavement can be relaxed by means of flow, and the low-temperature shrinkage cracking of bitumen pavement is reduced. As shown in Table 5, the J value of SBS-modified bitumen and original bitumen is 1.04 times and 1.55 times, respectively.

Table 4. Burgers viscoelastic parameters of different kinds of bitumen.

Bitumen Type — Parameters	Original Bitumen	SBS-Modified Bitumen	SBS/CRP-Modified Bitumen
E_1/MPa	9000.0	8000.0	7000.0
E_2/MPa	617.3	342.6	266.1
η_1/MPa·s	40,971.9	52,584.8	23,164.1
η_2/MPa·s	7282.8	6781.8	4605.7

Note: The values of E_1 were obtained using the empirical method: see Reference [28].

Table 5. Comprehensive creep compliance parameters of different kinds of bitumen.

Bitumen Type — Parameters	Original Bitumen	SBS-Modified Bitumen	SBS/CRP-Modified Bitumen
J_E/(1/MPa)	1.11 E-04	1.25 E-04	1.43 E-04
J_{De}/(1/MPa)	1.62 E-03	2.92 E-03	3.76 E-03
J_V/(1/MPa)	3.30 E-02	3.54 E-02	5.21 E-02
J/(1/MPa)	3.13 E-02	3.26 E-02	4.85 E-02

3.4. Infrared Spectrum Analyses of SBS/CRP-Modified Bitumen

Figure 6 shows the infrared spectra of different kinds of bitumen before and after aging. There are eight obvious characteristic absorption peaks in the infrared spectra of three kinds of bitumen before and after aging. The corresponding wave numbers are 2920 cm^{-1}, 2852 cm^{-1}, 1600 cm^{-1}, 1455 cm^{-1}, 1374 cm^{-1}, 850 cm^{-1}, 809 cm^{-1}, and 727 cm^{-1}, respectively. The strong absorption peaks of 2920 cm^{-1} and 2852 cm^{-1} are the reverse stretching peaks and the methylene symmetrical stretching peaks of aliphatic methylene C–H with acromion. The wave number 1600 cm^{-1} is the stretching vibration of the aromatic C=C skeleton. The wave numbers 1455 cm^{-1} and 1374 cm^{-1} are the bending vibration absorption peaks of methylene (–CH$_2$$^-$) and methyl (–CH$_3$), respectively, which prove the existence of long carbon chains. In the wave numbers 2920 cm^{-1} and 1455 cm^{-1} in three kinds of bitumen, C–H is the main chain. The wave number 650–900 cm^{-1} is called the benzene ring substituted region. The wave numbers 809 cm^{-1} and 727 cm^{-1} are the absorption peaks of C–H surface bending vibration on the aromatic benzene ring. As shown in Figure 6a, the absorption peaks of sulfoxide S=O at 1030 cm^{-1} and carbonyl at 1600 cm^{-1} are enhanced. The absorption magnitudes of 2920 cm^{-1}, 2852 cm^{-1}, and 1455 cm^{-1} are also enhanced. The transmittance of the sulfoxide group after PAV aging is stronger than that after TFOT aging. In addition, the characteristic peaks representing methylene 1261 cm^{-1} also appear after the PAV aging process.

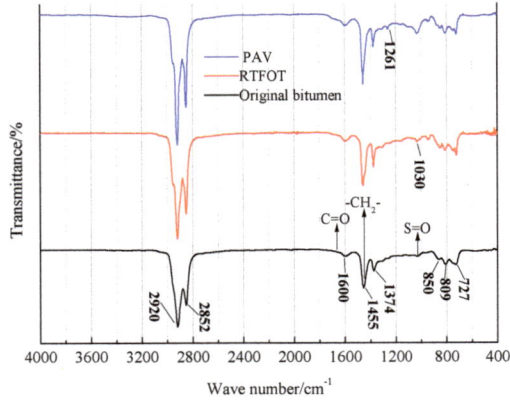

(**a**) Infrared spectrum of original bitumen before and after aging.

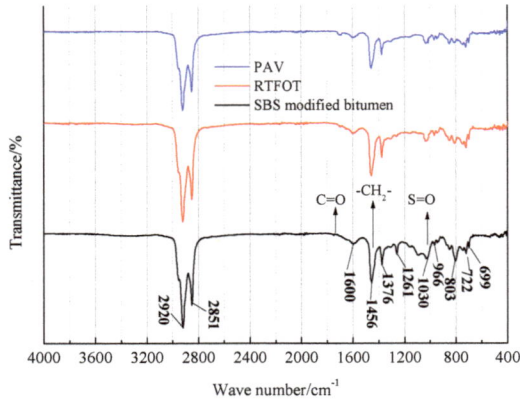

(**b**) Infrared spectrum of SBS-modified bitumen before and after aging.

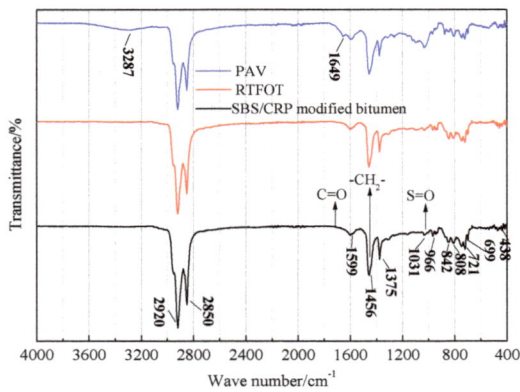

(**c**) Infrared spectrum of SBS/CRP-modified bitumen before and after aging.

Figure 6. Infrared spectra of three kinds of bitumen before and after aging.

Therefore, the aging process of bitumen is characterized by the increase in the methyl, methylene, sulfoxide, and carbonyl peaks, which indicates that the long chain of bitumen is broken; the short chain is dehydrogenated and condensed; and the active functional groups in the bitumen molecules combine with oxygen molecules in the air to form polar macromolecules such as sulfoxide and carbonyl. The sulfoxide group S=O increases obviously after long-term aging, and the sulfoxide group has poor thermal stability. Sulfur is added to the bitumen directly when the bitumen reacts with sulfur at 140 °C and it can decompose when the temperature rises. When the temperature rises to 180 °C, hydrogen sulfide and other substances are produced, which can join with sulfur to form macromolecules. When the temperature continues to rise, the dehydrogenation reaction occurs and the S=O is generated.

As shown in Figure 6b, in contrast to original bitumen, the absorption peak of 966 cm^{-1} of SBS-modified bitumen increases, which is the absorption peak corresponding to the C=C bond of butadiene in SBS-modified bitumen; and the absorption peak of 966 cm^{-1} goes down after aging. The results show that the unsaturated carbon-carbon double bond of the SBS modifier breaks after short-term aging and long-term aging; and the network structure of SBS is destroyed. With the increase in temperature, the interaction between the bitumen molecules becomes weaker; the softening point decreases, the composite shear modulus and stiffness modulus decrease slightly; and the hard segment of styrene is dispersed in the modified bitumen. In SBS-modified bitumen, the penetration decreases and the carbon-carbon double bonds break after long-term aging, tending to the variation of original bitumen. The composite shear modulus and stiffness modulus increase gradually, and the relaxation ability of SBS-modified bitumen decreases.

However, the characteristic peaks of SBS/CRP-modified bitumen do not change significantly compared with the SBS-modified bitumen, as shown in Figure 6c. After long-term aging, the absorption peaks of 3287 cm^{-1} and 1649 cm^{-1} appear. The wave number 3287 cm^{-1} can be generated by hydroxyl in air and 1649 cm^{-1} is generated by oxidized carbonyl.

4. Conclusions and Recommendations

In this study, SBS/CRP-modified bitumen was prepared. Temperature sensitivity performance, such as the high-temperature performance and low-temperature performance of original bitumen, SBS-modified bitumen and SBS/CRP-modified bitumen under different aging processes were evaluated and analyzed. Additionally, the relationship between macroscopic properties and microstructures of SBS/CRP-modified bitumen was discussed. Then, the following conclusions were drawn:

(1) The addition of modifiers can increase the softening point and viscosity and decrease the penetration of original bitumen. The ductility of three kinds of long-term aged bitumen decreases obviously and the viscosity increases. The softening point of aged original bitumen and aged SBS-modified bitumen increases slightly, but the softening point of SBS/CRP-modified bitumen does not decrease significantly. This shows that SBS/CRP-modified bitumen possesses a strong anti-aging ability in that its flexibility and structure remain in a good condition after long-term aging.

(2) In low-temperature surroundings, the stiffness modulus of SBS/CRP-modified bitumen is the lowest and the J value is the highest among three kinds of bitumen. That is to say, SBS/CRP-modified bitumen has excellent low-temperature relaxation ability and low-temperature crack resistance.

(3) In contrast to original bitumen and SBS-modified bitumen, SBS/CRP-modified bitumen possesses a stronger anti-aging ability. The composite shear modulus and phase angle fluctuation of SBS/CRP-modified bitumen with temperature change slightly, and the temperature sensitivity performance of SBS/CRP-modified bitumen decreases.

(4) The aging process of original bitumen is characterized by an increase in methyl, methylene, sulfoxide, and carbonyl peaks, and that of SBS-modified bitumen is close to that of original

bitumen. After long-term aging, the wave numbers 3287 cm^{-1} and 1649 cm^{-1}, the characteristic peaks of SBS/CRP-modified bitumen, increase, where 3287 cm^{-1} can be the hydroxyl group formed in the air and 1649 cm^{-1} is the carbonyl group formed by peroxidation. The physical blending effect is dominant and there is no evident chemical reaction between bitumen and crumb rubber powder.

(5) In contrast to original bitumen and SBS-modified bitumen, SBS/CRP-modified bitumen can be recommended for use in plateau areas. Of course, it is suggested that the low-temperature properties of SBS/CRP-modified bitumen under different temperatures be studied in future research.

Author Contributions: Conceptualization: R.H. and Z.W.; Data curation: R.H. and S.W.; Formal analysis: S.W., X.W. and H.C.; Investigation: S.W. and R.H.; Methodology: H.C.; Writing—original draft: Z.W.

Acknowledgments: This work is supported by the National Natural Science Foundation of China (No. 51508030), the Fundamental Research Funds for the Central Universities of China (Nos. 300102318401, 300102318402, 300102318501), the Basic Research Project in Qinghai Province (No. 2017-ZJ-715), and the Scientific Project of Guangxi Autonomous Region (No. AC16380112). The authors also wish to thank the reviewers for their valuable comments and suggestions concerning this manuscript.

Conflicts of Interest: There are no conflicts of interest regarding the publication of this paper.

References

1. Wu, S.; Pang, L.; Mo, L.; Chen, Y.; Zhu, G. Influence of aging on the evolution of structure, morphology and rheology of base and SBS modified bitumen. *Constr. Build. Mater.* **2009**, *23*, 1005–1010. [CrossRef]
2. Fu, H.; Xie, L.; Dou, D.; Li, L.; Yu, M.; Yao, S. Storage stability and compatibility of asphalt binder modified by SBS graft copolymer. *Constr. Build. Mater.* **2007**, *21*, 1528–1533. [CrossRef]
3. Zhang, F.; Yu, J.; Han, J. Effects of thermal oxidative ageing on dynamic viscosity, TG/DTG, DTA and FTIR of SBS- and SBS/sulfur-modified asphalts. *Constr. Build. Mater.* **2011**, *25*, 129–137. [CrossRef]
4. Zhang, F.; Hu, C. The research for thermal behaviour, creep properties and morphology of SBS-modified asphalt. *J. Therm. Anal. Calorim.* **2015**, *121*, 651–661. [CrossRef]
5. Chen, H.; Zhou, Y.; Wang, B. Dynamic mechanics performance of aged SBS modified-asphalt. *J. Chang'an Univ.* **2009**, *29*, 1–5.
6. Raqiqa, T.R.; Wang, S.; Zhang, Y.; Li, Y.; Zhang, G. Improving the aging resistance of SBS modified asphalt with the addition of highly reclaimed rubber. *Constr. Build. Mater.* **2017**, *145*, 126–134.
7. Liang, M.; Liang, P.; Fan, W.; Qian, C.; Xin, X.; Shi, J.; Nan, G. Thermo-rheological behavior and compatibility of modified asphalt with various styrene-butadiene structures in SBS copolymers. *Mater. Des.* **2015**, *88*, 177–185. [CrossRef]
8. Wang, T.; Xiao, F.; Amirkhanian, S. A review on low temperature performances of rubberized asphalt materials. *Constr. Build. Mater.* **2017**, *145*, 483–505. [CrossRef]
9. Kim, H.S.; Lee, S.J.; Amirkhanian, S. Rheology investigation of crumb rubber modified asphalt binders. *KSCE J. Civ. Eng.* **2010**, *14*, 839–843. [CrossRef]
10. Salim, M.N.; Hassan, A.A.; Rehan, K.M. A review on using crumb rubber in reinforcement of asphalt pavement. *Sci. World J.* **2014**, *1*, 1–22.
11. Ibrahim, I.M.; Fathy, E.S.; El-Shafie, M.; Elnaggar, M.Y. Impact of incorporated gamma irradiated crumb rubber on the short-term aging resistance and rheological properties of asphalt binder. *Constr. Build. Mater.* **2015**, *81*, 42–46. [CrossRef]
12. Lee, S.J.; Amirkhanian, S.N.; Shatanawi, K.; Kim, K.W. Short-term aging characterization of asphalt binders using gel permeation chromatography and selected Superpave binder tests. *Constr. Build. Mater.* **2008**, *22*, 2220–2227. [CrossRef]
13. Shen, J.; Amirkhanian, S.; Xiao, F.; Tang, B. Influence of surface area and size of crumb rubber on high temperature properties of crumb rubber modified binders. *Constr. Build. Mater.* **2009**, *23*, 304–310. [CrossRef]
14. Presti, D.L. Recycled tyre rubber modified bitumens for road asphalt mixtures: A literature review. *Constr. Build. Mater.* **2013**, *49*, 863–881. [CrossRef]

15. Cao, W. Study on properties of recycled tire rubber modified asphalt mixtures using dry process. *Constr. Build. Mater.* **2007**, *21*, 1011–1015. [CrossRef]
16. Guo, J.; Zhang, Bo. Research on the performance of SBS and rubber powder combined modified asphalt. *New Build. Mater.* **2010**, *5*, 66–69.
17. Liu, Z.; Wang, J.; Lu, F.; Tan, H. Experimental study on high-temperature performance of rubber powder and sbs compound modified asphalt. *West. China Commun. Sci. Technol.* **2016**, *4*, 6–9.
18. Chen, L. Research on behavior of SBS rubber compound modified asphalt. *North. Transp.* **2008**, *8*, 29–31.
19. Xiao, C. Experimental Study on compound modified technology of asphalt rubber based on high-temperature performance. *Mod. Transp. Technol.* **2016**, *13*, 5–9.
20. Li, G.; Wang, F.; Kuang, M.; Zhou, X. Performance of sbs/crumb rubber composite modified asphalt. *J. East China Univ. Sci. Technol. (Nat. Sci. Ed.)* **2016**, *42*, 21–27.
21. Tan, Y.; Guo, M.; Cao, L. Effects of common modifiers on viscoelastic properties of asphalt. *China J. Highw. Transp.* **2013**, *26*, 7–15.
22. Wang, Y.; Zhan, B.; Cheng, J. Study on preparation process of SBS/crumb rubber composite modified asphalt. *Adv. Mater. Res.* **2012**, *450–451*, 417–422. [CrossRef]
23. Rossi, C.O.; Ashimova, S.; Calandra, P.; De Santo, M.P.; Angelico, R. Mechanical resilience of modified bitumen at different cooling rates: A rheological and atomic force microscopy investigation. *Appl. Sci.* **2017**, *7*, 779. [CrossRef]
24. Rossi, C.O.; Caputo, P.; Ashimova, S.; Fabozzi, A.; D'Errico, G.; Angelico, R. Effects of natural antioxidant agents on the bitumen aging process: An epr and rheological investigation. *Appl. Sci.* **2018**, *8*, 1405. [CrossRef]
25. Katsuyuki, Y.; Iwao, S.; Seishi, M. Mechanism of asphalt binder aging by ultraviolet irradiation and aging resistance by adding carbon black. *J. Jpn. Pet. Inst.* **2004**, *47*, 266–273.
26. Xu, H. The analysis of the infrared spectrum for modified asphalt. *Hefei: Anhui Chem. Ind.* **2007**, *33*, 62–64.
27. Wei, D.; Wang, Y.; Li, S.; Qiu, X.; Ran, X.; Ji, X. Physical properties and microstructure of waste rubber powder and SBS complex modified asphalt. *J. Jilin Uni. (Eng. Technol. Ed.)* **2008**, *38*, 525–529.
28. Qin, X.T.; Zhu, S.Y.; Li, Z.Z.; Chen, S.F. High temperature performance of flame retardant asphalt mortar based on repeated creep test. *J. Highw. Transp. Res. Dev.* **2015**, *32*, 21–27.

materials

MDPI

Article

Influence of Water Solute Exposure on the Chemical Evolution and Rheological Properties of Asphalt

Ling Pang, Xuemei Zhang, Shaopeng Wu, Yong Ye * and Yuanyuan Li

State Key Laboratory of Silicate Materials for Architectures, Wuhan University of Technology, Wuhan 430070, China; lingpang@whut.edu.cn (L.P.); zhangxuemei@whut.edu.cn (X.Z.); wusp@whut.edu.cn (S.W.); liyuanyuan@whut.edu.cn (Y.L.)
* Correspondence: jhgs201309@126.com

Received: 19 April 2018; Accepted: 28 May 2018; Published: 11 June 2018

Abstract: The properties of asphalt pavement are damaged under the effects of moisture. The pH value and salt concentration of water are the key factors that affect the chemical and rheological properties of asphalt during moisture damage. Four kinds of water solutions, including distilled water, an acidic solution, alkaline solution and saline solution were used to investigate the effects of aqueous solute compositions on the chemical and rheological properties of asphalt. Thin-layer chromatography with flame ionization detection (TLC-FID), Fourier transform infrared (FTIR) spectroscopy and dynamic shear rheometer (DSR) were applied to investigate the components, chemistry and rheology characteristics of asphalt specimens before and after water solute exposure. The experimental results show that moisture damage of asphalt is not only associated with an oxidation process between asphalt with oxygen, but it is also highly dependent on some compounds of asphalt dissolving and being removed in the water solutions. In detail, after immersion in water solute, the fraction of saturates, aromatics and resins in asphalt binders decreased, while asphaltenes increased; an increase in the carbonyl and sulphoxide indices, and a decrease in the butadiene index were also found from the FTIR analyzer test. The rheological properties of asphalt are sensitive to water solute immersing. The addition of aqueous solutes causes more serious moisture damage on asphalt binders, with the pH11 solution presenting as the most destructive during water solute exposure.

Keywords: asphalt; water solute exposure; aqueous solute compositions; chemical evolutions; rheological properties

---

1. Introduction

Asphalt pavement, for its excellent driving performance and low noise advantages, is widely applied in the world [1]. During the service phase of asphalt pavement, moisture from the natural environment can diffuse into the asphalt binder, which results in the attenuation of the pavement performance of the asphalt mixture [2,3]. Recent studies suggest that the interaction between water oxygen molecules and asphalt can cause oxidation and aging of asphalt and increase its stiffness and viscosity [4,5]. Also the water may dissolve part of the components of asphalt, which softens the asphalt, reduces the adhesion between the asphalt binder and aggregate, and reduces the cohesion capability inside the asphalt binder. As a result, the mechanical properties of asphalt concrete and the binder are degraded [6–9].

Moisture damage is considered one of the most important factors affecting asphalt mixture durability [10–13]. Water is closely related to the moisture damage of asphalt pavement during its service phase, and the different aqueous solute compositions of water, such as in areas of acid rain, seaboards, or saline and alkaline land, may cause different moisture damage effects on an asphalt binder [14,15]. In salty and humid environments, the strength of asphalt mixture deteriorates easily because of moisture damage from water infiltration and salt chemical corrosion [16,17]. Following immersion in acid rain solutions, the pavement performance of the asphalt mixtures decreased with the decrease in the solution pH value [18]. In an acid rain area, where rainfall with pH < 3 occurs [19], and in saline and alkaline land areas where the pH value of soil may be over pH10 [20], the pavement is subjected to around 10% sodium chloride corrosion damage under snow melting [21]. The reduction in the properties of the asphalt mixture has been related to component changes under the effect of aqueous solute in the water solution. An important consideration for understanding the asphalt moisture damage process in different water solutions is that the chemical composition and rheological properties of asphalt may not be the same as at the asphalt-water surface where environmental factors have the greatest impact. Thus, understanding the process of moisture damage in asphalt under environmental exposure, especially the role and effect of the aqueous solute component in this process, is very important to the research and prevention of moisture damage to asphalt pavement.

However, there is little research on the effects of aqueous solute composition on the chemical and rheological properties of asphalt. Besides, due to the test conditions and material differences, the degree of influence of different aqueous solute composition factors on the chemical and rheological properties of asphalt is unclear. The main objective of this paper is to analyze the evolution of the chemical and rheological properties of asphalt at different pH values of water solution and saline solution conditions, hence, four kinds of water solution, including distilled water, acidic, alkaline and saline solution, were used to immerse the asphalt. The effects of water solute exposure on asphalt chemical composition, construction and rheology were investigated using four compositions which were analyzed using thin-layer chromatography with flame ionization detection (TLC-FID), Fourier transform infrared spectroscopy (FTIR) and dynamic shear rheometer (DSR).

2. Materials and Experimental Methods

2.1. Materials

The 70# asphalt and SBS modified asphalt were obtained from Hubei Guochuang Hi-tech Material Co., Ltd., Wuhan, China. Their basic properties are exhibited in Table 1. The ductility test temperatures of 70# asphalt and SBS modified asphalt were 10 °C and 5 °C, respectively.

Table 1. Basic properties of 70# asphalt and SBS modified asphalt.

Properties	Units	70# Asphalt	SBS Modified Asphalt	Test Specification
Penetration (25 °C, 10 g, 5 s)	0.1 mm	68	56	ASTM D5-61
Softening point	°C	47.2	74.0	ASTM D36-26
Ductility	cm	63.2	68.0	ASTM D113

2.2. Preparation of Solutions

In different regions, the aqueous solute composition of water solutions are different in nature. For example, the pH value of rainfall may be around 3 in acid rain areas, the pavement is subjected to around 10% sodium chloride corrosion damage under snow melting, or the pH value of the solution may be over pH10 in saline and alkaline land. So, pH3 acid solution, 10% NaCl salt solution and pH11 alkali solution were selected in this paper. The preparation methods of these are described below.

Artificial acid solution was prepared to simulate the natural ingredients in acid rain, such as the SO_4^{2-} and NO_3^- anions. It was prepared with excellent pure sulfuric acid and nitric acid by the serial dilution method [22], the molar ratio of sulfuric acid and nitric acid was 9:1, the pH value was 3. Alkaline solution was prepared with sodium hydroxide, and the pH value was 11. The water solute immersing tests of asphalt binders were conducted at 25 °C and normal pressure, the pH of the solution was measured with a precise pH test paper. Sodium chloride was diluted with distilled water, the concentration of which was 10%.

2.3. Water Solute Immersing Tests

First, a circular glass dish, with a diameter of 100 mm and height of 18 mm, was washed with deionized (DI) water. The selected 6 g of each asphalt binder was poured onto the glass dish and put into an oven at a constant temperature of 120 °C for 10 min to form a smooth asphalt film; the thickness of asphalt film was about 0.76 mm. After that, the samples were taken out and cooled to room temperature.

Then, 40 mL of the solutions were poured into the dish and immersed the asphalt films, the glass dish was capped. The temperature of the immersing water solute was 25 °C, the test times for 70[#] asphalt were designated as 7 days and 14 days, and 14 days and 28 days for the SBS modified asphalt. During the water solute immersion test, the solutions were replaced every week because the carboxylic acid of asphalt can be dissolved and reduce the pH value of the solution [23]. After the water solute immersion test, in order to make sure the water was fully evaporated from the surface of the asphalt samples, all the asphalt specimens were put into an oven at a constant temperature of 150 °C for 30 min. Subsequently, the asphalt specimens were separated from the glass dish with a spatula.

2.4. Characterization Methods

2.4.1. Experimental Program Plan

The experimental program plan is shown in Figure 1. At the first, two kinds of asphalt (70[#] asphalt and SBS modified asphalt) were prepared to be immersed in different water solutions (distilled water, acid solution, alkaline solution and sodium chloride solution). Secondly, the TLC-FID tests, FTIR tests and DSR tests were used to characterize the effect of immersion in water solute on the chemical evolution and rheological properties of asphalt. To increase the accuracy, two times replicate tests were carried out. Lastly, the mechanism of asphalt during moisture damage was summarized.

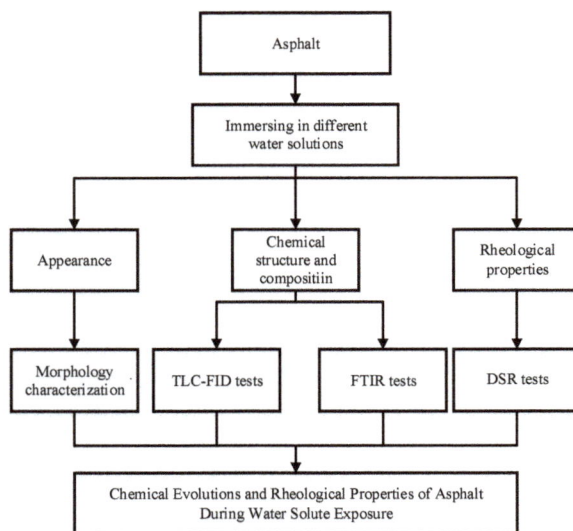

Figure 1. Experimental program plan.

2.4.2. Four Compositions Analysis

Asphalt consists of various molecular weights of hydrocarbons and its derivatives and based on the relative molecular size and polarity of asphalt, it can be divided into four components, namely saturates, aromatics, resins and asphaltenes [24,25]. In order to test the impact of aqueous solute composition on these four components of asphalt, TLC-FID (Iatron Laboratories Inc., Tokyo, Japan) was used to analyze the four components of asphalt before and after water solute exposure. Two percent (*w*/*v*) solutions of asphalt binders were prepared in dichloromethane, and 1 µL sample solution was spotted on chromarods. There was a three-stage process for the separation of asphalt fractions. The first stage was in n-heptane (70 mL) and expanded to 100 mm of the chromarods, the second stage in toluene/n-heptane (70 mL, 4/1 by volume) was developed to 50 mm of the chromarods, and the last development was in toluene/ethanol (70 mL, 11/9 by volume) and expanded to 25 mm of the chromarods. The solvent was dried in an oven at 80 °C after each stage. Then, the chromarods were scanned in the TLC-FID analyzer. Four chromarods were tested for each sample, and finally the average values were used as the results.

2.4.3. Fourier Transform Infrared (FTIR)

A Thermo Nicolet Nexus FTIR spectrophotometer (Thermo Fisher Scientific, Waltham, MA, USA) was used to test the chemical structure of asphalt binders before and after water solute exposure. The asphalt carbon disulfide solutions were prepared with a concentration of 5 wt % asphalt binders. All samples were obtained using a 0.1 mm path length KBr cell. Spectra were recorded using the following settings: number of scans 64; gain 1; apodization weak; and resolution 4. The change in chemical structure due to asphalt oxidation aging could be obtained by the calculation of functional and structural indices of some groups from the FTIR spectra. The oxidation aging of asphalt increases the area of carbonyl and the sulphoxide absorption peak, while greatly decreasing the area of butadiene double bonds absorption peak, the area of these absorption peaks are closely related with the degree of aging of asphalt, thus, carbonyl group C=O (centered around 1700 cm^{-1}), sulphoxide group S=O (centered around 1030 cm^{-1}) and chain segments of butadiene C=C (centered around 968 cm^{-1}) could be used to characterize the degree of aging of the asphalt [26–28]. The carbonyl index ($I_{C=O}$), sulphoxide index ($I_{S=O}$) and butadiene index (I_{SBS}) can be calculated according to Equations (1)–(3).

$$I_{C=O} = \frac{\text{Area of the carbonyl band centered around 1700 cm}^{-1}}{\sum \text{Area of the spectral bands between 2000 and 600 cm}^{-1}} \tag{1}$$

$$I_{S=O} = \frac{\text{Area of the sulphoxide band centered around 1030 cm}^{-1}}{\sum \text{Area of the spectral bands between 2000 and 600 cm}^{-1}} \tag{2}$$

$$I_{SBS} = \frac{\text{Area of the butadine band centered around 968 cm}^{-1}}{\sum \text{Area of the spectral bands between 2000 and 600 cm}^{-1}} \tag{3}$$

2.4.4. Dynamic Shear Rheometer (DSR) Test

The dynamic rheological properties of asphalt were investigated with a dynamic shear rheometer (MCR101, Anton Paar company, Graz, Austria) under a parallel plate configuration. A temperature sweep test from −10 °C to 30 °C with an increment of 2 °C/min was performed under a strain-controlled mode, the constant frequency was 10 rad/s. The specifications followed AASHTO T 315. The strain sweep test was carried out at −10 °C to determine whether 0.1% strain lies within the linear viscoelastic range of the aged binder in advance. The strain was maintained at 0.1% so that all testing would lie within the linear viscoelastic range. Moreover, the diameter of the plate was 8 mm, and the gap between the plates was 2 mm.

3. Results and Discussion

3.1. Water Solutions' Effect on the Appearance of Asphalt

Figure 2 displays the appearance of 70$^{\#}$ asphalt after 7 days of the water solute immersion test. As can be seen, the surface exhibits obvious differences after immersion in different water solutions.

Before water solute immersion, the surface of 70# asphalt is fairly even, however, after immersion in the four types of water solutions, the asphalt surface tends to be rougher and some micro bumps or pits can be observed. In detail, the roughness increases in the following order, pH11 alkaline solution > pH3 acidic solution > 10% NaCl saline solution > distilled water. The uneven surfaces may be caused by etching of the water solutions, that is, they are the results of moisture damage, also, the moisture damage effect of the pH11 alkaline solution immersion was the most serious.

Figure 3 shows the water solutions after 7 days of the water solute immersion test. It can be found from Figure 3 that some light-yellow oil patches exist on the surfaces of the four water solutions (marked with a white dashed line). The oil patches may derive from insoluble compounds of asphalt under the action of hydrostatic pressure, which are pushed off the asphalt surface and float on top of the water solutions. The changes in the features of the water solutions are consistent with the surface roughness of asphalt after being immersed in water solute, which indicates that the water solutions can dissolve and remove some compounds of asphalt and result in the damage of asphalt. In addition, the order of the oil patch size is pH11 alkaline solution > pH3 acidic solution > 10% NaCl saline solution > distilled water; therefore, the degree of damage caused by the pH11 alkaline solution, pH3 acidic solution and 10% NaCl saline solution are higher than that of distilled water. At the same time, the change in pH value of the alkaline solution after immersion for 7 days is shown in Figure 4, the pH value of the alkaline solution is about 11 before immersing the asphalt, however, the pH value is about 9 after immersing the asphalt, which means the alkaline solution is neutralized by some acidic composition of the asphalt, or that acid from the asphalt can be dissolved in the solution and reduce the pH value of alkaline solution.

Figure 2. 70# asphalt appearance before and after 7 days water solution immersion (**A**) original sample; (**B**) distilled water; (**C**) 10% NaCl salt solution; (**D**) pH3 acid solution; (**E**) pH11 alkali solution.

Figure 3. Water solutions' appearance after 7 days immersion (**A**) distilled water; (**B**) 10% NaCl salt solution; (**C**) pH3 acid solution; (**D**) pH11 alkali solution.

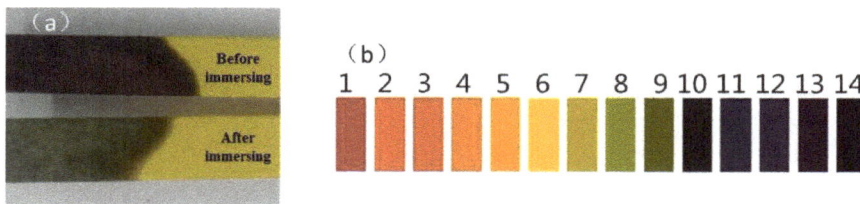

Figure 4. (**a**) The pH value change of alkaline solution. (**b**) The pH paper color swatches after 7 days immersing asphalt.

3.2. Four Component Fractions Analysis

The effect of the aqueous solute compositions on the component fractions of asphalt was studied with TLC-FID, the fractions of the four components of $70^{\#}$ asphalt before and after the water solute immersion test are shown in Table 2.

From Table 2, the fraction of asphaltenes increases, while the fractions of saturates, aromatics and resins tend to decline. For example, after 14 days moisture aging, the fraction of asphaltenes increases to 13.01% in distilled water, 14.68% in 10% NaCl saline solution, 18.89% in pH3 acidic solution, and 19.68% in pH11 alkaline solution, compared with 11.31% of original sample; the fraction of resins fluctuates in distilled water, and decreases observably from 34.46% to 32.70% in 10% NaCl saline solution, to 30.49% in pH3 acidic solution, and to 28.11% in pH11 alkaline solution. Similarly, saturates decline from 15.35% to 15.02% in distilled water, to 13.86 in 10% NaCl saline solution, to 12.29% in pH3 acidic solution, and slightly to 13.57% in pH11 alkaline solution as well; and the fraction of aromatics vary in a relatively narrow range from 38.64% to 37.56%. After water solute immersion, especially with the addition of the aqueous solute of acidic, alkaline and saline compositions, the components with relatively low molecular weight, including saturates and resins decrease while the asphaltene component increases.

The component changes of asphalt binders are caused by the absorption of oxygen, and by the dissolution of some hydrophilic groups and water-soluble substances of asphalt, these two actions occur simultaneously during immersion in water solute. The oxidation can be derived from the air-water-asphalt interaction and the oxygen in the water solutions [4]. Compared to the saturates and asphaltenes, aromatics and resins are more susceptible to oxidation and the polar or ionic chemical groups in the resins can also be solvated by water [23,29–31], therefore, there are more asphaltenes and less aromatics and resins after being immersed in water solute. With the water-soluble substances of asphalt solvated and displaced by water, some saturates presumably were seeped up, and pushed by hydrostatic pressure up to the surface of water and constitute a part of oil patch because it is water insoluble and less dense than water, resulting in a reduction of saturates. The decrease in the saturates of asphalt exposed to water solute was a little different compared to common oxidized aging in heat-oxidized and UV-oxidized aging. Thus, the oxidation of asphalt and the effect of dissolution and removal of some compounds of asphalt in water solutions seems to occur simultaneously, which changes the content and proportion of each component in asphalt. In addition, with an increase of water solute immersion time, the change in amplitude of the components of asphalt binders increases.

Table 2. Component fractions of 70# asphalt before and after water solute immersion test.

Sample		Saturates (%)	Aromatics (%)	Resins (%)	Asphaltenes (%)
Original sample		15.35	38.88	34.44	11.33
Distilled water	7 days	15.11	38.77	34.15	11.97
	14 days	15.02	37.79	34.68	12.51
10% NaCl salt solution	7 days	15.57	39.04	32.83	12.56
	14 days	15.36	37.56	33.02	14.06
pH3 acid solution	7 days	11.00	39.28	31.78	17.94
	14 days	12.29	38.32	30.49	18.89
pH11 alkali solution	7 days	16.73	39.03	27.70	16.54
	14 days	13.57	38.64	28.11	19.68

The colloidal structure of asphalt can be represented by the Gaestel index (Ic), and the Ic value is calculated according to the Equation (4) [28]. The increase in Ic is used to be characteristic of asphalt heat-oxidized and UV-oxidized aging usually, therefore the Ic value is also used to investigate the colloidal structure change of asphalt before and after moisture aging.

$$I_C = \frac{aromatics + resins}{saturates + asphaltenes} \tag{4}$$

The results are shown in Figure 5. As shown in Figure 5, the Gaestel index values of 70# asphalt increase over 7 days and 14 days water solute immersing, which means more gel-like asphalt has formed after water solute exposure, especially after immersion in pH11 alkaline solution.

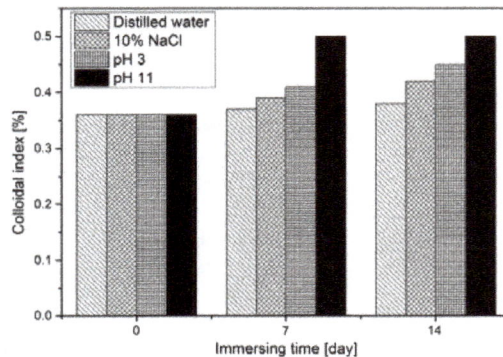

Figure 5. Gaestel index of the 70# asphalt before and after water solute immersion.

3.3. Chemical Structure Analysis

The chemical structures of asphalt binders were tested by the FTIR. The results of the FTIR spectrums of 70# asphalts and SBS modified asphalt are shown in Figures 6 and 7, respectively. Compared to the spectra of the origin asphalt, after immersing in water solute, the carbonyl and sulphoxide absorption peaks tend to be more obvious, while the butadiene absorption bands decrease obviously, which shows that more carbonyl and sulphoxide groups were generated, and the chain segments of butadiene decomposed due to immersing in distilled water.

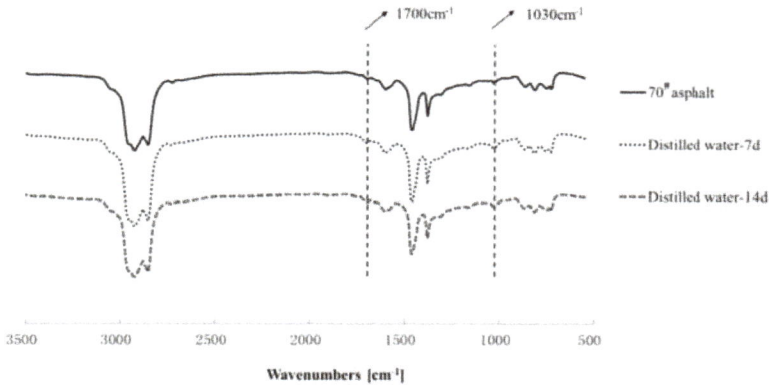

Figure 6. FTIR spectra of the 70# asphalt before and after distilled water aging.

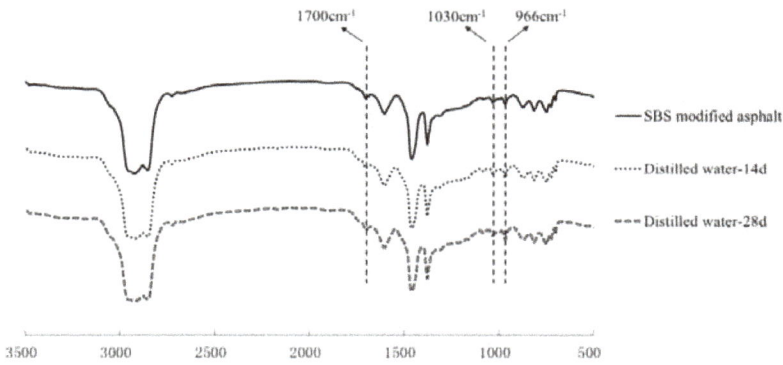

Figure 7. FTIR spectra of the SBS modified asphalt before and after distilled water aging.

The results of group indices of 70# asphalt and SBS modified asphalt before and after immersion in different water solutions are shown in Table 3.

From Table 3, both the carbonyl index and the sulphoxide index of asphalt increase continuously with extended solution immersion time, which shows the interaction between the water and asphalt can cause the oxidation of asphalt and form more asphaltene components. In addition, both the carbonyl index and sulphoxide index are the highest after immersion in pH11 alkaline solution, followed by the pH3 acid solution, 10% NaCl saline solution and distilled water, therefore, saline, alkaline and acidic components can increase the damage rate of asphalt binders during water solute immersion. The increase of the carbonyl and sulphoxide index were relatively faster when the immersion time is less than 7 days, however, there is a decrease in this trend when the immersion time is from 7 days to 14 days. It can be found from Table 4 that both the carbonyl index and sulphoxide index of SBS modified bitumen experience a similar change tendency as 70# asphalt. With respect to the butadiene index, it was found to reduce after moisture aging and decrease continuously with extended immersion time from 0 days to 28 days, which indicates that the chain segments of butadiene in SBS experience significant degradation.

The results of FTIR spectra demonstrate that oxidation of asphalt and degradation of SBS modifier occur simultaneously during water solute immersion. The aqueous solute in the solutions, that is, the saline, alkaline and acidic components promote the oxidation of asphalt, and the pH11 alkaline solution has the most severe effect during water solute immersion.

Table 3. Group indices of 70# asphalt before and after water solute immersion.

Results	Original Sample	Distilled Water		10% NaCl Saline Solution		pH3 Acid Solution		pH11 Alkaline Solution	
		7 days	14 days	7 days	14 days	7 days	14 days	7 days	14 days
Carbonyl index	0.0003	0.0009	0.0012	0.0009	0.0051	0.0106	0.0121	0.0148	0.0151
Sulphoxide index	0.0348	0.0387	0.0412	0.0567	0.0598	0.0651	0.0662	0.0717	0.0778

Table 4. Group indices of SBS modified asphalt before and after water solute immersion.

Results	Original Sample	Distilled Water		10% NaCl Saline Solution		pH3 Acid Solution		pH11 Alkaline Solution	
		14 days	28 days	14 days	28 days	14 days	28 days	14 days	28 days
Carbonyl index	0.0108	0.0109	0.0135	0.0110	0.0135	0.0117	0.0201	0.0129	0.0216
Sulphoxide index	0.0198	0.0202	0.0218	0.0220	0.0259	0.0223	0.0257	0.0233	0.0265
Butadiene index	0.0252	0.0245	0.0237	0.0229	0.0229	0.0228	0.0205	0.0220	0.0207

3.4. Rheological Properties Analysis

The above tests show there are some changes in chemical composition and structure can be observed during water solute immersing; the newly-formed material structure may cause certain effects on the rheological properties of asphalt. In this paper, the DSR was used to investigate the rheological properties of asphalt binders before and after being immersed in water solute. The complex modulus, phase angle, rutting parameter ($G*/\sin\delta$) and fatigue parameter ($G*\cdot\sin\delta$) of 70# asphalt from $-10\,°C$ to $30\,°C$ before and after 7 days of water solute immersion are shown in Figures 8–10.

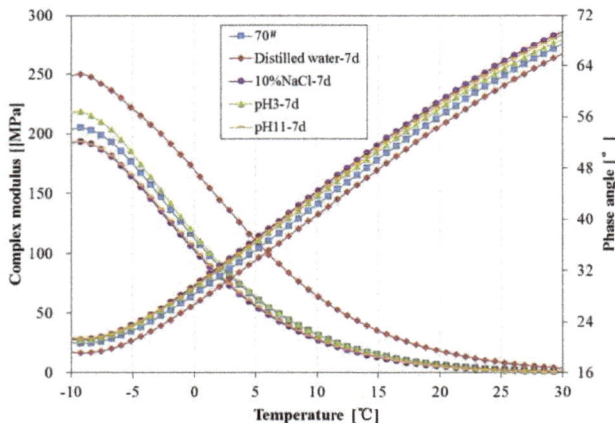

Figure 8. Complex modulus and phase angle of 70# asphalt before and after 7 days immersion.

From Figure 8, after 7 days immersion, the complex modulus and phase angle of 70[#] asphalt binders changed significantly. The complex modulus of 70[#] asphalt immersed in distilled water is higher than that of the original 70[#] asphalt, and the complex modulus of asphalt immersed in the pH11 alkaline solution and 10% NaCl saline solution are lower, while there is almost no difference between the complex modulus of asphalt binders immersed in the pH3 acidic solution and original asphalt. The phase angle of 70[#] asphalt immersed in distilled water is lower than that of the original 70[#] asphalt, and the phase angle of 70[#] asphalt immersed in the other three kinds of water solutions are higher. The G*/sinδ of 70[#] asphalt before and after 7 days immersion is shown in Figure 9, which demonstrates the permanent deformation resistance of asphalt under repeated loads. The rutting parameter of asphalt immersed in distilled water is apparently higher than that of the original asphalt, however, the rutting parameter of asphalt immersed in the pH11 alkaline solution and 10% NaCl saline solution are lower, and the rutting parameter of asphalt binders immersed in the pH3 acidic solution is close to the original asphalt. From Figure 10, after 7 days immersion, the G*·sinδ of 70[#] asphalt immersed in distilled water and pH3 acidic solution is apparently higher than that of the original asphalt, while the fatigue factor of asphalt immersed in the pH11 alkaline solution and 10% NaCl saline solution are slightly lower. The lower the G*·sinδ is, the better the fatigue cracking resistance of asphalt binder is. Therefore, distilled water and pH3 acidic solution can significantly decrease the resistant fatigue ability of 70[#] asphalt, and the effects of 7 days immersion in pH11 alkaline or 10% NaCl saline solution on the resistant fatigue ability of 70[#] asphalt are not obvious.

The change tendency of 70[#] asphalt in Figures 8–10 show that the addition of aqueous solute changes the variation trend of rheological properties of asphalt in the early phase of immersion. Previous studies showed that some asphalt components forming cohesion capability inside the asphalt are easily solvated and displaced by water [28]. This displacement should weaken the asphalt's inside bond and decrease the complex modulus, rutting parameter and fatigue factor, while increasing the phase angle. This process may be reinforced when the aqueous solute of acidic, alkaline or salt compositions exists in the water solutions.

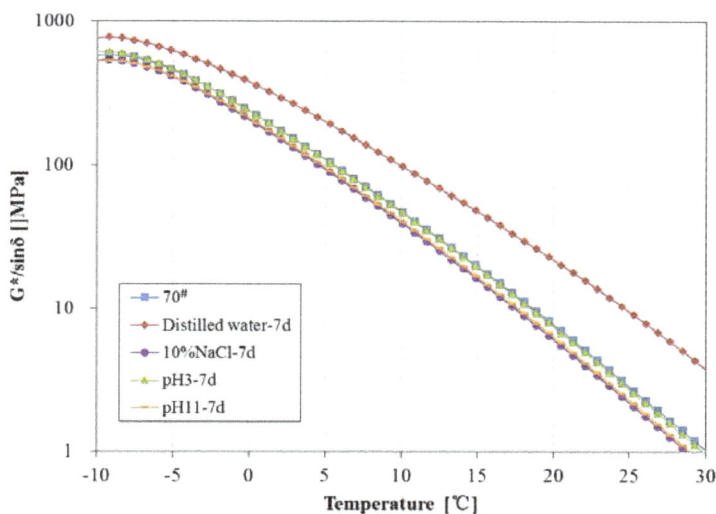

Figure 9. Rutting parameter of 70[#] asphalt before and after 7 days immersion.

The complex modulus, phase angle, rutting parameter (G*/sinδ) and fatigue parameter (G*·sinδ) of 70[#] asphalt before and after 14 days of water solute immersion are shown in Figures 11–13. From Figure 11, the complex modulus increases and phase angle clearly decreases after 14 days of immersion. From Figure 12, the G*/sinδ of 70# asphalt immersed in the pH11 alkaline solution, 10% NaCl saline solution and pH3 acidic solution are higher than that of the original 70# asphalt. In Figure 13, the G*·sinδ of 70# asphalt immersed in the pH11 alkaline solution and 10% NaCl saline solution are similar to that of the original 70# asphalt, and the G*·sinδ of asphalt immersed in the distilled water and pH3 acidic solution is higher than that of the original 70# asphalt. The effects of distilled water immersion on the rheological performance of 70# asphalt is most significant after 14 days immersion, followed by the pH3 solution on the complex modulus, phase angle and G*·sinδ, and the pH11 alkaline solution on G*/sinδ. Because of the displacement effect of water solutions with aqueous solute, more time is needed to observe the effect of the water solution, thus, the increase in the complex modulus, G*/sinδ and G*·sinδ, and the decrease of phase angle appear after 14 days immersion.

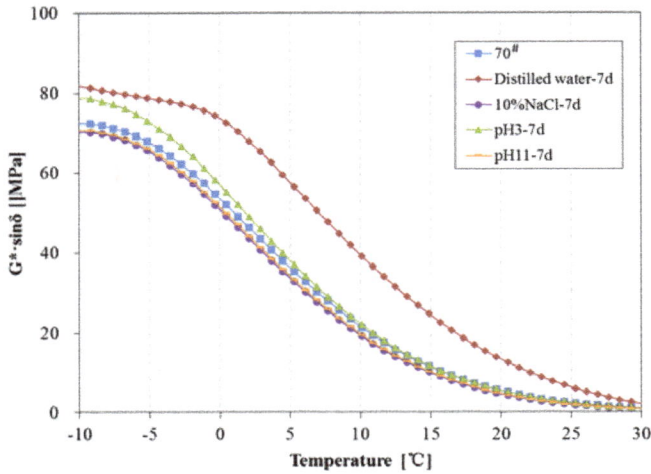

Figure 10. Fatigue factor of 70# asphalt before and after 7 days immersion.

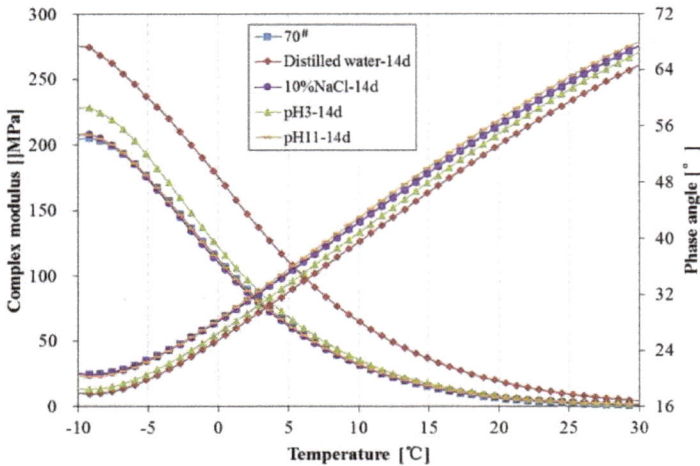

Figure 11. Complex modulus and phase angle of 70# asphalt before and after 14 days immersion.

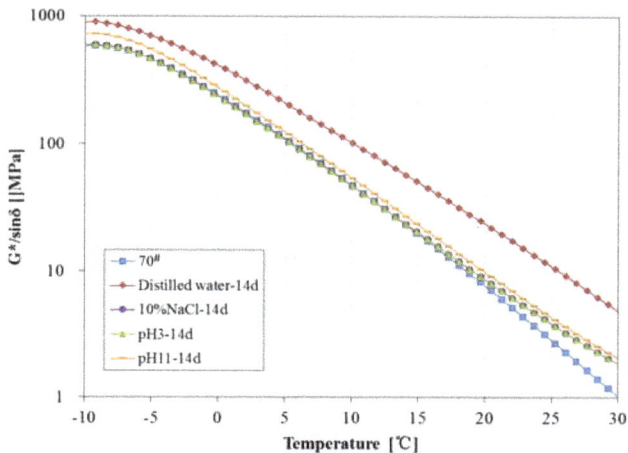

Figure 12. Rutting parameter of 70$^{\#}$ asphalt before and after 14 days immersion.

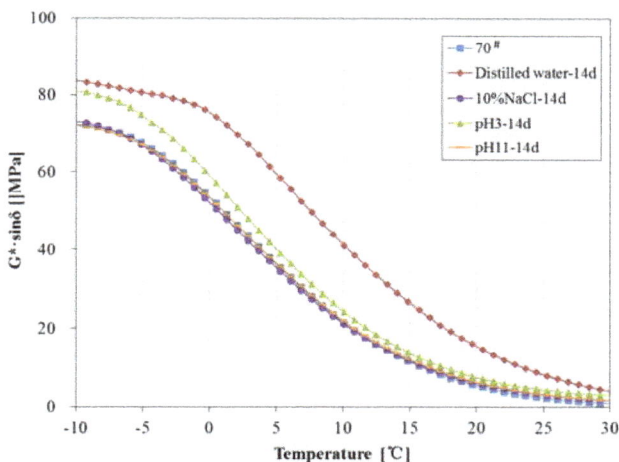

Figure 13. Fatigue factor of 70$^{\#}$ asphalt before and after 14 days immersion.

Figures 14 and 15 present the complex modulus and phase angle of SBS modified asphalt after 14 days and 28 days immersion, respectively. From Figures 14 and 15, the effects of immersion in water solutions on the complex modulus of SBS modified asphalt are similar to that of 70$^{\#}$ asphalt. However, there are some differences in the effects of the phase angle between SBS modified asphalt and 70$^{\#}$ asphalt; the phase angle of SBS modified asphalt after immersion is lower than the original asphalt, and the descent range is lower than that of 70$^{\#}$ asphalt. It seems that the modification of bitumen by SBS can lead to less change in the rheological properties when subjected to immersion in water solute. Additionally, among these different water solution immersion conditions, the effect of distilled water on the complex modulus and phase angle is the most remarkable. The asphalt become stiffer after being immersed in distilled water, especially for those samples immersed for 28 days. The results also indicated that to a certain extent, the rheological properties show a corresponding relationship to the change in chemical components. Asphaltene is the main component that affects the rheological property of asphalt [32]. With the prolonged water solute immersion time, the relative amount of asphaltene in the surface of the asphalt increases gradually, which results in the increase in the complex modulus.

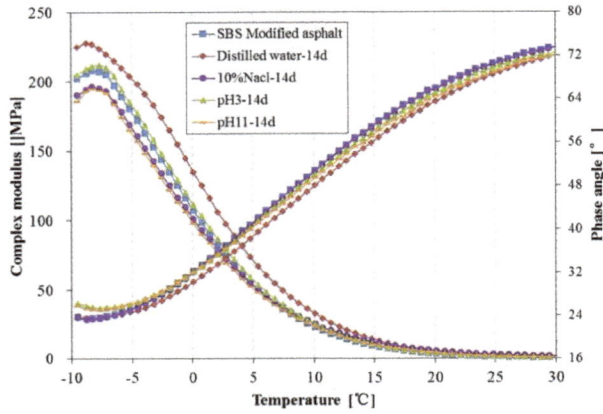

Figure 14. Complex modulus and phase angle of SBS modified asphalt before and after 14 days immersion.

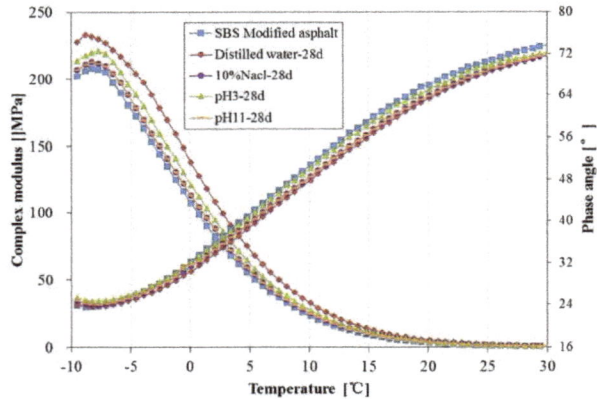

Figure 15. Complex modulus and phase angle of SBS modified asphalt before and after 28 days immersion.

Figures 16 and 17 present the rutting parameter of SBS modified asphalt after 14 days and 28 days water solute immersion, respectively. The rutting parameter of asphalt immersed in distilled water is distinctly higher than that of the original asphalt after 14 days, however, the rutting parameter of asphalt immersed in the water solution with the aqueous solute is close to the original asphalt over 14 days and are obviously higher than that of the original asphalt after 28 days.

Figures 18 and 19 present the $G^* \cdot \sin\delta$ of SBS modified asphalt after 14 days and 28 days water solute immersion, respectively. The fatigue factor of SBS modified asphalt immersed in distilled water and pH3 acidic solution is distinctly higher than that of original SBS modified asphalt over 14 days, however, the fatigue factor of asphalt immersed in pH11 alkaline solution and 10% NaCl saline solution is lower than the original SBS modified asphalt and are slightly higher than that of the original asphalt after 28 days.

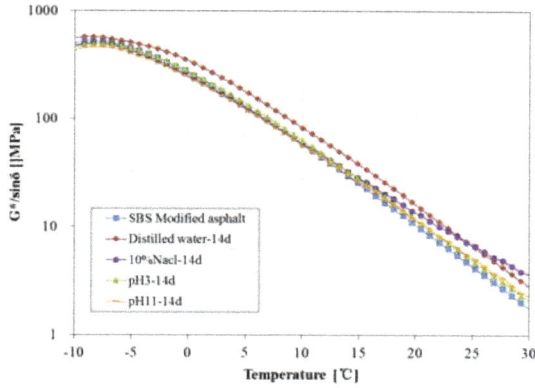

Figure 16. Rutting parameter of SBS modified asphalt before and after 14 days immersion.

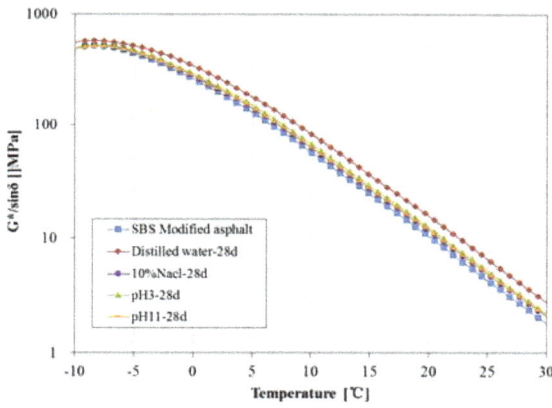

Figure 17. Rutting parameter of SBS modified asphalt before and after 28 days immersion.

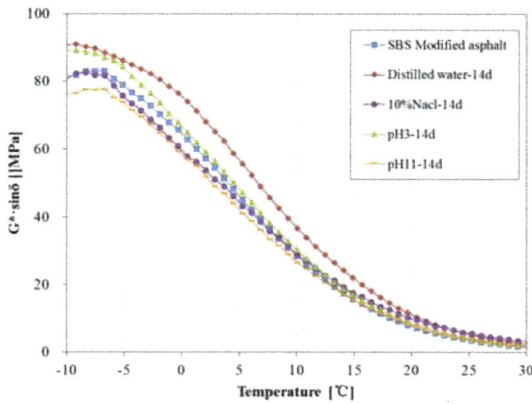

Figure 18. Fatigue factor of SBS modified asphalt before and after 14 days immersion.

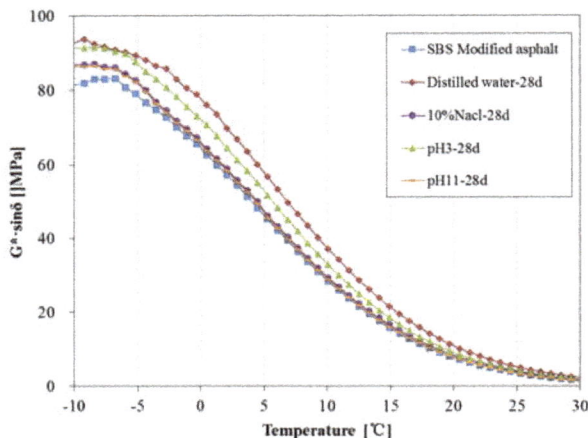

Figure 19. Fatigue factor of SBS modified asphalt before and after 28 days immersion.

All the results above demonstrate that all of these four kinds of water solutions can change the rheological performance of asphalt, the aqueous solute composition plays an important role in the properties of asphalt after water solute exposure, and the sum of these chemical processes leads to asphalt hardening and higher brittleness. The aging process, the dissolving and removal processes, is slow in distilled water, however the process is accelerated with the addition of aqueous solute compositions such as those found in acid rain areas, the seaboard, and in areas with saline and alkaline soil.

4. Conclusions

The effects of water solute exposure on asphalt chemistry and rheology were investigated by TLC-FID, FTIR spectra and DSR. Based on the analysis above, the following conclusions can be drawn:

(1) With the water-soluble substances of asphalt solvated and displaced by water, some water-insoluble light component fractions also can seep up and separate from the surface of the asphalt. Moisture damage is associated with dissolution and removes some polar or nonpolar compounds from the asphalt surface to water closely.

(2) Unlike common oxidized aging in heat-oxidized and UV-oxidized aging, saturate fractions may be removed onto the surface of water solutions and form a light-yellow oil patch, that reduces the content of saturates in asphalt after water solute immersion.

(3) Compared with distilled water, the addition of aqueous solute compositions can remarkably affect the chemical properties of asphalt during water solute exposure, and it was clear that the effects could be ordered as follows, pH11 alkaline solution > pH3 acidic solution > 10% NaCl saline solution. It may be that the addition of the aqueous solute compositions accelerates the dissolving and ionizing of some soluble compounds from the asphalt surface to water solutions and degrade the chemical properties of asphalt and the extent of moisture damage.

(4) Immersion in water solutions induced changes in the chemical composition and structure of asphalt. As a result, the newly-formed material structure caused certain effects on the rheological properties of asphalt. While the complex modulus, rutting parameter and fatigue factor increased and the phase angle decreased in distilled water, the adding of aqueous solute brought about the opposite effects in the early phases of immersion. With the immersion time prolonged, the relative amount of asphaltene in the surface of asphalt increased gradually, which resulted in an increase in the complex modulus and a decrease in the phase angle.

The changes in the chemical composition and rheological properties of asphalt binders during immersion in water solutions were investigated in this paper, however, the interactions between water solutions and asphalt and the influence on the micro-domains' mechanical and conventional physical properties of asphalt binders are still not clear. Investigations of the effects of exposure to water solutions on the micro-domains' mechanical properties and conventional physical properties such as penetration, softening point and ductility, will be carried out in our future research.

Author Contributions: Conceived and designed the experiments: L.P., X.Z. and Y.Y.; Performed the experiments: X.Z.; Attributed reagents/materials/analysis tools: L.P., S.W. and Y.Y.; Wrote the paper: L.P., X.Z. and Y.L.

Funding: This research was funded by the National Basic Research Program of China (973 Program No. 2014CB932104) and Technological Innovation Major Project of Hubei Province (2016AAA023).

Conflicts of Interest: The authors declare no conflict of interest.

References

1. National Asphalt Pavement Association. *Black and Green: Sustainable Asphalt, Now and Tomorrow*; Recycled Materials: Lanham, MD, USA, 2016.
2. Hung, A.M.; Goodwin, A.; Fini, E.H. Effects of water exposure on bitumen surface microstructure. *Constr. Build. Mater.* **2017**, *135*, 682–688. [CrossRef]
3. Behiry, E.M. Laboratory evaluation of resistance to moisture damage in asphalt mixtures. *Ain Shams Eng. J.* **2013**, *4*, 351–363. [CrossRef]
4. Noguera, J.A.H.; Quintana, H.A.R.; Gómez, W.D.F. The influence of water on the oxidation of asphalt cements. *Constr. Build. Mater.* **2014**, *71*, 451–455. [CrossRef]
5. Ma, T.; Huang, X.; Mahmoud, E.; Garibaldy, E. Effect of moisture on the aging behavior of asphalt binder. *Int. J. Miner. Metall. Mater.* **2011**, *18*, 460–466. [CrossRef]
6. Kang, A.H.; Zhou, X.; Mei-Ping, W.U.; Sun, L.J. Aging of TOR asphalt rubber in combination of environmental factors. *J. Nanjing Univ. Sci. Technol.* **2012**, *36*, 724–728.
7. Liu, K.; Deng, L.; Zheng, J. Nanoscale study on water damage for different warm mix asphalt binders. *Int. J. Pavement Res. Technol.* **2016**, *9*, 405–413. [CrossRef]
8. López-Montero, T.; Miró, R. Differences in cracking resistance of asphalt mixtures due to ageing and moisture damage. *Constr. Build. Mater.* **2016**, *112*, 299–306. [CrossRef]
9. Luo, Y.; Zhang, Z.; Cheng, G.; Zhang, K. The deterioration and performance improvement of long-term mechanical properties of warm-mix asphalt mixtures under special environmental conditions. *Constr. Build. Mater.* **2017**, *135*, 622–631. [CrossRef]
10. Yao, H.; Dai, Q.; You, Z. Chemo-physical analysis and molecular dynamics (MD) simulation of moisture susceptibility of nano hydrated lime modified asphalt mixtures. *Constr. Build. Mater.* **2015**, *101*, 536–547. [CrossRef]
11. Cheng, D.; Little, D.N.; Lytton, R.L.; Holste, J.C.; Davis, R.; Brown, S.; Hobson, K.; Dunning, M.; Mcdaniel, B.; Newman, K. Use of surface free energy properties of the asphalt-aggregate system to predict moisture damage potential. *Asph. Paving Technol. Assoc. Asph. Paving Technol. Proc. Tech. Sess.* **2002**, *71*, 59–88.
12. Huang, S.C.; Turner, T.F.; Pauli, A.T.; Miknis, F.P.; Branthaver, J.F.; Robertson, R.E. Evaluation of different techniques for adhesive properties of asphalt-filler systems at interfacial region. *J. ASTM Int.* **2005**, *2*, 1–15. [CrossRef]
13. Kakar, M.R.; Hamzah, M.O.; Valentin, J. A review on moisture damages of hot and warm mix asphalt and related investigations. *J. Clean. Prod.* **2015**, *99*, 39–58. [CrossRef]
14. Feng, D.; Yi, J.; Wang, D.; Chen, L. Impact of salt and freeze–thaw cycles on performance of asphalt mixtures in coastal frozen region of China. *Cold Reg. Sci. Technol.* **2010**, *62*, 34–41. [CrossRef]
15. Wang, Y.; Ye, J.; Liu, Y.; Qiang, X.; Feng, L. Influence of freeze–thaw cycles on properties of asphalt-modified epoxy repair materials. *Constr. Build. Mater.* **2013**, *41*, 580–585. [CrossRef]
16. Zhao, Q.; Zhao, J. Influence of coastal area salinity on road surface resistance to permanent deformation performance. *J. Dalian Jiaotong Univ.* **2016**, *37*, 69–72.
17. Ma, Q.; Wu, J.; Qin, K. Tests and analyses of the influence of chlorine salt on freezing-thawing splitting tensile strength of asphalt concrete. *J. Glaciol. Geocryol.* **2013**, *35*, 1202–1208.

18. Feng, X.J.; Tang, X.; Xiong, X. Study on influence of acid rain on pavement performances of asphalt mixtures. *J. Wuhan Univ. Technol.* **2015**, *37*, 39–43.

19. Zhang, Q.; Hu, J.R.; Peng, Y.H. Analysis of chemical mechanism for aggregates corroded by acid rain in asphalt mixtures. *Highway* **2004**, *2*, 104–106.

20. Dendooven, L.; Alcántara-Hernández, R.J.; Valenzuela-Encinas, C.; Luna-Guido, M.; Perez-Guevara, F.; Marsch, R. Dynamics of carbon and nitrogen in an extreme alkaline saline soil: A review. *Soil. Biol. Biochem.* **2010**, *42*, 865–877. [CrossRef]

21. Wang, Z.; Zhang, T.; Shao, M. Investigation on snow-melting performance of asphalt mixtures incorporating with salt-storage aggregates. *Constr. Build. Mater.* **2017**, *142*, 187–198. [CrossRef]

22. Zhang, Q.; Jia, X.J.; Li, Y.L.; Tong, S.J. the simulation study on effect of acid rain on the strength and void rate of asphalt mixture. *J. China Foreign Highw.* **2005**, *25*, 78–80.

23. Zhang, Q.; Meng, X.R.; Yan, W. Analysis of the chemical mechanism of influence of precipitation on components of asphalt binder. *J. Xian Univ. Archit. Technol.* **2010**, *42*, 669–673.

24. Liu, X.; Wu, S.; Liu, G.; Li, L. Effect of ultraviolet aging on rheology and chemistry of LDH-modified bitumen. *Materials* **2015**, *8*, 5238–5249. [CrossRef] [PubMed]

25. Huo, K.; Zhai, Y.; Liao, K.; Yang, P.; Yan, F.; Wei, Y. A study on change of family composition and properties of Liaoshu paving asphalt on aging. *Petrol. Sci. Technol.* **2001**, *19*, 651–660.

26. Ouyang, C.; Wang, S.; Zhang, Y.; Zhang, Y. Improving the aging resistance of asphalt by addition of Zinc dialkyldithiophosphate. *Fuel* **2006**, *85*, 1060–1066. [CrossRef]

27. Redelius, P.; Soenen, H. Relation between bitumen chemistry and performance. *Fuel* **2015**, *140*, 34–43. [CrossRef]

28. Yao, H.; Dai, Q.; You, Z. Fourier transform infrared spectroscopy characterization of aging-related properties of original and nano-modified asphalt binders. *Constr. Build. Mater.* **2015**, *101*, 1078–1087. [CrossRef]

29. Tarefder, R.A.; Arisa, I. Molecular dynamic simulations for determining change in thermodynamic properties of asphaltene and resin because of aging. *Energy Fuels* **2011**, *25*, 2211–2222. [CrossRef]

30. Lu, X.; Isacsson, U. Effect of ageing on bitumen chemistry and rheology. *Constr. Build. Mater.* **2002**, *16*, 15–22. [CrossRef]

31. Siddiqui, M.N.; Ali, M.F. Studies on the aging behavior of the arabian asphalts. *Fuel* **1999**, *78*, 1005–1015. [CrossRef]

32. Caro, S.; Diaz, A.; Rojas, D.; Nuñez, H. A Micromechanical model to evaluate the impact of air void content and connectivity in the oxidation of asphalt mixtures. *Constr. Build. Mater.* **2014**, *61*, 181–190. [CrossRef]

materials

MDPI

Article

Design Optimization of SBS-Modified Asphalt Mixture Reinforced with Eco-Friendly Basalt Fiber Based on Response Surface Methodology

Wensheng Wang [ORCID], Yongchun Cheng and Guojin Tan *[ORCID]

College of Transportation, Jilin University, Changchun 130025, China; wangws17@mails.jlu.edu.cn (W.W.); chengyc@jlu.edu.cn (Y.C.)
* Correspondence: tgj@jlu.edu.cn; Tel.: +86-0431-8509-5446

Received: 13 July 2018; Accepted: 27 July 2018; Published: 29 July 2018

Abstract: This paper investigates the effects of basalt fiber content, length and asphalt-aggregate ratio on the volumetric and strength properties of styrene-butadiene-styrene (SBS)-modified asphalt mixture reinforced with eco-friendly basalt fiber. An experimental scheme was designed to optimize three preparation parameters for the Marshall test indices based on response surface methodology (RSM). The results showed that basalt fiber content presents a more significant effect on air voids, voids in mineral aggregates and voids filled with asphalt. Basalt fiber length is more related to Marshall stability, and flow value exhibits a significant variation trend with asphalt-aggregate ratio. The optimization of preparation parameters is determined as follows: basalt fiber content is 0.34%, length is 6 mm, asphalt-aggregate ratio is 6.57%, which possesses favorable and reliable accuracy compared with experimental results. Furthermore, basalt fiber reinforced asphalt binder and mixture were also studied, and it was found that basalt fiber can enhance the performance of asphalt binder and mixture in terms of cone penetration, softening point, force ductility, as well as pavement performance tests.

Keywords: asphalt mixture; basalt fiber; response surface methodology; design optimization

---

1. Introduction

Asphalt mixture is a widely used pavement material around the world due to its superior performance. Stone matrix asphalt (SMA) is a gap-graded hot mix asphalt (HMA) mixture that was developed in Germany in the 1960s, and has more coarse aggregates and filler, higher asphalt content, modified asphalt and fibers compared with typical dense-graded asphalt mixtures [1]. Due to the high rutting resistance and durability, SMA has been widely used around the world, and SMA has been almost exclusively used for expressway as surface courses in China since 1992 [2]. As an indispensable part of SMA, fiber can stabilize asphalt binder to improve the performance of SMA. Previous studies have shown that fibers can improve rutting resistance, moisture susceptibility and prevent reflective cracks of asphalt mixtures [3–6].

Nowadays, various kinds of fibers are commonly applied to asphalt pavement, including lignin fiber, polyester fiber, glass fiber, and so on. As a new eco-friendly mineral fiber, basalt fiber is made of basalt rocks with high mechanical performance, low water absorption and appropriate temperature range [7]. The production of basalt fibers produces less waste and can be degraded directly in the environment without any harm after abandonment. Previously, a lot of research has been carried out to study the influences of basalt fiber on improving the performance of asphalt binder and its mixture by researchers and pavement engineers. Wang et al. [8] investigated the influences of basalt fiber on bitumen binder and mastic at low temperatures by using direct tension and fatigue tests. The results indicated that due to adding basalt fiber, the break stress of binder increased by about

4.5% and the stiffness modulus of mastic increased by 26%. Meanwhile, basalt fiber resulted in improving the fatigue-resistance property of binder and mastic to some extent. Gu et al. [9] analyzed the characteristics of basalt fiber and found that basalt fiber exhibits excellent strength, durability and suitability. Results revealed that the rheological performance of basalt fiber modified asphalt mastic are promoted and remarkable at higher temperature. Cheng et al. [10] indicated that basalt fiber can enhance the high- and low-temperature properties of asphalt mastics in terms of softening point, cone penetration, viscosity, force ductility, DSR and BBR tests. Zhang et al. [11,12] studied the rheological, viscoelastic characteristics of basalt fiber reinforced asphalt mastic and mortar based on 3D model and numerical analysis. Numerical analysis showed that basalt fiber leads to stress redistribution of asphalt mortar and decreases the stress. Moreover, the effects of basalt fiber content and aspect ratio on the shear and compressive properties of matrix were also discussed. Qin et al. [13] carried out an investigation on the properties of basalt fiber modified bitumen mastics with various fiber length and content. Basalt fiber with 6 mm length had much better bitumen adsorption and strength. Compared with lignin fiber and polyester fiber, basalt fiber-modified bitumen mastic presents the best comprehensive properties. In addition, Gao [14] investigated the performance of basalt fiber modified asphalt mixtures and showed that the low-temperature cracking resistance can be significantly improved. Morova [15] used the Marshall stability test to study the application of basalt fiber in asphalt mixture. The optimum asphalt content was determined, and a series of experiments was conducted with different fiber ratios.

However, not much study has been conducted on the design method of basalt fiber-reinforced asphalt mixture, based on the previous literature review. From the previous literature, it is worth noting that fiber content, fiber length and asphalt content play key roles for the performance of fiber-reinforced asphalt mixture [16]. Generally, there are two design methods for asphalt mixtures—the Marshall and Superpave methods—in which the volumetric properties are regarded as responses to design of more reliable asphalt mixtures. Response surface methodology (RSM) is more effectively used for analyzing and optimizing experimental responses, and is becoming more and more popular in the construction field [17]. Tan et al. [18] used the central composite design (CCD) method in conjunction with RSM for the optimal proportion of raw materials in terms of different test indices of asphalt binder. Kavussi et al. [19] investigated and analyzed the effects of preparation parameters including gradation, aggregate, etc., on indirect tensile strength (ITS) of warm mix asphalt (WMA) mixtures based on RSM. Hamzah et al. [20,21] investigated the influences of recycled aggregate content, compaction temperature and asphalt content on volumetric and mechanical properties of asphalt mixture for optimizing the binder content based on the CCD of RSM. Furthermore, Hamzah et al. [22] found that statistical models based on RSM can predict the effects of aging on binder viscosity behavior at high temperatures effectively. Khodaii et al. [23] employed RSM to evaluate the effects of lime content and gradation on ITS and its ratio of asphalt mixture under dry and saturated conditions. Haghshenas et al. [24] utilized RSM for optimizing the bitumen content and gradation of HMA mixtures based on tensile strength ratio. Based on the extensive literature, RSM can be successfully applied to study asphalt binder and mixture.

In this study, an experimental scheme was designed for styrene-butadiene-styrene (SBS)-modified asphalt mixture reinforced with eco-friendly basalt fiber based on RSM. The relationships between preparation parameters and Marshall test indices including volumetric and strength properties were analyzed to evaluate the effects of basalt fiber content, length and asphalt-aggregate ratio. A design optimization of basalt fiber and SBS-modified asphalt mixture was proposed and validated with experimental results. Meanwhile, the performance of basalt fiber-modified asphalt binder and mixture was also studied.

2. Materials and Methods

2.1. Raw Materials

In this study, SBS-modified asphalt was chosen, and its basic physical properties are listed in Table 1. The coarse and fine aggregates, as well as fillers, were obtained from a local quarry in Yitong of Jilin Province, China. Their physical properties are shown in Table 2. Basalt fiber with different lengths was used to modify the asphalt binder and mixture, as shown in Figure 1, and its basic properties are summarized in Table 3.

Table 1. Basic physical properties of SBS-modified asphalt.

Properties	Measurement	Technical Criterion
Penetration @ 25 °C, 100 g, 5 s (0.1 mm)	57.9	40~60
Ductility @ 5 °C, 5 cm/min (cm)	24.9	≥20
Softening point (°C)	64.4	≥60
Flash point (°C)	286	≥230
Elastic recovery @ 25 °C (%)	88	≥75
Solubility (trichloroethylene, %)	100.2	≥99
RTFOT		
Mass loss (%)	0.22	±1.0
Penetration ratio @ 25 °C (%)	85	≥65
Ductility @ 5 °C (cm)	18.1	≥15

Table 2. Physical properties of aggregates and filler.

Index	Coarse Aggregate				Fine Aggregate	Filler
	13.2	9.5	4.75	2.36	2.36~0.075	<0.075
Crushed stone value (%)	22.45	23.13	23.62	24.16	–	–
Los Angeles abrasion loss (%)	26.47	25.99	25.78	26.14	–	–
Apparent specific gravity (g/cm^3)	2.687	2.675	2.659	2.689	2.691	2.738
Water absorption (%)	0.99	1.13	1.57	1.76	–	–

Table 3. Physical properties of basalt fiber.

Index	Length	Diameter	Specific Gravity	Tensile Strength	Elongation at Break
Units	mm	μm	g/cm^3	MPa	%
Value	3/6/9	13	2.55~2.65	≥3000	32

Figure 1. Morphology of basalt fiber with length of 6 mm in this study.

2.2. Sample Preparation

Basalt fibers with different lengths of 3, 6 and 9 mm were used in this study. According to the previous literature [9,10,15,25], the percentage of basalt fiber to asphalt binder should not exceed 5%. Therefore, in order to investigate the influence of basalt fiber at the level of asphalt binder, their proportions added into SBS-modified asphalt were 0%, 1%, 2%, 2.5%, 3%, 3.5% and 4% by mass of SBS-modified asphalt, respectively. Based on previous research [10,14,26], the detailed preparation procedures of SBS-modified asphalt binder reinforced with basalt fiber are as follows: (i) basalt fiber with different lengths and SBS-modified asphalt were heated in an oven at 170 °C until the constant weight and stable state; (ii) three kinds of basalt fiber with different lengths were added into SBS-modified asphalt at seven different proportions, and twenty-one experimental groups could be obtained; (iii) in order to ensure that basalt fibers can be distributed uniformly in asphalt binder, the mixture of basalt fiber and asphalt was placed in a shear homogenizer (KRH-I, Shanghai Konmix Mechanical & Electrical Equipment Technology Co. Ltd., Shanghai, China) after a preliminary manual blending. Then, shearing temperature and speed were set as 170 °C and 6000 rev/min, respectively. After mixing of asphalt with basalt fibers for one hour, SBS-modified asphalt containing basalt fiber was prepared.

Asphalt mixture specimens were produced to investigate the design optimization of SBS-modified asphalt mixture reinforced with basalt fiber. Figure 2 illustrates the gradation of SMA with a nominal maximum size of 13.2 mm. According to the Chinese specification JTG E20-2011 [27], the Marshall specimens of asphalt mixture with height of 63.5 mm and diameter of 101 mm were made by Marshall procedures, which were used for laboratory tests and optimization analysis based on RSM. The detailed preparation procedures of SBS-modified asphalt mixture reinforced with basalt fiber are as follows: (i) the aggregates and fillers were weighted and placed in an oven at 180 °C for two hours and SBS-modified asphalt was heated to 170 °C; (ii) the weighted aggregates and basalt fiber were blended together in a mixing pot and then asphalt was poured and mixed at 165 °C until the aggregates were coated; (iii) the weighted fillers were added and mixed well at 165 °C; (iv) asphalt mixtures were compacted with 50 blows of Marshall hammer per side for the target of 4% air void content.

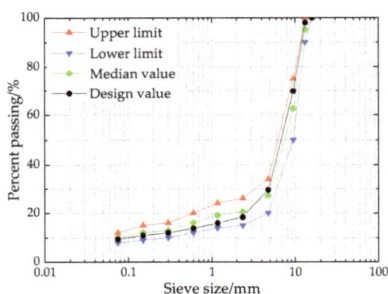

Figure 2. Gradation of stone matrix asphalt (SMA)-13 used in this study.

2.3. Testing Procedure

2.3.1. Cone Penetration Test

Cone penetration test, as shown in Figure 3, is a method to characterize the shearing resistance of fiber modified asphalt, which was developed by Chen [28]. In the cone penetration test, basalt fiber-modified asphalt binder samples were prepared after melting and cooling under controlled conditions. Afterwards, a cone of penetration instrument penetrated into an asphalt binder sample from the sample surface and the sink depth in a stable state can be measured and obtained. It should be noted that penetration depth would be smaller than the sample height. By using the equilibrium equation of force, the shear stress (τ) of SBS-modified asphalt reinforced with basalt fiber could be calculated as follows:

$$\tau = [981Q\cos^2(\alpha/2)]/[\pi h^2 \tan(\alpha/2)], \tag{1}$$

where Q is the cone mass (150 g), h is the sink depth (0.1 mm), α is the cone angle (30°). Three replicate samples were used for each experimental group of modified asphalt binders.

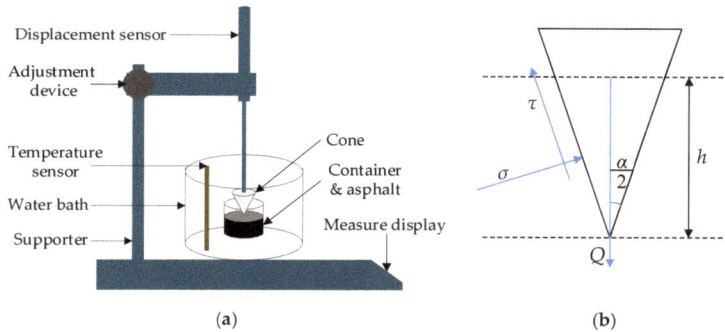

Figure 3. Cone penetration test [28]: (**a**) diagram of cone penetration test; (**b**) cone structure.

2.3.2. Softening Point Test

Softening point of asphalt is defined as the temperature at which asphalt can't bear a steel ball weighing 3.5 g. Therefore, the softening point test is the basic method for the high-temperature susceptibility evaluation of asphalt. According to JTG E20-2011 [27], two replicate samples were prepared and measured for each group of SBS-modified asphalt reinforced with basalt fiber by using the ring and ball apparatus.

2.3.3. Force Ductility Test

The force ductility test refers to the force measurement during the elongation of asphalt at low temperatures, which was firstly introduced by Anderson and Wiley since 1976 [29]. Previous research has shown that the force ductility test is an effective modified ductility method to indicate the low-temperature performance and tensile property of asphalt [30]. In this study, three replicate asphalt samples were prepared in ductility molds in a water bath of 5 °C and then force ductility testing was carried out at a deformation rate of 50 mm/min. Based on the recorded force-elongation curve, strain energy, also called deformation energy, can be calculated by measuring the area under the recorded curve between 20 and 40 mm.

2.3.4. Marshall Test Method

SBS-modified asphalt mixture specimens containing basalt fiber were prepared using the Marshall test method in this study, which is a very popular design method of asphalt mixture due to the relative simplicity, economical equipment and procedure. Then, different parameters of compacted asphalt specimens can be measured, including Marshall stability (*MS*), flow value (*FV*), air voids (*VA*) and voids in mineral aggregates (*VMA*), as well as voids filled with asphalt (*VFA*). Before testing, specimens were immersed in water at 60 °C for 0.5 h. Then, following the Chinese specification [27], a constant compressive loading was applied on cylindrical asphalt specimens at a speed of 50 mm/min until failure occurs. The maximum loading and deformation are defined as *MS* and *FV*, respectively. *VA*, *VMA* and *VFA* can be obtained by the given calculation equations below:

$$VA = [1 - \gamma_f/\gamma_{TMD}] \times 100, \tag{2}$$

$$VMA = [1 - \gamma_f \times P_s/\gamma_{sb}] \times 100, \tag{3}$$

$$VFA = [(VMA - VA)/VMA] \times 100, \tag{4}$$

where γ_f is the bulk specific gravity, γ_{TMD} is the theoretical maximum specific density, P_s is the aggregate content percent by weight of mixture, γ_{sb} is the bulk specific gravity of aggregates.

2.3.5. Pavement Performance Test

Based on the optimal modified asphalt mixture design, pavement performance testing was carried out to evaluate the engineering properties of the optimal modified asphalt mixture in this study, including the high-temperature rutting resistance by wheel tracking test [27], the low-temperature cracking resistance by indirect tensile stiffness modulus (ITSM) test [6] and the moisture stability by immersion Marshall and freeze-thaw splitting tests [27]. The detailed experimental processes of the wheel tracking test, immersion Marshall and freeze-thaw splitting tests have been described in previous studies [31,32]. The ITSM test was conducted to characterize the low-temperature properties of asphalt mixture, which is a widely used experimental method of measuring tensile properties. In this study, the Marshall specimens were prepared and then put in a chamber for at least 6 h at 0 °C before tested. Then a servo-pneumatic universal testing machine (NU-14, Cooper Research Technology, Ltd., Ripley, UK) was employed for ITSM test at a loading speed of 1 mm/min. The target horizontal deformation was set as 5 μm, and the peak value of loading could be obtained at the target deformation. The stiffness modulus can be calculated by using the following equation:

$$S_m = [F \times (\mu + 0.27)] / (h \times Z), \tag{5}$$

where S_m is the stiffness modulus, F is the peak load, μ is the Poisson ratio, h is the specimen height, Z is the measured horizontal deformation.

2.4. Response Surface Methodology

RSM is a statistical method for optimizing random experimental processes, which is used to explore the quantitative relationship between independent variables and response variables [17]. A response surface model is established, and a suitable fitting model based on test data can be chosen to determine the optimum experimental conditions and procedure. In general, CCD is a common experimental design method in RSM, which is a fractional factorial experiment design. Then, the face-centered central composite design (FCCD) was adopted to investigate the relationship between independent variables and response variables for determining a suitable experimental formulation [33]. A three-factor layout for the FCCD is shown in Figure 4, which is a cube with axial points on the face centers. The number of experimental samples can be given by ($2^k + 2k + n$), in which k represents the number of factors, n is the number of center points.

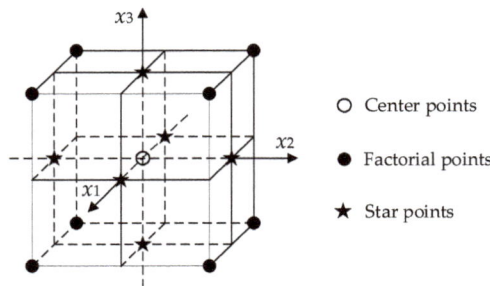

Figure 4. A three-factor layout for face-centered central composite design (FCCD).

The experimental design in this study required nineteen experimental runs, which was composed of eight factorial points, five center points and six star points at three experimental levels ($k = 3$, $n = 5$). Center points were set as five because some replication should be required to estimate the experimental error. These three independent variable factors are basalt fiber content (X_1), basalt fiber length (X_2) and asphalt-aggregate ratio (X_3), which are abbreviated as *BFC*, *BFL* and *AAR*, respectively. The quantitative relationship between the amount of added basalt fibers in asphalt binders and mixtures is not clear, but the approximate percentage range of basalt fiber to asphalt is basically consistent in asphalt binder and mixture. From the previous literature [15,16,25,34], the appropriate basalt fiber content should exceed 0.5%. Additionally, based on the previous research [15,35–38], the optimum asphalt content of modified SMA mixtures can reach up to approximately 6.7%. In view of the added basalt fibers in this study varying from 3 to 9 mm, the basalt fiber content and asphalt-aggregate ratio were chosen for appropriate ranges accordingly. Table 4 lists three independent factors at three experimental levels, in which coded levels are the normalized levels and "−1, 0 and +1" stand for the low, medium and high levels of independent factors, respectively. Thus, the experimental processes can be optimized for *MS* (Y_1), *FV* (Y_2), *VA* (Y_3), *VMA* (Y_4) and *VFA* (Y_5).

Table 4. Experimental design for face-centered central composite design (FCCD).

Factors		Units	Levels: Actual (Coded)		
			Low (−1)	Medium (0)	High (+1)
X_1	BFC	%	0.25	0.40	0.55
X_2	BFL	mm	3	6	9
X_3	AAR	%	6.4	6.6	6.8

The Design-Expert 8.0 software (Stat-Ease, Inc., Minneapolis, MN, US) was employed for the experimental design, response surface modeling, statistical regression analysis and process optimization. Based on the experimental data, it is well suited for fitting a quadratic surface model using the following equation:

$$y = \beta_0 + \sum_{i=1}^{k} \beta_i x_i + \sum_{i=1}^{k} \beta_{ii} x_i^2 + \sum_{i<j}^{k} \beta_{ij} x_i x_j + \varepsilon, \tag{6}$$

where y is the response, x_i and x_j are the coded independent variables, β_0 is the mean value of response constant coefficient, β_i is the linear effect of independent variable x_i, β_{ii} is the secondary effects of x_i, β_{ij} is the linear interaction between x_i and x_j, ε is the random error.

3. Results and Discussion

3.1. Test Results of SBS-Modified Asphalt Binder Reinforced with Basalt Fiber

Figure 5 shows the cone penetration test result at 30 °C, the softening point result and the strain energy result of SBS-modified asphalt binder reinforced with basalt fiber. It can be clearly seen that these test results for modified bitumen binder varied with the fiber length and content. Furthermore, the fiber length had a slight influence on the softening point and strain energy results of bitumen binder when the basalt fiber content was lower. With the increment of basalt fiber content, the changes of modified asphalt binder became more and more significant.

As seen in Figure 5a, basalt fiber can improve the shear resistance of asphalt binder compared to the asphalt binder without fiber. With the addition of basalt fiber, the shear strength of modified asphalt binder gradually increased. This is because basalt fiber can lead to a spatial networking structure in the asphalt binder. On the other hand, fibers can also absorb the light components in asphalt, which saturates it and increases its viscosity [28,38]. Thus, the addition of basalt fiber resulted in a smaller cone penetration and a higher shear strength. However, it should be noted that the shear resistance of modified asphalt binder decreases under the conditions of higher basalt fiber content and larger fiber length, which could be attributed to the uneven distribution of basalt fiber in the asphalt binder. Consequently, it is believed that suitable basalt fiber content and fiber length should be crucial for asphalt properties and its production process.

Figure 5b illustrates that the softening point results for modified asphalt binders increased with increasing fiber content and length. Generally, the softening point is used to evaluate the temperature susceptibility of asphalt materials, and a higher softening point means a lower temperature susceptibility. As a result, it is evident that basalt fiber can well improve the high-temperature susceptibility of asphalt binder. This is because the addition of basalt fiber into asphalt leads to less bitumen, higher stiffness, and bitumen absorption, resulting, to a certain extent, in a reinforcement mechanism, filler-like action, and increased viscosity.

Asphalt material is a typical viscoelastic material, and asphalt exhibits elastic characteristics at low temperatures. The work done by the external force on asphalt at low temperatures is stored as elastic strain and is converted into surface energy when fracture occurs. Normally, the higher the strain energy is, the better the tensile properties are [30]. As demonstrated in Figure 5c, the strain energy results of the modified asphalt binders clearly increased with the increase of basalt fiber content and length. It is worth noting that the variation trend of strain energy was especially significant when the basalt fiber content exceeded 3%. This indicates that adding more basalt fiber could greatly improve the low-temperature tensile properties of asphalt binder.

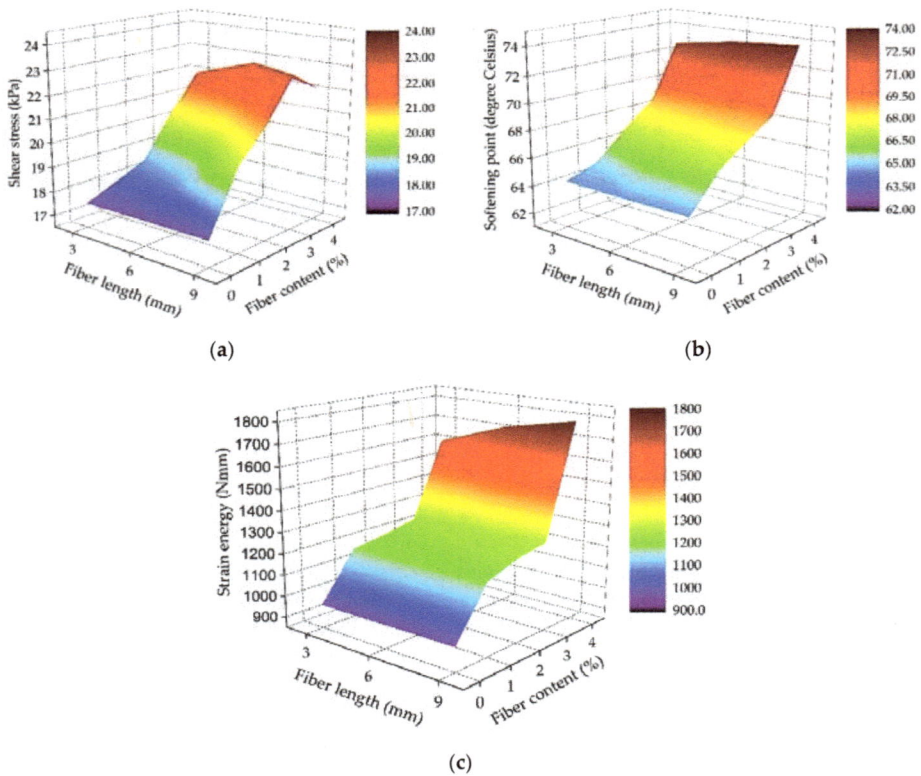

(a)

(b)

(c)

Figure 5. Test results of modified asphalt binders in this study: (a) Cone penetration test; (b) Softening point test; (c) Force ductility test.

3.2. Analysis and Optimization of SBS-Modified Asphalt Mixture Reinforced with Basalt Fiber Using RSM

3.2.1. Experimental Design and Test Results Based on FCCD

RSM was adopted to investigate the influences of basalt fiber content, fiber length and asphalt-aggregate ratio on the Marshall test indices of modified asphalt mixtures. By using FCCD, an experimental design was established with 19 experimental groups, and the total number of experimental groups could be greatly reduced. Table 5 details the experimental design and Marshall test results by using the FCCD. Preparation parameters including *BFC*, *BFL* and *AAR* were considered to be independent variables, and *MS*, *FV*, *VA*, *VMA* and *VFA* were responses or dependent variables.

Table 5. Experimental design and test outcomes by face-centered central composite design (FCCD).

No.	Preparation Parameters			Responses				
	BFC X_1 (%)	*BFL* X_2 (mm)	*AAR* X_3 (%)	*MS* Y_1 (kN)	*FV* Y_2 (mm)	*VA* Y_3 (%)	*VMA* Y_4 (%)	*VFA* Y_5 (%)
1	0.40	6	6.6	9.98	3.052	4.69	18.35	74.45
2	0.25	3	6.4	7.48	3.053	2.80	16.79	83.32
3	0.55	3	6.8	6.23	3.328	7.19	20.47	64.88
4	0.55	6	6.6	9.92	2.778	5.99	19.46	69.22
5	0.40	6	6.8	9.33	4.153	5.89	19.36	69.58
6	0.40	6	6.6	10.56	3.054	4.78	18.42	74.05
7	0.55	9	6.8	7.19	3.053	7.55	20.78	63.67
8	0.25	6	6.6	10.21	3.190	2.75	16.69	83.52
9	0.40	3	6.6	8.72	3.191	4.37	18.08	75.83
10	0.40	6	6.6	10.48	3.053	4.69	18.35	74.44
11	0.40	6	6.6	10.29	3.051	4.82	18.46	73.89
12	0.55	3	6.4	6.63	2.641	6.05	19.57	69.09
13	0.25	9	6.8	7.82	4.291	4.28	17.97	76.18
14	0.25	9	6.4	8.41	2.916	3.13	17.07	81.66
15	0.25	3	6.8	6.89	4.428	3.96	17.70	77.63
16	0.40	6	6.6	10.38	3.053	4.65	18.32	74.62
17	0.40	6	6.4	9.61	2.916	5.07	18.73	72.93
18	0.40	9	6.6	9.04	2.915	5.02	18.63	73.05
19	0.55	9	6.4	8.11	2.503	6.37	19.85	67.91

3.2.2. Statistical Analysis and Discussion

Analysis of Variance (ANOVA) Results

According to the above experimental design and test outcomes using FCCD, the superior regression model was suggested and chosen for the further analysis. The superior model was determined based on *R*-squared, Adjusted *R*-squared (Adj. *R*-squared), Adeq. precision, Fisher's test value (*F*-value) and the probability "Prob > *F*-value" (*p*-value) among different responses, including linear, two-factor interaction (2FI), quadratic, and cubic polynomials. The analysis of variance (ANOVA) was adopted to evaluate the statistical significance of independent variables (i.e., preparation parameters) and their interactions on the responses. The statistical significance level was chosen as 0.05; that is, models and factors can be considered significant when the *p*-value falls below 0.05. The Design-Expert 8.0 software indicated that the quadratic model looks best, and these terms are significant. Corresponding ANOVA results for quadratic models and independent variables were obtained and are listed in Tables 6 and 7, respectively.

Table 6. ANOVA results for quadratic models of modified asphalt mixture.

Responses		R-Squared	Adj. R-Squared	Adeq. Precision	F-Value	p-Value	Significant
Y_1	MS	0.9834	0.9668	21.472	59.31	<0.0001	Yes
Y_2	FV	0.9796	0.9591	25.279	47.94	<0.0001	Yes
Y_3	VA	0.9966	0.9932	59.932	291.94	<0.0001	Yes
Y_4	VMA	0.9968	0.9935	61.691	308.72	<0.0001	Yes
Y_5	VFA	0.9953	0.9905	50.560	210.31	<0.0001	Yes

Table 7. ANOVA results for independent variables.

Responses	Factors	Sum of Squares	Degree of Freedom	Mean Square	F-Value	p-Value	Significant
MS	BFC	0.75	1	0.75	10.89	0.0092	**
	BFL	2.13	1	2.13	31.18	0.0003	***
	AAR	0.77	1	0.77	11.29	0.0084	**
	BFC × BFL	0.042	1	0.042	0.61	0.4533	–
	BFC × AAR	0.002	1	0.002	0.036	0.8541	–
	BFL × AAR	0.034	1	0.034	0.49	0.5000	–
	$(BFC)^2$	0.48	1	0.48	7.02	0.0265	*
	$(BFL)^2$	7.03	1	7.03	102.73	<0.0001	****
	$(AAR)^2$	2.81	1	2.81	41.06	0.0001	***
FV	BFC	1.28	1	1.28	112.77	<0.0001	****
	BFL	0.093	1	0.093	8.18	0.0188	*
	AAR	2.73	1	2.73	240.79	<0.0001	****
	BFC × BFL	0.002	1	0.002	0.21	0.6553	–
	BFC × AAR	0.29	1	0.29	25.25	0.0007	***
	BFL × AAR	0.002	1	0.002	0.21	0.6599	–
	$(BFC)^2$	0.053	1	0.053	4.66	0.0592	–
	$(BFL)^2$	0.013	1	0.013	1.18	0.3054	–
	$(AAR)^2$	0.46	1	46.27	40.82	0.0001	***
VA	BFC	26.34	1	26.34	2183.93	<0.0001	****
	BFL	0.39	1	0.39	32.50	0.0003	***
	AAR	2.97	1	2.97	246.26	<0.0001	****
	BFC × BFL	0.0001	1	0.0001	0.009	0.9252	–
	BFC × AAR	0.00001	1	0.00001	0.001	0.9750	–
	BFL × AAR	0.0001	1	0.0001	0.009	0.9252	–
	$(BFC)^2$	0.30	1	0.30	24.49	0.0008	***
	$(BFL)^2$	0.00003	1	0.00003	0.003	0.9555	–
	$(AAR)^2$	1.67	1	1.67	138.25	<0.0001	****
VMA	BFC	19.35	1	19.35	2337.08	<0.0001	****
	BFL	0.29	1	0.29	34.50	0.0002	***
	AAR	1.82	1	1.82	220.23	<0.0001	****
	BFC × BFL	0.0002	1	0.0002	0.024	0.8799	–
	BFC × AAR	0.00005	1	0.00005	0.006	0.9398	–
	BFL × AAR	0.00005	1	0.00005	0.006	0.9398	–
	$(BFC)^2$	0.22	1	0.22	26.41	0.0006	***
	$(BFL)^2$	0.00002	1	0.00002	0.003	0.9593	–
	$(AAR)^2$	1.29	1	1.29	155.82	<0.0001	****

Table 7. *Cont.*

Responses	Factors	Sum of Squares	Degree of Freedom	Mean Square	F-Value	p-Value	Significant
VFA	BFC	456.17	1	456.17	1572.14	<0.0001	****
	BFL	6.86	1	6.86	23.63	0.0009	***
	AAR	52.76	1	52.76	181.84	<0.0001	****
	BFC × BFL	0.065	1	0.065	0.22	0.6478	–
	BFC × AAR	0.92	1	0.92	3.19	0.1079	–
	BFL × AAR	0.004	1	0.004	0.014	0.9085	–
	$(BFC)^2$	10.03	1	10.03	34.56	0.0002	***
	$(BFL)^2$	0.0005	1	0.0005	0.002	0.9661	–
	$(AAR)^2$	27.97	1	27.97	96.38	<0.0001	****

Note: "****" $p < 0.0001$; "***" $0.0001 \leq p < 0.001$; "**" $0.001 \leq p < 0.01$; "*" $0.01 \leq p < 0.05$; "–" $p \geq 0.05$.

Analysis of Marshall Stability (*MS*)

The ANOVA results of *MS* (Y_1) are listed in the first row of Table 6. The results illustrate that the quadratic model of *MS* possessed satisfactory fitting levels, in which *R*-squared is 0.9834, Adj. *R*-squared is 0.9668, and they are close to one. Additionally, Adeq. precision stands for the signal-to-noise ratio, and it is desirable when this ratio exceeds 4. The Adeq. precision of *MS* is 21.472, and thus greater than 4, which indicates an adequate signal, meaning that this model can be used to navigate the design space.

Based on the ANOVA results for independent variables in Table 7, the factors of the quadratic model of *MS* were demonstrated in detail. The *p*-values quantify the significance of these factors in the quadratic model of *MS*, in which the significance level was chosen as 0.05. A factor with *p*-value less than 0.05 means a statistically significant relationship between this factor and *MS*. Therefore, the significant factors in the quadratic model of *MS* include X_1, X_2, X_3, $(X_1)^2$, $(X_2)^2$ and $(X_3)^2$. Based on the least squares method, the regression coefficients of factors can be determined. Then, by leaving out the insignificant factors, the reasonable second-order polynomial equation in terms of actual factors for *MS* can be established as:

$$Y_1 = -1099.95 + 15.97X_1 + 2.94X_2 + 334.19X_3 - 18.63(X_1)^2 - 0.17(X_2)^2 - 25.36(X_3)^2, \qquad (7)$$

The diagnostics of the statistical model in Figure 6 presents an approximately linear set of data points, indicating a higher significance. The ANOVA results indicate that the linear terms and quadratic terms of basalt fiber content, fiber length and asphalt-aggregate ratio are significant model terms, in which the quadratic term of basalt fiber length has the most significant effect on *MS* of modified asphalt mixture, with *p*-value < 0.0001. Subsequently, Figure 7 illustrates the three-dimensional (3D) response surface and two-dimensional (2D) contour plots for *MS*, which are plotted by the fitting quadratic polynomial equation to reveal the relationship between preparation parameters and *MS*, as well as the interaction of preparation parameters.

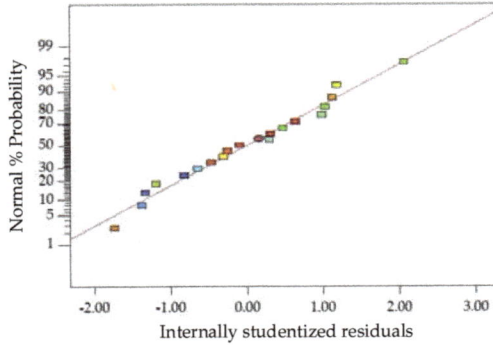

Figure 6. Diagnostics of statistical model: normal plot of residuals for *MS*.

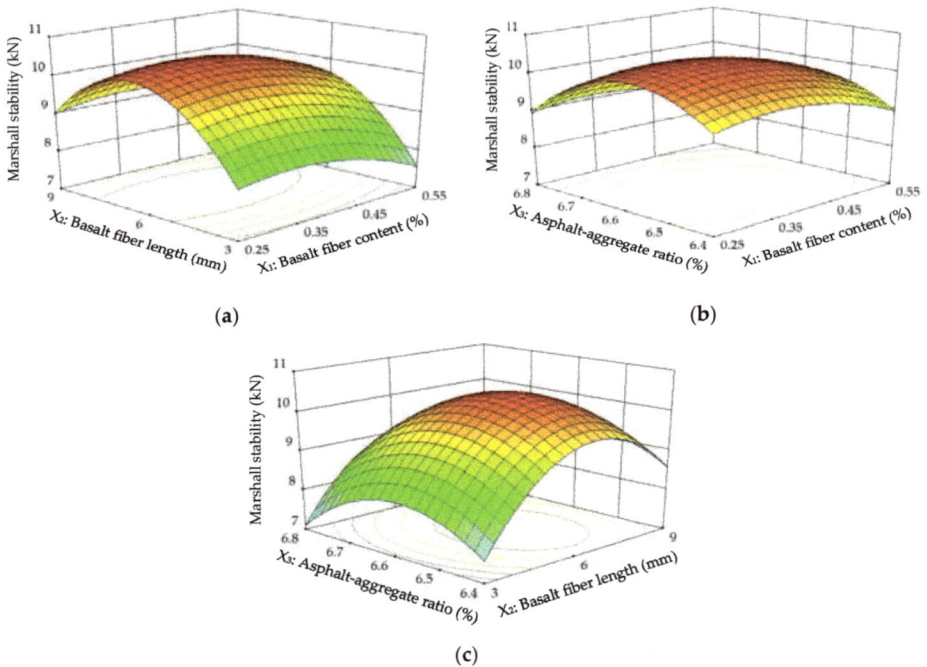

(a)

(b)

(c)

Figure 7. Response surface plots between *MS* and factors: (**a**) Factors: *BFC* and *BFL* at *AAR* = 6.6%; (**b**) Factors: *BFC* and *AAR* at *BFL* = 6 mm; (**c**) Factors: *BFL* and *AAR* at *BFC* = 0.4%.

As illustrated in Figure 7, the *MS*s of modified asphalt mixture firstly presented increasing trends, and then decreased when the basalt fiber content or fiber length or asphalt-aggregate ratio increased continuously. These results could be attributed to a spatial networking structure and fiber's adhesion in asphalt mixture by the addition of basalt fiber, in which fiber has a reinforcement effect on asphalt material for a networking structure. However, *MS*s decreased at higher basalt fiber content and larger fiber length, which is because it would be difficult for basalt fibers to disperse uniformly in asphalt, and more asphalt is needed to wrap around the surfaces of fiber and aggregates to form effective interface adhesions. In other words, due to the higher basalt fiber content and larger fiber length, fibers in asphalt easily coagulated together, resulting in weak points. Meanwhile, a lower asphalt-aggregate ratio led to an SMA structure without sufficient filling asphalt, meaning that the structure became looser, while too much asphalt may cause it not to form a dense interlocking structure.

Analysis of Flow Value (*FV*)

The ANOVA results of *FV* (Y_2) are listed in the second row of Table 6. *R*-squared is 0.9796, Adj. *R*-squared is 0.9591 and Adeq. precision of *FV* is 25.279, illustrating that the quadratic model of *FV* also possessed satisfactory fitting levels. As listed in Table 7, the factors of the quadratic model of *FV* have been demonstrated in detail. Based on the *p*-values, the significant factors in the quadratic model of *FV* were obtained, i.e., X_1, X_2, X_3, X_1X_3 and $(X_3)^2$. Based on the least squares method, the regression coefficients of factors can be determined, and the reasonable second-order polynomial equation in terms of actual factors for *FV* can be established as:

$$Y_2 = 415.99 + 44.40X_1 + 0.27X_2 - 130.49X_3 - 6.30X_1X_3 + 10.29(X_3)^2, \tag{8}$$

The diagnostics of the statistical model in Figure 8 presents an approximately linear set of data points, indicating a higher significance. The ANOVA results indicate that the linear terms of basalt fiber content, fiber length and asphalt-aggregate ratio, the quadratic term of the asphalt-aggregate ratio and the interaction terms between basalt fiber content and asphalt-aggregate ratio are significant model terms, in which the quadratic terms of basalt fiber length and asphalt-aggregate ratio have the most significant effects on the *FV* of modified asphalt mixture, with *p*-value < 0.0001. Subsequently, Figure 9 illustrates the 3D response surface and 2D contour plots for *FV*, which are plotted by fitting the quadratic polynomial equation to reveal the relationship between preparation parameters and *FV*, as well as the interaction of preparation parameters.

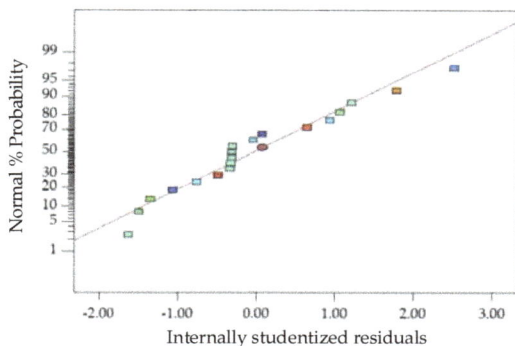

Figure 8. Diagnostics of statistical model: normal plot of residuals for *FV*.

As demonstrated in Figure 9, the *FV*s of the modified asphalt mixture significantly increased with increasing asphalt-aggregate ratio. Generally, *FV* depends on asphalt content, and asphalt mixtures with higher asphalt content usually have a larger *FV* [16]. Accordingly, *FV* presented a significant variation trend with asphalt-aggregate ratio. On the other hand, *FV*s remained approximately unchanged or slightly decreased with increasing basalt fiber content and length. These trends are consistent with the previous study [5]. Based on the comparative analysis in Figure 9, as well as the ANOVA results, it is evident that the asphalt-aggregate ratio has the most significant effect on *FV*.

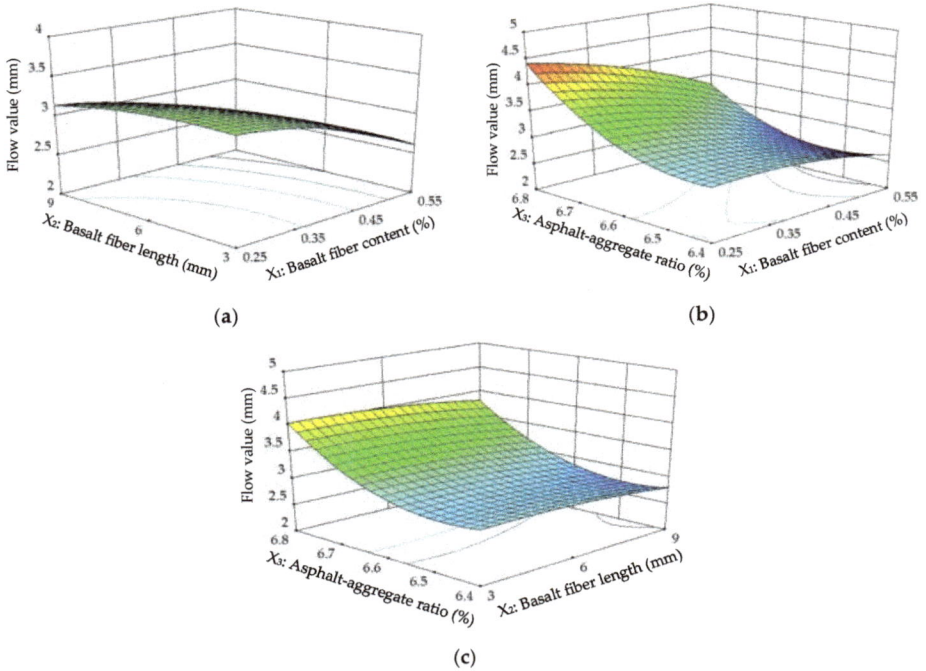

Figure 9. Response surface plots between *FV* and factors: (a) Factors: *BFC* and *BFL* at *AAR* = 6.6%; (b) Factors: *BFC* and *AAR* at *BFL* = 6 mm; (c) Factors: *BFL* and *AAR* at *BFC* = 0.4%

Analysis of Air Voids (*VA*)

The ANOVA results of *VA* (Y_3) are listed in the third row of Table 6. *R*-squared is 0.9966, Adj. *R*-squared is 0.9932 and Adeq. precision of *VA* is 59.932, illustrating that the quadratic model of *VA* also possessed satisfactory fitting levels. As seen in Table 7, the factors of the quadratic model of *VA* were demonstrated in detail. Based on the *p*-values, the significant factors in the quadratic model of *VA* were obtained, i.e., X_1, X_2, X_3, X_1X_3, $(X_1)^2$ and $(X_3)^2$. Based on the least squares method, the regression coefficients of factors can be determined, and the reasonable second-order polynomial equation in terms of actual factors for *VA* can be established as:

$$Y_3 = 830.74 + 22.19X_1 + 0.03X_2 - 255.12X_3 - 14.61(X_1)^2 + 19.53(X_3)^2, \tag{9}$$

The diagnostics of the statistical model in Figure 10 present an approximately linear set of data points, indicating a higher significance. The ANOVA results indicate that the linear terms of basalt fiber content, fiber length and asphalt-aggregate ratio and the quadratic terms of basalt fiber content and asphalt-aggregate ratio are significant model terms, of which the linear terms of basalt fiber content and asphalt-aggregate ratio, as well as the quadratic term of asphalt-aggregate ratio, have the most significant effects on the *VA* of modified asphalt mixture, with *p*-value < 0.0001. Subsequently, Figure 11 illustrates the 3D response surface and 2D contour plots for *VA*, which are plotted by fitting the quadratic polynomial equation to reveal the relationship between preparation parameters and *VA*, as well as the interaction of preparation parameters.

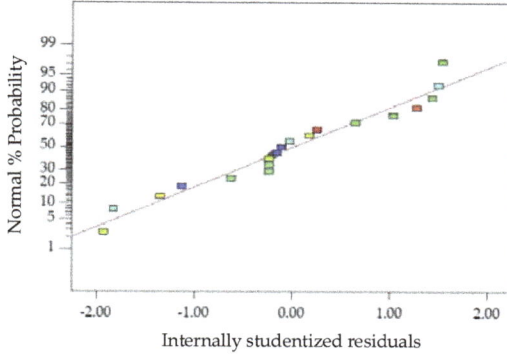

Figure 10. Diagnostics of statistical model: normal plot of residuals for *VA*.

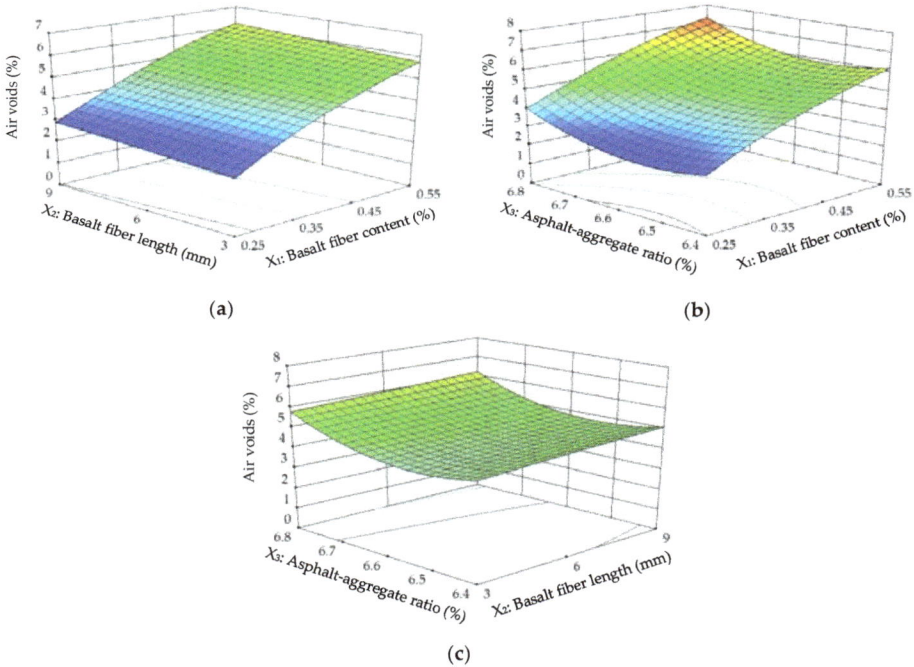

(a)

(b)

(c)

Figure 11. Response surface plots between *VA* and factors: (**a**) Factors: *BFC* and *BFL* at *AAR* = 6.6%; (**b**) Factors: *BFC* and *AAR* at *BFL* = 6 mm; (**c**) Factors: *BFL* and *AAR* at *BFC* = 0.4%.

As shown in Figure 11, the *VA*s of the modified asphalt mixture exhibited a significant increasing trend with increasing basalt fiber content, and *VA*s remained approximately unchanged at different basalt fiber lengths. With respect to asphalt-aggregate ratio, *VA*s firstly slightly decreased and then increased gradually with increasing ratio. It is evident from the comparative analysis that basalt fiber content and asphalt-aggregate ratio have more significant effects on *VA* than basalt fiber length. Compared with mineral aggregates, basalt fiber has the lowest specific gravity (i.e., 2.55~2.65 < 2.738 as shown in Tables 2 and 3), which could cause the bulk specific gravity (γ_f) of modified asphalt mixture to decrease after adding basalt fiber. Then, combined with Equation (2), these findings can explain the higher *VA* at higher basalt fiber contents. In addition, asphalt mixtures with optimum asphalt content have much lower *VAs*, as discussed previously [16,31]. Hence, these trends in Figure 11 are consistent with the previous research, and the optimum asphalt content of modified asphalt mixture is between 6.4% and 6.8%.

Analysis of Voids in Mineral Aggregates (*VMA*)

The ANOVA results of *VMA* (Y_4) are listed in the fourth row of Table 6. *R*-squared is 0.9968, Adj. *R*-squared is 0.9935 and Adeq. precision of *VMA* is 61.691, illustrating that the quadratic model of *VMA* also possessed satisfactory fitting levels. As shown in Table 7, the factors of the quadratic model of *VMA* were demonstrated in detail. Based on the *p*-values, the significant factors in the quadratic model of *VMA* were X_1, X_2, X_3, X_1X_3, $(X_1)^2$ and $(X_3)^2$. Then, the regression coefficients of factors can be determined by the least squares method, and the reasonable second-order polynomial equation in terms of actual factors for *VMA* can be established as:

$$Y_4 = 746.89 + 18.71X_1 + 0.03X_2 - 224.67X_3 - 12.57(X_1)^2 + 17.18(X_3)^2, \tag{10}$$

The diagnostics of the statistical model in Figure 12 presents an approximately linear set of data points, indicating a higher significance. The ANOVA results indicate that the linear terms of basalt fiber content, fiber length and asphalt-aggregate ratio and the quadratic terms of basalt fiber content and asphalt-aggregate ratio are significant model terms, in which the linear terms of basalt fiber content and asphalt-aggregate ratio, as well as the quadratic term of asphalt-aggregate ratio, have the most significant effects on the *VMA* of modified asphalt mixture, with *p*-value < 0.0001. Subsequently, Figure 13 illustrates the 3D response surface and 2D contour plots for *VMA*, which are plotted by fitting the quadratic polynomial equation to reveal the relationship between preparation parameters and *VMA*, as well as the interaction of preparation parameters.

As shown in Figure 13, the *VMAs* of modified asphalt mixture exhibited approximately similar variation trends to *VA*. This is expected, due to the decrease of bulk specific gravity (γ_f) of the modified asphalt mixture after adding basalt fiber, as discussed above. The results can also be explained by Equation (3). As a result, the bulk specific gravity influenced by basalt fiber content can be considered a significant factor on *VMA*.

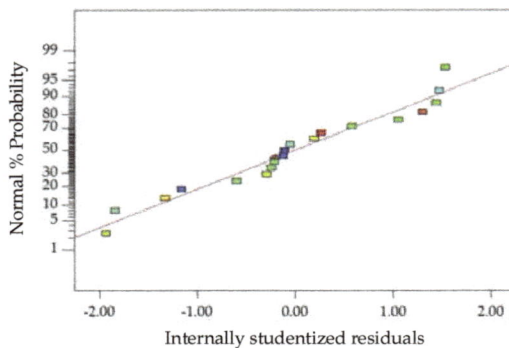

Figure 12. Diagnostics of statistical model: normal plot of residuals for *VMA*.

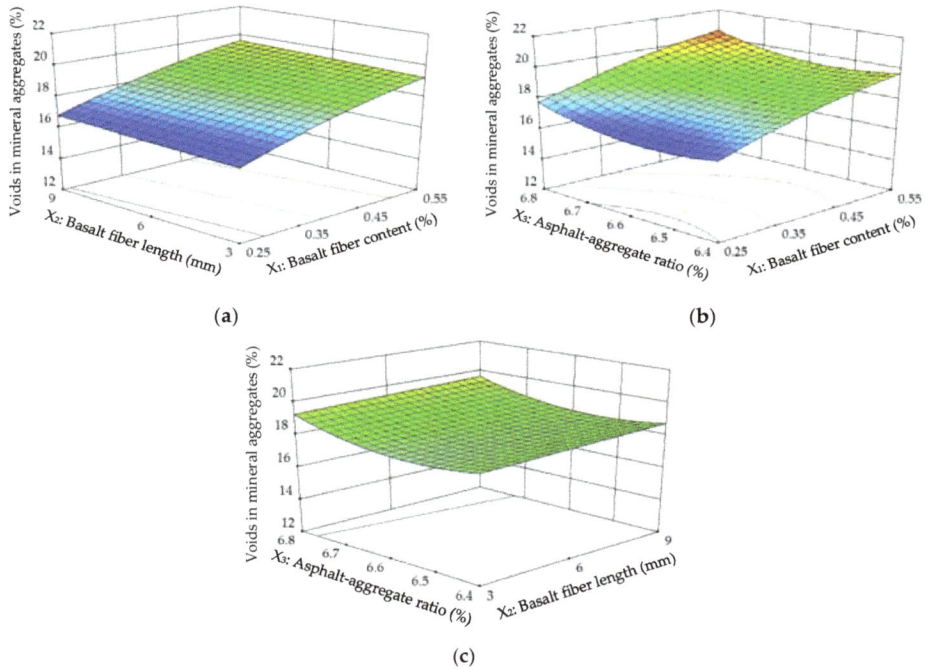

(a)

(b)

(c)

Figure 13. Response surface plots between *VMA* and factors: (a) Factors: *BFC* and *BFL* at *AAR* = 6.6%; (b) Factors: *BFC* and *AAR* at *BFL* = 6 mm; (c) Factors: *BFL* and *AAR* at *BFC* = 0.4%.

Analysis of Voids Filled with Asphalt (*VFA*)

The ANOVA results of *VFA* (Y_5) are listed in the fifth row of Table 6. R-squared is 0.9953, Adj. R-squared is 0.9905 and Adeq. precision of *VFA* is 50.560, illustrating that the quadratic model of *VFA* also possessed satisfactory fitting levels. As listed in Table 7, the factors of the quadratic model of *VFA* were demonstrated in detail. Based on the *p*-values, the significant factors in the quadratic model of *VFA* included X_1, X_2, X_3, X_1X_3, $(X_1)^2$ and $(X_3)^2$. Then, the regression coefficients of factors can be determined by the least squares method, and the reasonable second-order polynomial equation in terms of actual factors for *VFA* can be established as:

$$Y_5 = -3268.68 - 189.14X_1 - 0.58X_2 + 1039.50X_3 + 85.15(X_1)^2 - 79.98(X_3)^2, \tag{11}$$

The diagnostics of the statistical model in Figure 14 presents an approximately linear set of data points, indicating a higher significance. The ANOVA results indicate that the linear terms of basalt fiber content, fiber length and asphalt-aggregate ratio and the quadratic terms of basalt fiber content and asphalt-aggregate ratio are significant model terms, in which the linear terms of basalt fiber content and asphalt-aggregate ratio, as well as the quadratic term of asphalt-aggregate ratio, have the most significant effects on the *VFA* of modified asphalt mixture, with *p*-value < 0.0001. Subsequently, Figure 15 illustrates the 3D response surface and 2D contour plots for *VFA*, which are plotted by fitting the quadratic polynomial equation to reveal the relationship between preparation parameters and *VFA*, as well as the interaction of preparation parameters.

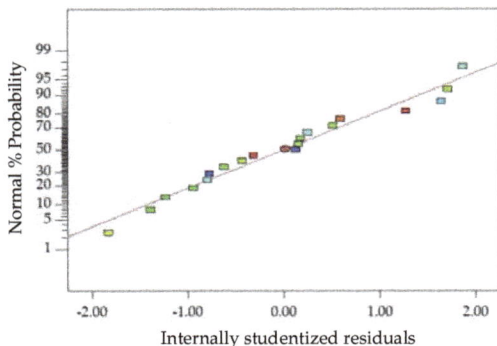

Figure 14. Diagnostics of statistical model: normal plot of residuals for *VFA*.

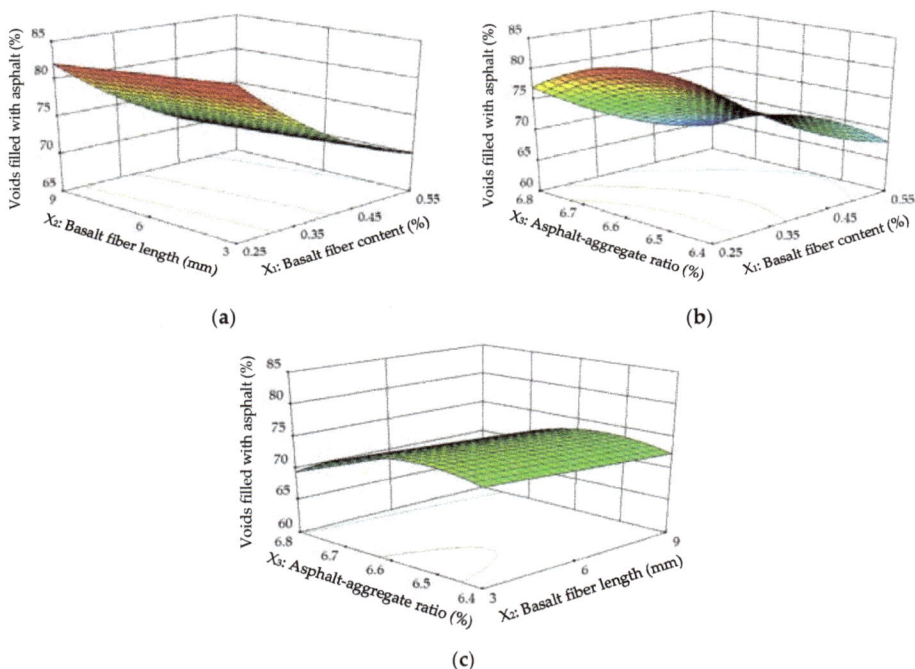

Figure 15. Response surface plots between *VFA* and factors: (**a**) Factors: *BFC* and *BFL* at *AAR* = 6.6%; (**b**) Factors: *BFC* and *AAR* at *BFL* = 6 mm; (**c**) Factors: *BFL* and *AAR* at *BFC* = 0.4%.

As illustrated in Figure 15, the *VFA*s of modified asphalt mixture remained approximately unchanged or slightly decreased with increasing basalt fiber length and asphalt-aggregate ratio, whereas *VFA*s significantly decreased with increasing basalt fiber content. This is evidence that basalt fiber content has the most significant effect on *VFA*. Additionally, a lower *VFA* indicates a thinner asphalt film between aggregates, resulting in an unstable interface adhesion. Therefore, higher basalt fiber content is not recommended for avoiding a much lower *VFA*.

3.2.3. Optimization of Preparation Parameters and Model Verification

As discussed above, the preparation parameters exhibit various effects on the response variables of modified asphalt mixture, and the corresponding influence significance levels are also different from each other. In order to determine the optimal combination of preparation parameters, a multiple response optimization was carried out based on the best fitted response surface model. The target values were selected in accordance with JTG F40-2004 [39], which are presented in Table 8. Then, the optimal combination of preparation parameters can be obtained based on RSM by the Design-Expert 8.0 software, and Table 9 lists the optimization of preparation parameters, as well as the corresponding responses.

Table 8. Target values of response variables.

Response	MS (Y_1)	FV (Y_2)	VA (Y_3)	VMA (Y_4)	VFA (Y_5)
Units	kN	mm	%	%	%
Target value	Maximize	2~5	3~4	≥17	75~85

Table 9. Optimal preparation parameters and prediction vs. experiment.

Response		BFC X_1 (%)	BFL X_2 (mm)	AAR X_3 (%)	MS Y_1 (kN)	FV Y_2 (mm)	VA Y_3 (%)	VMA Y_4 (%)	VFA Y_5 (%)
Prediction		0.34	6.41	6.57	10.49	3.113	4	17.76	77.43
	1	0.34	6	6.57	10.35	3.118	3.99	17.66	77.41
Experiment	2	0.34	6	6.57	10.33	3.117	4.01	17.71	77.36
	3	0.34	6	6.57	10.39	3.113	4	17.67	77.36
	Mean	0.34	6	6.57	10.36	3.116	4	17.68	77.38
Relative error (%)		—	—	—	−1.24	0.10	0	−0.45	−0.06

In view of the experimental conditions, the optimal combination of preparation parameters was chosen as follows: *BFC* is 0.34%, *BFL* is 6 mm, *AAR* is 6.57%. Three replicate samples were prepared and tested by the Marshall test method. The experimental results are summarized in Table 9. According to the relative error results, a very good agreement between the prediction and experiment can be observed, and the relative errors are less than 2%. This reveals that the design optimization of the preparation parameters of the modified asphalt mixture by RSM possesses favorable accuracy for Marshall test indices.

3.3. Comparative Analysis of Pavement Performance

In order to evaluate the pavement performance of optimal basalt fiber-modified asphalt mixture by RSM, wheel tracking, indirect tensile stiffness modulus, immersion Marshall and freeze-thaw splitting tests were conducted, and these pavement performance results can be used for comparison with the previous work [26]. Lignin fiber is a commonly used fiber in road engineering. In the previous work, SBS-modified asphalt mixture containing lignin fiber was prepared with a lignin fiber content of 0.4%, fiber length of 1.1 mm and asphalt-aggregate ratio of 6.8%. Then, the pavement performance of the lignin fiber-modified asphalt mixture was obtained. Figure 16 illustrates the pavement performance results, including dynamic stability, indirect tensile stiffness modulus, residual Marshall stability and tensile strength ratio.

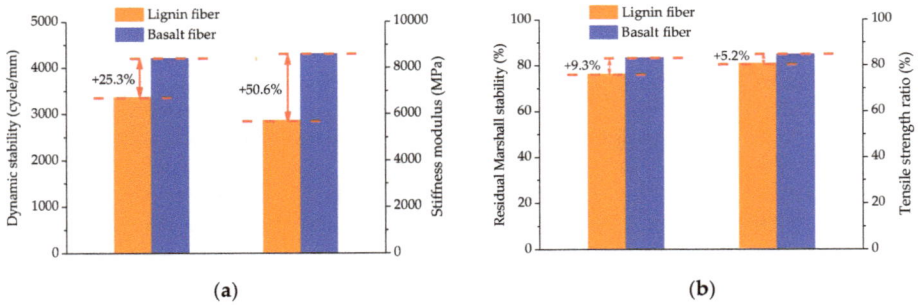

Figure 16. Pavement performance test results: (**a**) High-temperature dynamic stability and low-temperature cracking resistance; (**b**) Moisture stability.

As shown in Figure 16a, it can be observed from the comparative analysis that the dynamic stability of basalt fiber-modified asphalt mixture was improved by approximately 25.3%, and the indirect tensile stiffness modulus can be improved by up to 50.6% compared to lignin fiber-modified asphalt mixture. In general, a higher dynamic stability indicates a better high-temperature rutting resistance and a higher indirect tensile stiffness modulus is preferable for low-temperature cracking resistance. Therefore, this implies that the high-temperature rutting resistance and low-temperature cracking resistance can be greatly improved by using basalt fiber.

As illustrated in Figure 16b, the residual Marshall stability and tensile strength ratio of basalt fiber-modified asphalt mixture were improved by 9.3% and 5.2%, respectively. Higher residual Marshall stability and tensile strength ratio are usually desirable for better moisture stability. As a result, the moisture stability is improved by the addition of basalt fiber.

4. Conclusions

This study optimized the design of SBS-modified asphalt mixture reinforced with eco-friendly basalt fiber based on response surface methodology. The effects of the preparation parameters on Marshall test indices were discussed and analyzed. Meanwhile, the performance of basalt fiber-modified asphalt binder and mixture were also studied. The following conclusions can be drawn:

- A design optimization of basalt fiber and SBS-modified asphalt mixture is proposed based on response surface methodology—i.e., basalt fiber content: 0.34%, fiber length: 6 mm and asphalt-aggregate ratio: 6.57%—which possesses favorable and reliable accuracy compared with experimental results.
- Basalt fiber length has a more significant effect on Marshall stability than basalt fiber content and asphalt-aggregate ratio.
- Flow value presents a significant variation trend with asphalt-aggregate ratio and a larger flow value depends on a higher asphalt content.
- Basalt fiber content exhibits the most significant effect on air voids due to the lower specific gravity of basalt fiber. Additionally, voids in mineral aggregates also exhibit an approximately similar variation trend to air voids.
- Voids filled with asphalt are also more related to basalt fiber content; hence, a higher basalt fiber content should not be recommended in view of the interface adhesion.
- The spatial networking structure and absorption of light components in asphalt by basalt fiber can lead to better shear resistance and tensile properties, as well as lower temperature susceptibility of the asphalt binder. Meanwhile, the high-temperature stability, low-temperature cracking resistance and moisture stability can also be improved by the addition of basalt fiber.

- In comparing the cost of asphalt mixtures modified by basalt fiber and lignin fiber, the cost mainly depends on the fiber prices. The basalt fiber price is about 2~3 times that of the lignin fiber, while the performance of basalt fiber-modified asphalt mixture is better than that of lignin fiber-modified asphalt mixture. Therefore, in the long run, basalt fiber modified asphalt mixture would produce good economic benefits.

Author Contributions: Conceptualization, W.W. and Y.C.; Methodology, W.W. and G.T.; Validation, Y.C.; Formal Analysis, G.T.; Investigation, W.W.; Writing-Original Draft Preparation, W.W.; Writing-Review & Editing, Y.C. and G.T.; Project Administration, Y.C.; Funding Acquisition, Y.C.

Funding: This research was funded by [National Natural Science Foundation of China] grant number [51678271] and [Science Technology Development Program of Jilin Province] grant number [20160204008SF].

Acknowledgments: The authors would like to appreciate anonymous reviewers for their constructive suggestions and comments to improve the quality of the paper.

Conflicts of Interest: The authors declare no conflict of interest.

References

1. Nejad, F.M.; Aflaki, E.; Mohammadi, M.A. Fatigue behavior of SMA and HMA mixtures. *Constr. Build. Mater.* **2010**, *24*, 1158–1165. [CrossRef]
2. Cao, W.D.; Liu, S.T.; Feng, Z.G. Comparison of performance of stone matrix asphalt mixtures using basalt and limestone aggregates. *Constr. Build. Mater.* **2013**, *41*, 474–479. [CrossRef]
3. Chen, H.X.; Li, N.L.; Hu, C.S.; Zhang, Z.Q. Mechanical performance of fibers-reinforced asphalt mixture. *J. Chan. Univ.* **2004**, *24*, 1–6. (In Chinese)
4. Putman, B.J.; Amirkhanian, S.N. Utilization of waste fibers in stone matrix asphalt mixtures. *Resour. Conserv. Recy.* **2004**, *42*, 265–274. [CrossRef]
5. Tapkın, S. The effect of polypropylene fibers on asphalt performance. *Build. Environ.* **2008**, *43*, 1065–1071. [CrossRef]
6. Guo, Q.L.; Li, L.L.; Cheng, Y.C.; Jiao, Y.B.; Xu, C. Laboratory evaluation on performance of diatomite and glass fiber compound modified asphalt mixture. *Mater. Des.* **2015**, *66*, 51–59. [CrossRef]
7. Zhang, X.Y.; Gu, X.Y.; Lv, J.X. Effect of basalt fiber distribution on the flexural–tensile rheological performance of asphalt mortar. *Constr. Build. Mater.* **2018**, *179*, 307–314. [CrossRef]
8. Wang, D.; Wang, L.B.; Gu, X.Y.; Zhou, G.Q. Effect of Basalt Fiber on the Asphalt Binder and Mastic at Low Temperature. *J. Mater. Civ. Eng.* **2013**, *25*, 355–364. [CrossRef]
9. Gu, X.Y.; Xu, T.T.; Ni, F.J. Rheological behavior of basalt fiber reinforced asphalt mastic. *J. Wuhan Univ. Technol.* **2014**, *29*, 950–955. [CrossRef]
10. Cheng, Y.C.; Zhu, C.F.; Tan, G.J.; Lv, Z.H.; Yang, J.S.; Ma, J.S. Laboratory Study on Properties of Diatomite and Basalt Fiber Compound Modified Asphalt Mastic. *Adv. Mater. Sci. Eng.* **2017**, *2017*, 1–10. [CrossRef]
11. Zhang, X.Y.; Gu, X.Y.; Lv, J.X.; Zhu, Z.K.; Zou, X.Y. Numerical analysis of the rheological behaviors of basalt fiber reinforced asphalt mortar using ABAQUS. *Constr. Build. Mater.* **2017**, *157*, 392–401. [CrossRef]
12. Zhang, X.Y.; Gu, X.Y.; Lv, J.X.; Zou, X.Y. 3D numerical model to investigate the rheological properties of basalt fiber reinforced asphalt-like materials. *Constr. Build. Mater.* **2017**, *138*, 185–194. [CrossRef]
13. Qin, X.; Shen, A.Q.; Guo, Y.C.; Li, Z.N.; Lv, Z.H. Characterization of asphalt mastics reinforced with basalt fibers. *Constr. Build. Mater.* **2018**, *159*, 508–516. [CrossRef]
14. Gao, C.M. Microcosmic Analysis and Performance Research of Basalt Fiber Asphalt Concrete. Doctor's Thesis, Jilin University, Changchun, China, 2012.
15. Morova, N. Investigation of usability of basalt fibers in hot mix asphalt concrete. *Constr. Build. Mater.* **2013**, *47*, 175–180. [CrossRef]
16. Chen, H.X.; Xu, Q.W.; Chen, S.F.; Zhang, Z.Q. Evaluation and design of fiber-reinforced asphalt mixtures. *Mater. Des.* **2009**, *30*, 2595–2603. [CrossRef]
17. Omranian, S.R.; Hamzah, M.O.; Valentin, J.; Hasan, M.R.M. Determination of optimal mix from the standpoint of short term aging based on asphalt mixture fracture properties using response surface method. *Constr. Build. Mater.* **2018**, *179*, 35–48. [CrossRef]

18. Tan, Y.Q.; Guo, M.; Zhang, L.; Cao, L.P. Performance Optimization Method of Composite Modified Asphalt Sealant. *J. Highway Transport. Res. Dev.* **2012**, *7*, 1–7. (In Chinese) [CrossRef]
19. Kavussi, A.; Qorbani, M.; Khodaii, A.; Haghshenas, H.F. Moisture susceptibility of warm mix asphalt: A statistical analysis of the laboratory testing results. *Constr. Build. Mater.* **2014**, *52*, 511–517. [CrossRef]
20. Hamzah, M.O.; Golchin, B.; Tye, C.T. Determination of the optimum binder content of warm mix asphalt incorporating Rediset using response surface method. *Constr. Build. Mater.* **2013**, *47*, 1328–1336. [CrossRef]
21. Hamzah, M.O.; Gungat, L.; Golchin, B. Estimation of optimum binder content of recycled asphalt incorporating a wax warm additive using response surface method. *Int. J. Pavement Eng.* **2017**, *2017*, 1–11. [CrossRef]
22. Hamzah, M.O.; Omranian, S.R. Effects of extended short-term aging duration on asphalt binder behaviour at high temperatures. *Balt. J. Road Bridge Eng.* **2016**, *11*, 302–312. [CrossRef]
23. Khodaii, A.; Haghshenas, H.F.; Tehrani, H.K. Effect of grading and lime content on HMA stripping using statistical methodology. *Constr. Build. Mater.* **2012**, *34*, 131–135. [CrossRef]
24. Haghshenas, H.F.; Khodaii, A.; Khedmati, M.; Tapkin, S. A mathematical model for predicting stripping potential of Hot Mix Asphalt. *Constr. Build. Mater.* **2015**, *75*, 488–495. [CrossRef]
25. Chen, Y.Z.; Li, Z.X. Study of Road Property of Basalt Fiber Asphalt Concrete. *Appl. Mech. Mater.* **2012**, *238*, 22–25. [CrossRef]
26. Ni, P. Study on Design and Pavement Performance Experiment of Basalt Fiber SMA-13 Mixture. Master's Thesis, Jilin University, Changchun, China, 2017.
27. *JTG E20-2011. Standard Test Methods of Bitumen and Bituminous Mixtures for Highway Engineering*; Ministry of Transport of the People's Republic of China: Beijing, China, 2011. (In Chinese)
28. Chen, H.X.; Xu, Q.W. Experimental study of fibers in stabilizing and reinforcing asphalt binder. *Fuel* **2010**, *89*, 1616–1622. [CrossRef]
29. Edwards, Y.; Tasdemir, Y.; Isacsson, U. Rheological effects of commercial waxes and polyphosphoric acid in bitumen 160/220-low temperature performance. *Fuel* **2006**, *85*, 989–997. [CrossRef]
30. Cheng, Y.C.; Tao, J.L.; Jiao, Y.B.; Tan, G.J.; Guo, Q.L.; Wang, S.R.; Ni, P. Influence of the properties of filler on high and medium temperature performances of asphalt mastic. *Constr. Build. Mater.* **2016**, *118*, 268–275. [CrossRef]
31. Wang, W.S.; Cheng, Y.C.; Tan, G.J.; Shi, C.L. Pavement Performance Evaluation of Asphalt Mixtures Containing Oil Shale Waste. *Road Mater. Pavement* **2018**. [CrossRef]
32. Cheng, Y.C.; Wang, W.S.; Tan, G.J.; Shi, C.L. Assessing High- and Low-Temperature Properties of Asphalt Pavements Incorporating Waste Oil Shale as an Alternative Material in Jilin Province, China. *Sustainability* **2018**, *10*, 2179. [CrossRef]
33. Balachandran, M.; Devanathan, S.; Muraleekrishnan, R.; Bhagawan, S.S. Optimizing properties of nanoclay-nitrile rubber (NBR) composites using Face Centred Central Composite Design. *Mater. Des.* **2012**, *35*, 854–862. [CrossRef]
34. Zhao, L.H.; Chen, J.Y.; Wang, S.W. Using Mineral Fibers to Improve Asphalt and Asphalt Mixture Behavior. In Proceedings of the 7th International Conference on Traffic and Transportation Studies, Kunming, China, 3–5 August 2010. [CrossRef]
35. Kordi, Z.; Shafabakhsh, G. Evaluating mechanical properties of stone mastic asphalt modified with Nano Fe_2O_3. *Constr. Build. Mater.* **2017**, *134*, 530–539. [CrossRef]
36. Khedmati, M.; Khodaii, A.; Haghshenas, H.F. A study on moisture susceptibility of stone matrix warm mix asphalt. *Constr. Build. Mater.* **2017**, *144*, 42–49. [CrossRef]
37. Ameri, M.; Mohammadi, R.; Vamegh, M.; Molayem, M. Evaluation the effects of nanoclay on permanent deformation behavior of stone mastic asphalt mixtures. *Constr. Build. Mater.* **2017**, *156*, 107–113. [CrossRef]
38. Xiong, R.; Fang, J.H.; Xu, A.H.; Guan, B.W.; Liu, Z.Z. Laboratory investigation on the brucite fiber reinforced asphalt binder and asphalt concrete. *Constr. Build. Mater.* **2015**, *83*, 44–52. [CrossRef]
39. *JTG F40-2004. Technical Specifications for Construction of Highway Asphalt Pavements*; Ministry of Transport of the People's Republic of China: Beijing, China, 2004. (In Chinese)

materials

MDPI

Article

Laboratory Evaluation on Performance of Eco-Friendly Basalt Fiber and Diatomite Compound Modified Asphalt Mixture

Yongchun Cheng [1], Di Yu [1], Yafeng Gong [1,*], Chunfeng Zhu [2], Jinglin Tao [3] and Wensheng Wang [1]

1   College of Transportation, Jilin University, Changchun 130025, China; chengyc@jlu.edu.cn (Y.C.);
    yudi16@mails.jlu.edu.cn (D.Y.); wangws17@mails.jlu.edu.cn (W.W.)
2   College of Transportation, Jilin Jianzhu University, Changchun 130021, China; zcf-mine@163.com
3   Jiangxi Transportation Institute, Nanchang 330200, China; taojinglinok@163.com
*   Correspondence: gongyf@jlu.edu.cn; Tel.: +86-0431-4186-5446

Received: 13 November 2018; Accepted: 26 November 2018; Published: 28 November 2018

Abstract: This study proposed an asphalt mixture modified by basalt fiber and diatomite. Performance of diatomite modified asphalt mixture (DAM), basalt fiber modified asphalt mixture (BFAM), diatomite and basalt fiber compound modified asphalt mixture (DBFAM), and control asphalt mixture (AM) were investigated by experimental methods. The wheel tracking test, low-temperature indirect tensile test, moisture susceptibility test, fatigue test and freeze–thaw cycles test of four kinds of asphalt mixtures were carried out. The results show that the addition of basalt fiber and diatomite can improve the pavement performance. Diatomite has a significant effect on the high temperature stability, moisture susceptibility and resistance to moisture and frost damage under freeze–thaw cycles of asphalt mixture. Basalt fiber has a significant effect on low-temperature cracking resistance of asphalt mixture. Composed modified asphalt mixture has obvious advantages on performance compared to the control asphalt mixture. It will provide a reference for the design of asphalt mixture in seasonal frozen regions.

Keywords: laboratory evaluation; diatomite; basalt fiber; compound modify; asphalt mixture

1. Introduction

Asphalt pavement has been widely used in the world, but it is affected by the action of vehicle loading and complex natural environment factors [1,2]. The seasonal frozen regions account for 53.5% total area of China and has complex climatic conditions [3–5], including high temperatures in the summer, low temperatures in the winter, freeze–thaw (F–T) cycles in the autumn, winter and spring. Asphalt mixture, as a kind of viscoelastic material, suffers rutting deformation at high temperature and pavement cracking at low temperature. What's more, asphalt mixture has short service life under F–T cycles and suffers fatigue damage due to the action of vehicle loading. Therefore, to address common distresses of asphalt pavement, researchers have been trying to modify asphalt mixture with additives [6–8].

Diatomite is a mineral filler with low cost, relative abundance, large specific surface area, and high absorptive capacity [9,10]. What's more, as an inorganic modified material, diatomite is more environmentally friendly than polymer modifier. Diatomite has been widely used to modify asphalt binder and mixture. Additives can modify the property of asphalt binder, which finally influence the performance of asphalt mixture [11–13]. Cong et al. [14] investigated the effects of diatomite on the chemical properties of modified asphalt by Fourier Transform Infrared Spectroscopy test. The results revealed that no chemical reaction occurred between diatomite and asphalt. Cheng et al. [15] indicated that the diatomite modified asphalt had stronger anti-ageing capacity than pure one. Bao [16] studied the pavement performance of diatomite modified asphalt mixture. The results showed that diatomite-modified asphalt mixture had higher Marshall stability, moisture sensitivity, and high temperature stability than pure one. Yang et al. [17] evaluated the modification mechanisms and performance of diatomite modified asphalt mixture. The results showed that the improvement mechanism of diatomite modified asphalt was physical absorption. The bitumen absorption by diatomite brought better resistance in the permanent deformation and moisture damage for asphalt mixture. While tensile strength ratio (TSR) was used as an index to evaluate the moisture susceptibility, the moisture susceptibility had a certain relationship with the resistance to F–T cycles. The addition of diatomite is a possible solution to improve the mechanical properties and F–T resistance for asphalt mixture in the seasonal frozen regions.

Basalt fiber is an eco-friendly mineral fiber with high strength, and low water absorption. In addition, basalt fiber has stable chemical properties when subjected to acid, alkali, and high temperature [18]. Basalt fiber has been widely used to improve the performance of asphalt binder and mixture. Wang et al. [19] investigated the tensile and fatigue property of basalt fiber modified asphalt binder. The results showed that the tensile strength and fatigue life of basalt fiber modified asphalt binder were significantly improved than pure one. Gu et al. [20] studied the effect of basalt fiber on rheological characteristics of asphalt mastic with dynamic shear rheological tests. The results showed that the addition of basalt fiber improved the elasticity characteristics and shear modulus of the asphalt mastic, which meant better high temperature stability. Zhang et al. [21,22] studied viscoelastic characteristics of basalt fiber modified asphalt mortar with a new three dimensional fiber distribution model. The results indicated that the basalt fiber caused stress redistribution and reduced the stress of the asphalt mortar. Zhao [23] and Gao [24] studied the effect of basalt fiber on asphalt mixture, and the results showed that the resistance to low-temperature crack of the mixture was remarkably improved. However, the compound modification effect of the diatomite and basalt fiber on the performance of asphalt mixture is still unclear. Cheng et al. [25] used basalt fiber and diatomite to modify asphalt mastic and found that the high and low temperature properties were improved. Davar et al. [26] studied the properties of basalt fiber–diatomite asphalt mixtures through fatigue test and indirect tensile (IDT) test. But the other mechanical performance and the durability are still unknown. If the addition of diatomite and basalt fibers can give full play to their respective advantages in different performance of asphalt mixtures, and be suitable for the complex environments, then the basalt fiber–diatomite modified asphalt mixture can be well-used to address the common distresses of asphalt pavement in seasonal frozen regions.

In this study, the effect of basalt fiber and diatomite on performance of asphalt mixture were tested and evaluated. A diatomite modified asphalt mixture (DAM), basalt fiber modified asphalt mixture (BFAM), and diatomite–basalt fiber-modified asphalt mixture (DBFAM) and control asphalt mixture (AM) were prepared. Laboratory experiments were carried out to investigate the properties of modified asphalt mixtures. The tests adopted in this research included the wheel tracking test, indirect tensile test at low-temperature, moisture susceptibility test, fatigue test, and F–T cycles test. The pavement performance of modified asphalt mixtures were discussed. It will provide a reference for the design of asphalt mixture in seasonal frozen regions.

2. Materials and Methods

2.1. Raw Materials

The asphalt AH-90 in this study was obtained from Panjin petrochemical industry, Panjin in Liaoning province, China, and its basic properties are given in Table 1. Basalts were chosen as coarse and fine aggregates, obtained from Wantong Road Building Materials Co., Ltd., Yitong in Jilin Province, China. Their physical properties are shown in Table 2. Limestone powder was chosen as the filler. The selected gradation of asphalt mixture is shown in Figure 1.

Diatomite was used to modify the asphalt mixture, as shown in Figure 2, and its chemical composition are summarized in Table 3. Its pH value is 7.

Basalt fiber was used to modify the asphalt mixture, as shown in Figure 3, and its basic properties are listed in Table 4.

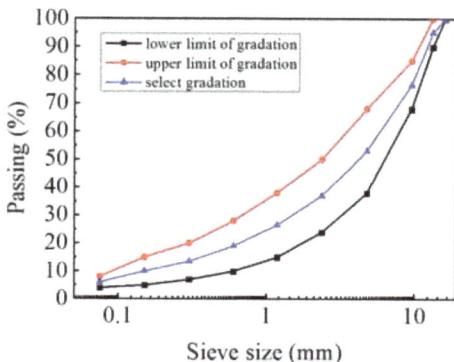

Figure 1. The Grading Curve aggregates used in this study.

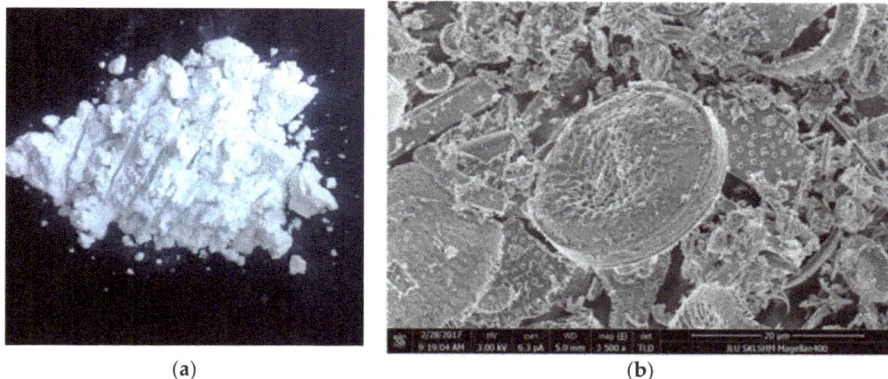

(a)　　　　　　　　　　　　　　　　　　(b)

Figure 2. Images of diatomite. (**a**) Morphology of diatomite; (**b**) SEM images of diatomite.

Table 1. Basic properties of AH-90 asphalt.

Property	Value
Density (15 °C, g/cm^3)	1.018
Penetration (25 °C, 0.1 mm)	92.3
Softening point (°C)	46.9
Ductility (25 °C, cm)	>150

Table 2. Properties of aggregate.

Sieve Size (mm)	13.2	9.5	4.75	2.36	1.18	0.6	0.3	0.15	0.075
Crushed stone value (%)	14.9	13.6	16.0	15.8	-	-	-	-	-
Flakiness index (%)	8.8	9.2	9.1	8.3	-	-	-	-	-
Los Angeles abrasion loss (%)	19.0	18.5	19.0	15.7	-	-	-	-	-
Apparent density(g/cm^3)	2.811	2.805	2.815	2.817	2.808	2.805	2.778	2.777	2.768
Absorption coefficient of water (%)	0.33	0.44	0.54	0.75	-	-	-	-	-

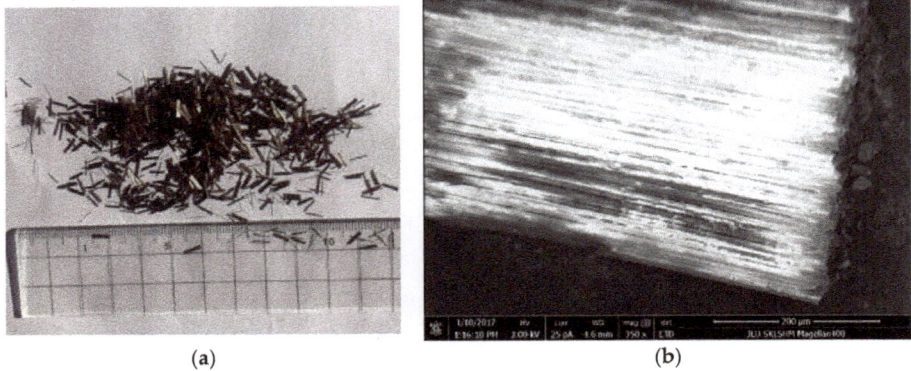

(a)   (b)

Figure 3. Images of basalt fiber. (a) Morphology of basalt fiber; (b) SEM images of basalt fiber.

Table 3. Diatomite chemical composition.

Chemical Composition	SiO_2	Al_2O_3	Fe_2O_3	CaO	MgO	TiO_2	K_2O	Loss on Ignition
Content (%)	85.60	4.50	1.50	0.52	0.45	0.30	0.67	4.61

Table 4. Technical properties of basalt fiber.

Property	Test Result	Standard Requirements
Diameter (μm)	10–13	—
Length (mm)	6	—
Moisture content (%)	0.030	≤ 0.2
Combustible content (%)	0.56	—
Linear density (Tex)	2398	2400 ± 120
Fracture strength (N/Tex)	0.55	≥ 0.40
Tensile strength (MPa)	2320	≥ 2000
Tensile modulus of elasticity (GPa)	86.3	≥ 85
Elongation at break (%)	2.84	≥ 2.5

2.2. Specimen Preparation

In this study, AM, DAM, BFAM, and DBFAM were produced at the optimum asphalt content (OAC). The contents of modifiers and OAC were determined according to our previous researches [27,28]. The experimental design of these four kinds of asphalt mixtures are listed in Table 5. Modifiers are generally added into asphalt mixture by two ways. One is adding modifier into asphalt, making it dispersed evenly with high-speed shear method, and preparing modified asphalt mixture with modified asphalt. The other is adding the modifier in the mixing process of asphalt mixture. The second method is more convenient. In this research, two methods were tested to find the appropriate method. The prepared modified asphalt by first method and modified asphalt mixture by second method were observed by electron microscopy (Phenom Scientific Instrument, Shanghai, China), as shown in Figure 4. It can be found that both methods can produce evenly dispersed modified materials. Basalt fiber has been dispersed from bundles to single strains of fiber. So the both methods were found to be reasonable. Considering that the progress is simple and can be easily used in the practical engineering, the second method is finally used to prepare the samples of composite modified asphalt mixture.

Table 5. Experimental design of asphalt mixtures. AM: control asphalt mixture; DAM: diatomite modified asphalt mixture; BFAM: basalt fiber modified asphalt mixture; DBAFM: diatomite and basalt fiber compound modified asphalt mixture.

Group	Diatomite [1] (%)	Basalt Fiber [2] (%)	OAC (%)
AM	0	0	4.78
DAM	0	6.5	5.12
BFAM	0.25	0	5.09
DBFAM	0.25	6.5	5.22

[1] Diatomite content is the replace volume ratio of the diatomite to-entire filler; [2] Basalt fiber content is the weight ratio of the basalt fiber to asphalt mixture.

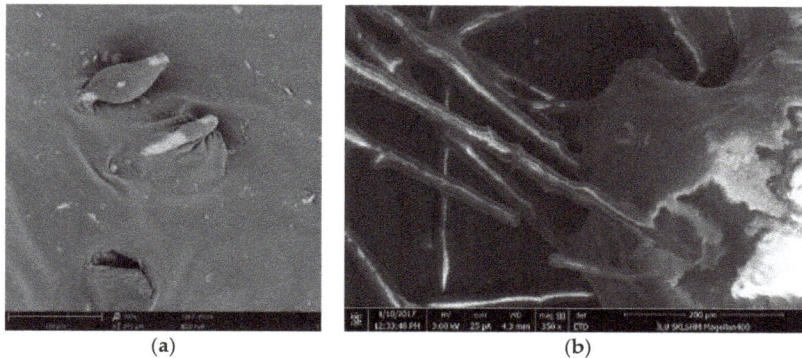

(a)   (b)

Figure 4. Images of modified asphalt and asphalt mixture. (a) SEM images of modified asphalt; (b) SEM images of modified asphalt mixture.

The cylinder specimens (diameter of 101.6 ± 0.2 mm, height of 63.5 mm ± 1.3 mm) were prepared in accordance to the compaction method of Chinese national standard (JTG E20-2011) [29], which were compacted in a Marshall Compactor (San Yu Lu Tong Instrument Co., Ltd., Beijing, China) with 75 blows on each side. Three identical samples were prepared for each kind of asphalt mixture in each test. The preparation procedures of pure asphalt mixture were conducted as the standard regulates. In order to disperse the basalt fiber into asphalt mixture, the aggregates and the basalt fiber were mixed at temperature of 160 °C for 60 s, then the asphalt was added and mixed for 90 s. The diatomite was added with mineral filler and mixed for 90 s.

2.3. Experimental Procedures

2.3.1. Wheel Tracking Test

The pavement in the seasonal frozen regions suffers a high temperature environment in the summer, so adequate high temperature performance of the mixture is required. The wheel tracking test is a widely adopted method to evaluate the high-temperature stability of asphalt mixture [30]. The square slab specimens (length and width of 300 mm, thickness of 50 mm) made of four kinds of asphalt mixtures were prepared in accordance to the Chinese standard [29], which were compacted by rolling compactor for 24 times. Three identical samples were used for each kind of asphalt mixture. The loading time of solid rubber tire applied on the specimen was 60 min with a pressure of 0.7 MPa and running speed of 42 cycle/min. The temperature of the test was 60 °C. The Dynamic stability (*DS*) was used to analysis the high temperature performance of modified asphalt mixture. What's more, permanent deformation was also used in this paper to evaluate the total deformation in whole process. The *DS* can be calculated by the following equation:

$$DS = \frac{(t_2 - t_1) \times N}{d_2 - d_1} \times C_1 \times C_2 \tag{1}$$

where *DS* is the dynamic stability, in times/mm; d_1 and d_2 are the deformation at time of t_1 (45 min) and t_2 (60 min), in mm; C_1 is the equipment coefficient, which is 1.0; C_2 is the specimen coefficient, which is 1.0; *N* is the round trip speed of wheel, which is 42 times/min.

2.3.2. The low Temperature Indirect Tensile Test

The pavement in the seasonal frozen regions suffers a low temperature environment in the winter, so the low temperature performance of the mixture is required. The low-temperature indirect tensile test was a widely used method to evaluate the crack resistance of modified asphalt mixture [31,32]. To control the temperature, the specimens were placed at −10 °C for 5 h, and an environmental chamber was used in the test, as shown in Figure 5. The loading rate was 1 mm/min. The load and vertical deformation on the top surface of the specimen were recorded. The IDT strength and the IDT failure strain were used to evaluate the low-temperature performance of modified asphalt mixture, which can be calculated by following equations:

$$R_T = 0.006287 P_T / h \tag{2}$$

$$\varepsilon_T = X_T \times (0.0307 + 0.0936\mu) / (1.35 + 5\mu) \tag{3}$$

$$X_T = Y_T \times (0.135 + 0.5\mu) / (1.794 - 0.0314\mu) \tag{4}$$

where R_T is the IDT strength, in MPa; ε_T is the IDT failure strain; P_T is the IDT failure load, in N; Y_T is the vertical deformation, in mm; X_T is the horizontal deformation, in mm; μ is the Poisson ratio, which is 0.25 at −10 °C; and *h* is the height of specimen, in mm.

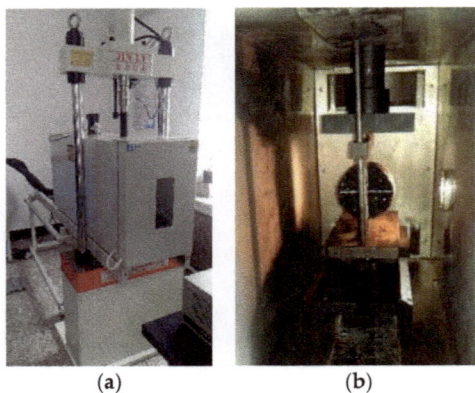

(a) (b)

Figure 5. The low temperature indirect tensile test. (a) Testing equipment. (b) Testing progress.

2.3.3. Moisture Susceptibility Test

The immersion Marshall test and the freeze–thaw splitting test were the widely adopted methods to evaluate the moisture susceptibility of asphalt mixture [33]. The specimens in treated group of immersion Marshall test were immersed in water for 48 h at 60 °C. The specimens in controlled group were only immersed in water for 30 min at 60 °C. The Marshall stability of all specimens were tested, and the residual stability (MS_0) was calculated and used as evaluation index of moisture susceptibility.

The freeze–thaw splitting test was similar as immersion Marshall test. The specimens in treated group were vacuumed in water, kept at $-18\ ^\circ$C for 16 h and thawed in water for 24 h at 60 $^\circ$C. After that, all specimens in two groups were placed in water bath for 2 h at 25 $^\circ$C. The tensile strength ratio (TSR) were calculated as index of moisture susceptibility.

2.3.4. Fatigue Test

Fatigue property of asphalt mixture can reflect the service life of asphalt pavement under vehicle loading. Four-point bending fatigue test was widely used to evaluate the performance of modified asphalt mixture [33,34]. This test was conducted according to EN 12697-24D standard. The NU-14 fatigue testing machine produced by Cooper Research Technology Ltd., Ripley, United Kingdom was used. Slab of $400 \times 300 \times 75$ mm was molded and then compacted under a vibration rolling compactor (Earth produce China Ltd., Guangzhou, China). The slabs were cut into beams of $380 \times 63 \times 50$ mm with the cutting machine (Zhongyu Shunfeng Stone Market Co., Ltd., Changchun, China). Fatigue testing machine and vibration rolling compactor are shown in Figure 6. The samples are shown in Figure 7. Strain control loading mode was adopted to study the fatigue life of the asphalt mixture under a microstrain of 800 μm at 20 $^\circ$C. In addition, sine wave-shaped was applied at frequency of 10 Hz. The stiffness were recorded in the progress of the test. The total number of loading cycles was regarded as the fatigue life when stiffness reaches the 50% of initial stiffness. The initial stiffness was calculated at the 50th loading cycles after the sine wave reach the 800 μm lever.

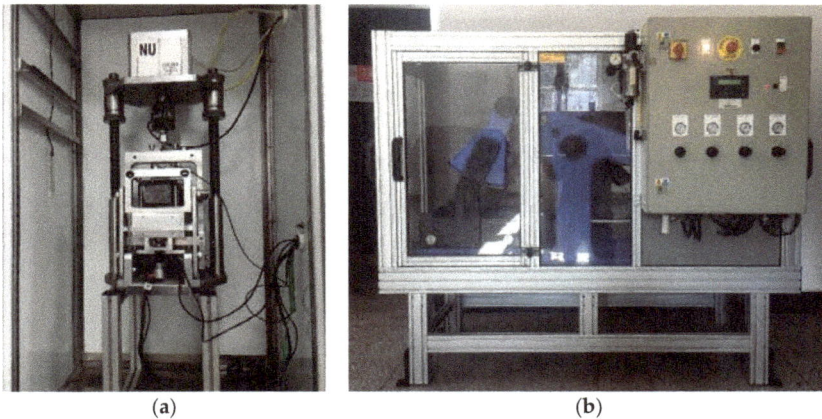

Figure 6. Equipment of Fatigue test. (**a**) Fatigue testing machine; (**b**) Vibration rolling compactor.

Figure 7. Samples of Four-point Bending Fatigue test.

2.3.5. Freeze–Thaw Cycles Test

Asphalt pavement suffers severe F–T damage in seasonal freeze regions. The water in the pores of the asphalt mixture will produce frost heave when it freezes and flow into new micro-cracks when ice thaws. The repeated F–T cycles can cause significant damage to the pavement performance [35,36]. The moisture susceptibility test can't directly reflect the damage caused by F–T cycles. So the F–T cycles test were used in this paper to simulate the environmental condition in seasonal frozen regions. To compare with the moisture susceptibility test, the Marshall stability and Indirect tensile (IDT)strength were also conducted under F–T cycles. The F–T condition was choose as vacuumed in water, freeze for 16h at −18 °C and thaw in water for 8 h at 60 °C. After 3, 6, 9, 12, 15 F–T cycles, the residual stability and tensile strength ratio were calculated, which are similar to moisture susceptibility test. However, the freeze–thaw conditions applied in this test was different from the moisture susceptibility test, in order to reflect the performance of asphalt mixture under the effect of freeze–thaw factor more significantly than moisture factor. So the F–T cycles test was more suitable to simulate the environment of seasonal frozen regions.

3. Results and Discussion

3.1. Wheel Tracking Test Results

The DS and permanent deformation of four kinds of asphalt mixture are shown in Figures 8 and 9.

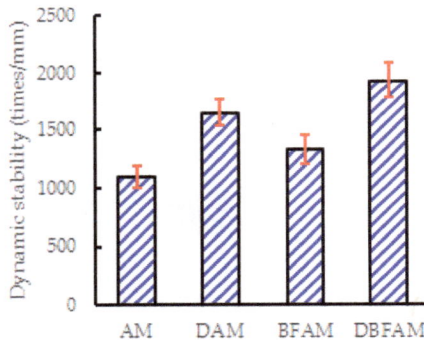

Figure 8. Dynamic stability of asphalt mixtures.

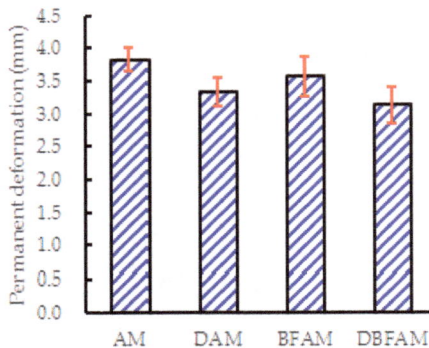

Figure 9. Permanent deformation of asphalt mixtures.

As can be seen from Figures 8 and 9, the addition of diatomite and basalt fiber both improve the *DS* and decrease the permanent deformation. The *DS* of DBFAM is higher than DAM, BFAM and AM, which are increased by 15.2%, 42.4% and 72.7%, respectively, and the permanent deformation of DBFAM is the smallest. The high temperature stability of asphalt mixtures is often affected by many factors. Asphalt is a continuous phase in the mixture and plays an important role in the high temperature performance of the mixture. Diatomite has large specific surface area and high absorptive capacity, which can absorb free asphalt, increase the proportion of structural asphalt and improve the adhesiveness between asphalt and aggregate in the mixture. So diatomite can better improve the high temperature performance of the mixture. The basalt fiber has the effects in improving cohesiveness, toughening, crack arrest, bridging macro-cracks, and asphalt adsorption on asphalt mixture. The dispersed basalt fiber in the asphalt mixture can form a three-dimensional network structure and play a reinforcing role, which improve the high temperature performance of asphalt mixture to a certain extent. What's more, the void of the Marshall specimens of AM, DAM, BFAM, DBFAM are 4.03%, 4.35%, 3.97%, and 4.29% respectively. The reinforcing role of basalt fiber for forming three-dimensional network structure is much obvious than the disadvantage for the increasing void. Under the combined action of basalt fiber and diatomite, the high temperature deformation resistance of the composite modified asphalt mixture is significantly enhanced. Asphalt mixture is greatly affected by the performance of binder. After adding modified materials, the shear viscosity and glass transition temperature of asphalt may also change, which will affect the high and low temperature performance of asphalt mixture [11–14]. In the previous study, the performance of asphalt mastic has been tested, and the high temperature performance has been significantly improved. The wheel tracking test results of modified asphalt mixture can mutually corroborated with the results of modified asphalt mastic in previous study [25].

3.2. The Low-Temperature Indirect Tensile Test Results

The IDT strength and failure strain at -10 °C of four kinds of asphalt mixture are shown in Figures 10 and 11. The addition of basalt fiber improve in the IDT strength and failure strain. The addition of diatomite improved the IDT strength and decreased the failure strain. The strength and failure strain of DBFAM were higher than that of AM. It is clearly that the basalt fiber can improve the low-temperature performance. But the low temperature performance of DAM seems contradictory. The increase in IDT strength may due to the hardening effects of diatomite on asphalt. Diatomite is a porous material with a large specific surface area. After it is added to the asphalt mixture, the bonding ability of the asphalt mortar is improved due to the hardening effects of adsorption between diatomite and asphalt. As a result, the mechanical properties of the mixture will increase. This result meet well with the study of Yang which shows that the indirect tensile (IDT) strength at 20 °C of diatomite modified asphalt mixture has also increased [6]. However, different evaluation indexes for low temperature performance may obtain different conclusions. The deformability at low temperature is also an important low temperature performance. Due to the hardening effects of adsorption between diatomite and asphalt, diatomite also increase the stiffness modulus of the mixture, and its failure strain is not significantly improved, which is not conducive to the low–temperature crack resistance. Considering the indexes in this paper, it can be concluded that the effect of diatomite on the low temperature performance of asphalt mixture is not obvious.

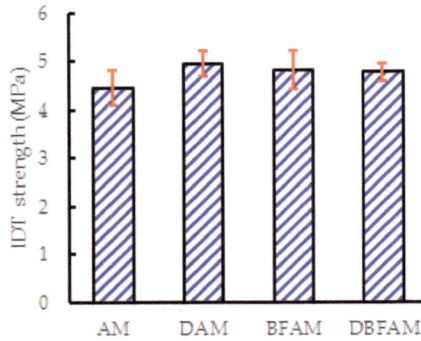

Figure 10. Indirect tensile strength of asphalt mixtures at −10 °C.

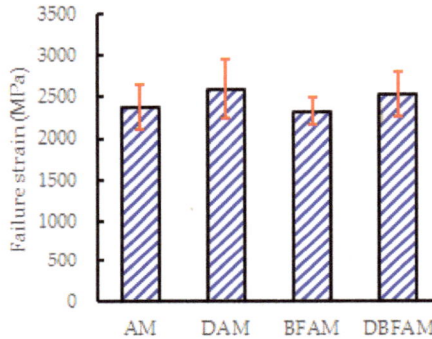

Figure 11. IDT failure strain of asphalt mixtures.

However, it is important to find a kind of modified material that can effectively improve the low temperature performance of asphalt mixture in seasonal frozen regions of China. For example, Jilin province of China is facing severe low-temperature environment, the lowest temperature can reach −30 °C. Basalt fiber can improve the low temperature performance better because of its toughening and preventing cracks and redistribution of stress in asphalt mixture. Therefore, diatomite has little effect on the improvement of low temperature performance than other modified materials. Basalt fiber can be used to improve the low temperature performance of mixture in seasonal frozen regions. The result of DBFAM reflects that adding basalt fiber can solve the low temperature problem of diatomite-modified asphalt mixture. Under the combined action of basalt fiber and diatomite, the low-temperature performance of the composite modified asphalt mixture is significantly enhanced.

3.3. Moisture Susceptibility Test Results

The MS_0 and TSR for each kinds of asphalt mixture are shown in Figures 12 and 13. The addition of diatomite improve the TSR and MS_0, which means a clear improvement in resistance to the moisture susceptibility. But the addition of basalt fiber decreases the TSR and MS_0. It seems that adding basalt fiber would decrease the resistance to moisture susceptibility. It is noted that this problem has also appeared in the research of other fibers modified asphalt mixtures, such as glass fiber [37], polypropylene and aramid fibers [38]. The results may due to the reason that the asphalt mixtures become harder to compact after adding hard fiber, and the void of asphalt mixtures would also increase. Higher voids make BFAM more susceptible to moisture damage than AM. However, DBFAM has a high resistance to rutting and moisture susceptibility than AM. The reinforced adhesion effects of diatomite in asphalt can reduce the moisture damage in the bounding between asphalt mortar and aggregate, and can solve the moisture susceptibility problem of BFAM.

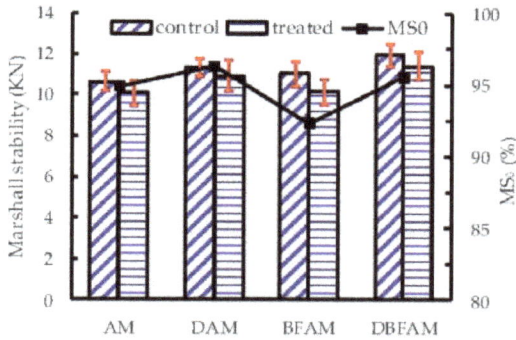

Figure 12. MS_0 of asphalt mixtures.

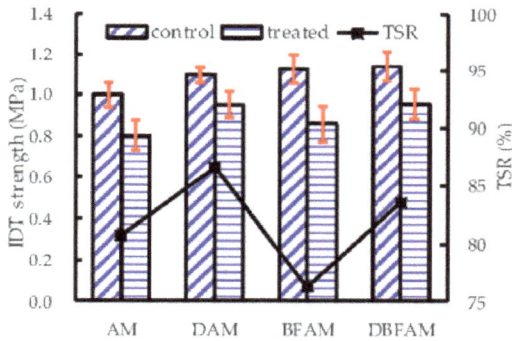

Figure 13. Tensile strength ratio of asphalt mixtures.

3.4. Fatigue Performance Test Results

Fatigue test can reflect the performance degradation under repeated loading and show the effects of basalt fiber and diatomite on fatigue life. The changes of bending stiffness modulus under loading cycles is shown in the Figure 14.

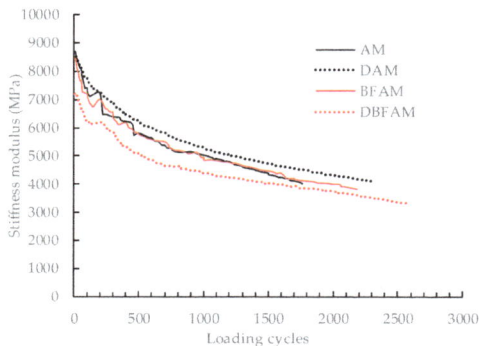

Figure 14. Stiffness modulus under fatigue test.

Based on the data in Figure 14, DBFAM has the highest fatigue life of 2582 cycles, followed by DAM of 2300 cycles, BFAM of 2182 cycles, while the AM has the lowest one of 1762 cycles. The fatigue life of DBFAM is 46.5% longer than AM. The fatigue lives of asphalt mixtures in this research are lower than other researches by the high strain levels and shape of the wave [6,26,33]. Because the sine wave-shaped loading was applied in this study. The fatigue damage caused by sine wave-shaped was more serious than that of half-sine wave-shape loading. The bending stiffness modulus decreases under the loading cycles. It can be clearly seen from the figures that the fatigue process of asphalt mixture mainly goes through two stages before the 50% initial stiffness modulus. In the first stage, the bending stiffness modulus of the four materials decrease sharply. In the second stage, the modulus of flexural and tensile stiffness decreased slowly and approach linearity. In the first stage, the sharp decrease of bending stiffness modulus is mainly due to the material restructuring in the test specimens under alternating loads. The second stage is the main stage of fatigue damage of the materials. The fatigue damage of the specimens under alternating loads occurs during the initiation and development of micro-cracks. The diatomite and basalt fiber can both enhance the bounding force and adhesive strength between asphalt and aggregates. So the two kinds of modifier can both improve the fatigue performance of mixture at high strain levels.

3.5. Freeze–Thaw Cycles Test Results

The F–T cycles test is similar to the moisture susceptibility test. But this method is more reasonable to reflect the damage caused by the repeated action of water icing and water erosion under the F–T cycle. The changes of stability and IDT strength for different asphalt mixtures after F–T cycles are shown in Figures 15–18.

It can be seen from Figures 15–18 that the Marshall stability and IDT strength of each type of asphalt mixture is continuously reducing, and the loss ratio of them are continuously increasing under the F–T cycles. The MS and IDT strength of BFAM, DAM, DBFAM are still high than that of AM under F–T cycles. The loss of MS for the DAM and DBFAM are obviously smaller than that of AM. As for the BFAM, the loss is slightly less than AM. The results show that the addition of diatomite can effectively improve the resistance to F–T cycles, while the basalt fiber also has some improvement but not obviously. The results of IDT strength under F–T cycles are the same as that of MS.

The change of performance can show the damage caused by F–T cycles. As the number of F–T cycles increases, the performance of all asphalt mixtures decreases. At the beginning of the F–T cycle, the performance decreases rapidly. After 6 F–T cycles, the performance degradation rate slows down. At the beginning of the F–T cycle, under the effect of freezing, the void of the specimen increases, and the mixture becomes loose. The moisture gradually invades into the internal void and the asphalt film, and the mechanical properties are significantly reduced. As the F–T cycles continues to increase, the voids are gradually connected, which creates communication channel. The pressure generated during the freezing of the void water can be released through this channel. Therefore, the increasing rate of damage is slowed down. The addition of diatomite can reinforce adhesion between asphalt and aggregate, which reduce the moisture damage in the bounding between asphalt mortar and aggregate. It is more difficult for moisture to intrude into internal voids and asphalt film. So diatomite can effectively improve the resistance to F–T cycles. The basalt fiber may also has a little improving due to the bridging and crack arrest effect on limiting the crack propagation during the freezing process. For the effect of basalt fiber on air voids, the freezing effect will increase, but for the toughening and crack resistance effect of basalt fibers on inhibiting the expansion of cracks, and ultimately the resistance to F–T cycles does not decrease. In addition, the diatomite can improve the resistance to F–T cycles. Therefore, the method of composite modification is reasonable. However, The DBFAM has the high resistance to F–T cycles and can be well used in seasonal frozen regions.

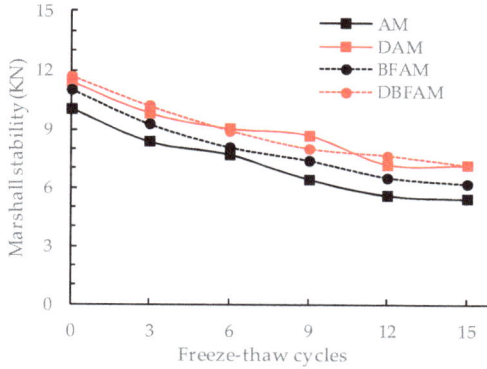

Figure 15. Marshall stability under freeze–thawcycles.

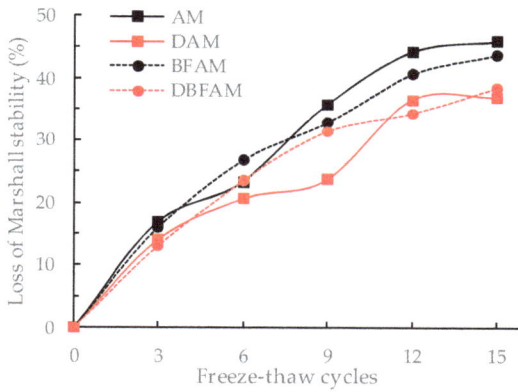

Figure 16. Loss of Marshall stability under F–T cycles.

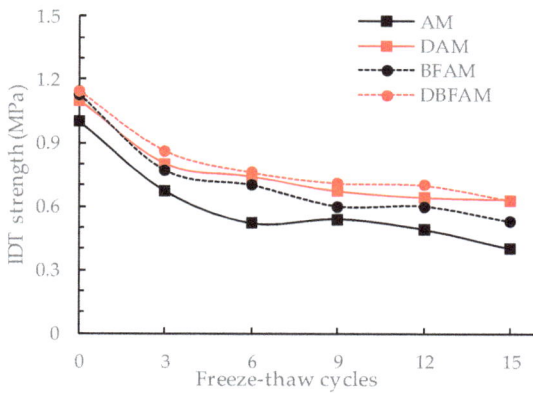

Figure 17. IDT strength under F–T cycles.

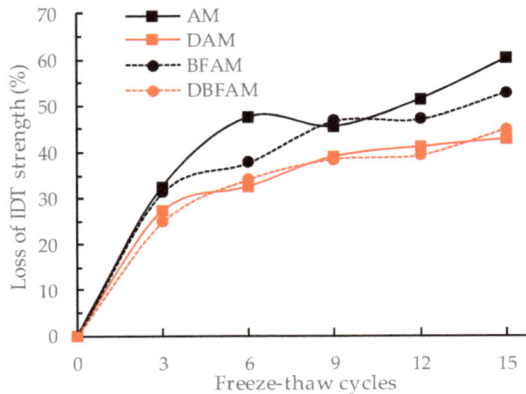

Figure 18. Loss of IDT strength under F–T cycles.

4. Conclusions

This paper investigated the performance of asphalt mixtures modified by basalt fiber and diatomite. The effect of diatomite and basalt fiber on the performance of asphalt mixture were studied by means of wheel tracking test, low temperature IDT test, moisture susceptibility test, four-point bending fatigue test and F–T cycles test. Based on the above analysis, the following conclusions could be drawn:

- The addition of basalt fiber and diatomite can both improve the resistance to rutting than AM. DBFAM has highest dynamic stability, which means DBFAM can better resist rutting or permanent deformation at high temperature in asphalt pavements.
- The addition of basalt fiber can obviously improve the resistance to low-temperature cracking than AM while diatomite is not obviously. DBFAM has high IDT strength and IDT failure strain, which means DBFAM can better resist cracking at low temperature in seasonal frozen regions.
- The addition of diatomite can increase the resistance to moisture damage while basalt fiber can't. DBFAM have a higher resistance to moisture susceptibility than AM, which means DBFAM can better resist moisture damage in asphalt pavements.
- The addition of basalt fiber and diatomite can both improve the fatigue cracking resistance than AM. DBFAM has longest fatigue life than others under high strain lever, which means DBFAM has longer service life under the repeated vehicle loading in asphalt pavements.
- The addition of diatomite can obviously increase the resistance to F–T cycles while basalt fiber is not obviously. DBFAM have a higher resistance to F–T cycles than AM, which means DBFAM can better resist the freeze–thaw cycles damage in seasonal frozen regions.

Author Contributions: Conceptualization, Y.C. and D.Y.; Methodology, D.Y. and Y.G.; Validation, Y.C.; Formal Analysis, J.T. and C.Z.; Investigation, D.Y.; Writing-Original Draft Preparation, D.Y.; Writing-Review and Editing, J.T., W.W. and C.Z.; Project Administration, Y.C.; and Funding Acquisition, Y.C. and Y.G.

Funding: This research was funded by National Natural Science Foundation of China (grant number 51678271), Science Technology Development Program of Jilin Province (grant number 20160204008SF), and Transportation Science & Technology Project of Jilin Province (grant number 2015-1-13).

Acknowledgments: This research was funded by National Natural Science Foundation of China, Science Technology Development Program of Jilin Province, and Transportation Science & Technology Project of Jilin Province. This financial support is gratefully acknowledged.

Conflicts of Interest: The authors declare no conflict of interest.

References

1. De Almeida, A.J.; Momm, L.; Triches, G.; Shinohara, K.J. Evaluation of the influence of water and temperature on the rheological behavior and resistance to fatigue of asphalt mixtures. *Constr. Build. Mater.* **2018**, *158*, 401–409. [CrossRef]
2. Cui, P.; Xiao, Y.; Fang, M.; Chen, Z.; Yi, M.; Li, M. Residual fatigue properties of asphalt pavement after long-term field service. *Materials* **2018**, *11*, 892. [CrossRef] [PubMed]
3. Yu, W.; Han, F.; Yi, X.; Liu, W.; Hu, D. Cut-slope icing prevention: Case study of the seasonal frozen area of western China. *J. Cold Reg. Eng.* **2016**, *30*. [CrossRef]
4. Wei, H.; Li, Z.; Jiao, Y. Effects of diatomite and SBS on freeze–thaw resistance of crumb rubber modified asphalt mixture. *Adv. Mater. Sci. Eng.* **2017**, *2017*, 7802035. [CrossRef]
5. Ma, H.; Wang, D.; Zhou, C.; Feng, D. Calibration on MEPDG low temperature cracking model and recommendation on asphalt pavement structures in seasonal frozen region of China. *Adv. Mater. Sci. Eng.* **2015**, *19*, 259–272. [CrossRef]
6. Moreno-Navarro, F.; Sol-Sánchez, M.; Tomás-Fortún, E.; Rubio-Gámez, M.C. High-modulus asphalt mixtures modified with acrylic fibers for their use in pavements under severe climate conditions. *J. Cold Reg. Eng.* **2016**, *30*. [CrossRef]
7. Ye, Q.; Wu, S.; Li, N. Investigation of the dynamic and fatigue properties of fiber-modified asphalt mixtures. *Int. J. Fatigue* **2009**, *31*, 1598–1602. [CrossRef]
8. Chomiczkowalska, A. Laboratory testing of low temperature asphalt concrete produced in foamed bitumen technology with fiber reinforcement. *Bull. Pol. Acad. Sci.-Tech.* **2017**, *65*, 779–790.
9. Sun, Y.S.; Chen, X.L.; Han, Y.X.; Zhang, B. Research on Performance of the Modified Asphalt by Diatomite-Cellulose Composite. *Adv. Mater. Res.* **2010**, *158*, 211–218. [CrossRef]
10. Jiang, L.; Liu, Q.L. Application of Diatomite Modified Asphalt. *Appl. Mech. Mater.* **2013**, *477–478*, 959–963. [CrossRef]
11. Fallah, F.; Khabaz, F.; Kim, Y.-R.; Kommidi, S.R.; Haghshenas, H. Molecular dynamics modeling and simulation of bituminous binder chemical aging due to variation of oxidation level and saturate-aromaticresin-asphaltene fraction. *Fuel* **2019**, *237*, 71–80. [CrossRef]
12. Khabaz, F.; Khare, R. Glass transition and molecular mobility in styrene–butadiene rubber modified asphalt. *J. Phys. Chem. B* **2015**, *119*, 14261–14269. [CrossRef] [PubMed]
13. Khabaz, F.; Khare, R. Molecular simulations of asphalt rheology: Application of time–temperature superposition principle. *J. Rheol.* **2018**, *62*, 941–954. [CrossRef]
14. Cong, P.L.; Chen, S.F.; Chen, H.X. Effects of diatomite on the properties of asphalt binder. *Constr. Build. Mater.* **2012**, *30*, 495–499. [CrossRef]
15. Cheng, Y.C.; Tao, J.L.; Jiao, Y.B.; Guo, Q.L.; Li, C. Influence of Diatomite and Mineral Powder on Thermal Oxidative Ageing Properties of Asphalt. *Adv. Mater. Sci. Eng.* **2015**, *2015*, 947834. [CrossRef]
16. Bao, Y.N. Study of Asphalt Mixture Modified by Diatomite. Master's Thesis, Chang'an University, Xi'an, China, 2005.
17. Yang, C.; Xie, J.; Zhou, X.; Liu, Q.; Pang, L. Performance Evaluation and Improving Mechanisms of Diatomite-Modified Asphalt Mixture. *Materials* **2018**, *11*, 686. [CrossRef] [PubMed]
18. Wang, W.S.; Cheng, Y.C.; Tan, G.J. Design Optimization of SBS-Modified Asphalt Mixture Reinforced with Eco-Friendly Basalt Fiber Based on Response Surface Methodology. *Materials* **2018**, *11*, 1311. [CrossRef] [PubMed]
19. Wang, D.; Wang, L.B.; Gu, X.Y.; Zhou, G.Q. Effect of Basalt Fiber on the Asphalt Binder and Mastic at Low Temperature. *J. Mater. Civ. Eng.* **2013**, *25*, 355–364. [CrossRef]
20. Gu, X.Y.; Xu, T.T.; Ni, F.J. Rheological behavior of basalt fiber reinforced asphalt mastic. *J. Wuhan Univ. Technol.* **2014**, *29*, 950–955. [CrossRef]
21. Zhang, X.Y.; Gu, X.Y.; Lv, J.X.; Zhu, Z.K.; Zou, X.Y. Numerical analysis of the rheological behaviors of basalt fiber reinforced asphalt mortar using ABAQUS. *Constr. Build. Mater.* **2017**, *157*, 392–401. [CrossRef]
22. Zhang, X.; Gu, X.Y.; Lv, J.X.; Zou, X.Y. 3D numerical model to investigate the rheological properties of basalt fiber reinforced asphalt-like materials. *Constr. Build. Mater.* **2017**, *138*, 185–194. [CrossRef]
23. Gao, C.M. Microcosmic Analysis and Performance Research of Basalt Fiber Asphalt Concrete. Ph.D. Thesis, Jilin University, Changchun, China, 2012.

24. Zhao, L.H. Study on the Influence Mechanism of Basalt Fiber on Asphalt Mixture Property. Ph.D. Thesis, Dalian University of Technology, Dalian, China, 2013.

25. Cheng, Y.C.; Zhu, C.F.; Tan, G.J.; Lv, Z.H.; Yang, J.S.; Ma, J.S. Laboratory study on properties of diatomite and basalt fiber compound modified asphalt mastic. *Adv. Mater. Sci. Eng.* **2017**, *3*, 4175167. [CrossRef]

26. Davar, A.; Tanzadeh, J.; Fadaee, O. Experimental evaluation of the basalt fibers and diatomite powder compound on enhanced fatigue life and tensile strength of hot mix asphalt at low temperatures. *Constr. Build. Mater.* **2017**, *153*, 238–246. [CrossRef]

27. Zhu, C.F. Research on Road Performance and Mechanical Properties of Diatomite-Basalt Fiber Compound Modified Asphalt Mixture. Ph.D. Thesis, Jilin University, Changchun, China, 2018.

28. Cheng, Y.C.; Di, Y.; Tan, G.J.; Zhu, C.F. Low-Temperature Performance and Damage Constitutive Model of Eco-Friendly Basalt Fiber–Diatomite-Modified Asphalt Mixture under Freeze–thaw Cycles. *Materials* **2018**, *11*, 2148. [CrossRef] [PubMed]

29. JTG E20-2011. *Standard Test Methods of Asphalt and Asphalt Mixtures for Highway Engineering*; Ministry of Transport: Beijing, China, 2011. (In Chinese)

30. Ma, T.; Zhang, D.; Zhang, Y.; Wang, S.; Huang, X. Simulation of wheel tracking test for asphalt mixture using discrete element modelling. *Road Mater. Pavement* **2018**, *19*, 367–384. [CrossRef]

31. Guo, Q.L.; Li, L.L.; Cheng, Y.C.; Jiao, Y.B.; Xu, C. Laboratory evaluation on performance of diatomite and glass fiber compound modified asphalt mixture. *Mater. Des.* **2015**, *66*, 51–59. [CrossRef]

32. Krcmarik, M.; Varma, S.; Kutay, M.E.; Jamrah, A. Development of predictive models for low-temperature indirect tensile strength of asphalt mixtures. *J. Mater. Civ. Eng.* **2016**, *28*. [CrossRef]

33. Cai, L.; Shi, X.; Xue, J. Laboratory evaluation of composed modified asphalt binder and mixture containing nano-silica/rock asphalt/SBS. *Constr. Build. Mater.* **2018**, *172*, 204–211. [CrossRef]

34. Omrani, H.; Tanakizadeh, A.; Ghanizadeh, A.R.; Fakhri, M. Investigating different approaches for evaluation of fatigue performance of warm mix asphalt mixtures. *Mater. Struct.* **2017**, *50*, 149. [CrossRef]

35. Özgan, E.; Serin, S. Investigation of certain engineering characteristics of asphalt concrete exposed to freeze–thaw cycles. *Cold Reg. Sci. Technol.* **2013**, *85*, 131–136. [CrossRef]

36. Yan, K.Z.; Ge, D.D.; You, L.Y.; Wang, X.L. Laboratory investigation of the characteristics of SMA mixtures under freeze–thaw cycles. *Cold Reg. Sci. Technol.* **2015**, *119*, 68–74. [CrossRef]

37. Fakhri, M.; Hosseini, S.A. Laboratory evaluation of rutting and moisture damage resistance of glass fiber modified warm mix asphalt incorporating high rap proportion. *Constr. Build. Mater.* **2017**, *134*, 626–640. [CrossRef]

38. Klinsky, L.M.G.; Kaloush, K.E.; Faria, V.C.; Bardini, V.S.S. Performance characteristics of fiber modified hot mix asphalt. *Constr. Build. Mater.* **2018**, *176*, 747–752. [CrossRef]

materials

MDPI

Article

Performance Evaluation and Improving Mechanisms of Diatomite-Modified Asphalt Mixture

Chao Yang, Jun Xie *, Xiaojun Zhou, Quantao Liu and Ling Pang

State Key Laboratory of Silicate Materials for Architectures, Wuhan University of Technology,
Wuhan 430070, China; hbyangc@whut.edn.cn (C.Y.); zhouxj25@whut.edu.cn (X.Z.);
liuqt@whut.edu.cn (Q.L.); lingpang@whut.edu.cn (L.P.)
* Correspondence: xiejun3970@whut.edu.cn

Received: 3 April 2018; Accepted: 25 April 2018; Published: 27 April 2018

Abstract: Diatomite is an inorganic natural resource in large reserve. This study consists of two phases to evaluate the effects of diatomite on asphalt mixtures. In the first phase, we characterized the diatomite in terms of mineralogical properties, chemical compositions, particle size distribution, mesoporous distribution, morphology, and IR spectra. In the second phase, road performances, referring to the permanent deformation, crack, fatigue, and moisture resistance, of asphalt mixtures with diatomite were investigated. The characterization of diatomite exhibits that it is a porous material with high SiO_2 content and large specific surface area. It contributes to asphalt absorption and therefore leads to bonding enhancement between asphalt and aggregate. However, physical absorption instead of chemical reaction occurs according to the results of FTIR. The resistance of asphalt mixtures with diatomite to permanent deformation and moisture are superior to those of the control mixtures. But, the addition of diatomite does not help to improve the crack and fatigue resistance of asphalt mixture.

Keywords: diatomite; styrene–butadiene–styrene (SBS) modified bitumen; diatomite-modified asphalt mixture

1. Introduction

Due to its good driving comfort, fast construction speed, convenient maintenance, and easy recycling, asphalt pavement prevails in highway engineering [1,2]. The Chinese government has been committed to developing fully the transportation industry in the past few decades. By the end of 2016, the mileage of expressways in China exceeded 130,000 km, of which more than 90% is asphalt pavement [3].

But during service periods, major damage inevitably occurs in the asphalt pavement, including rutting, cracking, and permanent deformation [4]. It is caused by the degradation of asphalt, including bonding strength breaking, high-temperature softening, low-temperature embrittlement, and heat aging [5].

In order to mitigate pavement damage, it is essential to improve the full temperature range performance of asphalt during the service period [6]. In recent years, a variety of modifiers, including organic and inorganic materials have been introduced. Researchers have conducted various studies to investigate their effects on the improvement of road performance. Amir [7] investigated the effect of temperature on the toughness index and fatigue properties of styrene–butadiene–styrene (SBS), a styrene-butadiene block copolymer-modified asphalt mixture created by a Universal Test Machine (UTM) apparatus. The results suggest that the SBS can increase the indirect tensile strength of an asphalt mixture at high temperatures. Taher [8] evaluated the permanent deformation characteristics of polyethylene terephthalate (PET)-modified asphalt mixtures. The results indicate that mixtures with PET modification have better resistance against permanent deformation. However, its price is too high to promote. Mahyar [4] investigated the effects of rice husk ash (RHA) as an asphalt modifier on binders and mixtures. The results suggest that the properties of the binders and mixtures were enhanced remarkably with the addition of RHA; although, the preparation process of RHA-modified asphalt is quite complex. Paravita [9] investigated the effect of crumb rubber on the properties of asphalt mixtures. The crumb rubber-modified asphalt mixture exhibited better mechanical properties. But, the modified mixture showed uncontrolled volume properties, which may affect durability. Erol [10] evaluated the effect of nano-clay materials on the enhancement of the mechanical properties of an asphalt mixture. The mixtures with nano-clay modification exhibit acceptable water damage resistance and rutting resistance except for its fatigue performance.

Nowadays, the SBS-modified bitumen is widely used. But, as for the other modifiers, they are either too expensive or show fewer improving effects according to the literature. New modifiers with low prices, easy modification procedures, and good modification effects are still in urgent demand.

As a non-metallic mineral, diatomite is an inorganic natural resource in large reserve [11,12]. Researchers have tried to introduce it into asphalt mixtures for its rough surface, high hardness, acid and alkali resistance, wear resistance, anti-skidding, porous structure, unique component activity, stable properties, etc. [13,14]. Alejandra's [15] research indicates that the fatigue resistance of a binder with 4% diatomite content is improved. Cong [12] investigated the physical properties, dynamic rheological behaviors, storage stability, and aging properties of different contents of modified asphalt binders. The results suggest that both viscosity and complex modulus of binders increase rapidly at high temperatures with the addition of diatomite. Compared with base asphalt binders, the resistance of modified asphalt binders to high-temperature deformation and low-temperature cracking has been greatly improved.

Meanwhile, the pavement performances of diatomite-modified asphalt mixtures have been studied. Zhang [16] and Tan [17] evaluated the effect of diatomite on the low-temperature performance of asphalt mixtures. The results indicate that the bending strain energy density of a mixture increases with the addition of diatomite. Chen's [18] research shows that the dynamic stability of an SBS-modified asphalt mixture is the greatest, followed by the diatomite-modified asphalt mixture and the controlled asphalt mixture. Wei [19] stated that the anti-icing performance of diatomite-modified asphalt mixtures was improved. Chen [20] suggested that the fatigue life of modified asphalt mixtures with diatomite was certainly improved under the same stress levels. Bao [21] indicates that diatomite can improve the stability and splitting strength of an asphalt mixture.

Based on the findings mentioned above, it can be found that diatomite can improve the performance of asphalt mixtures with respect to rutting resistance at high temperatures and splitting or crack resistance at low temperatures. Nevertheless, the improving mechanism necessitates further systematical research.

In this paper, X-ray Diffraction (XRD), X-ray Fluorescence (XRF), particle-size and pore-size analyzer, Scanning Electron Microscope (SEM), and Fourier-Transform Infrared Spectrometer (FTIR) tests were employed to evaluate the characteristics of diatomite. The effects of diatomite on the pavement performance of a modified asphalt mixture were also investigated. In particular, low-temperature properties were given much importance, because inorganic fillers seldom have significant effects on the low-temperature aspects. Based on the tests, we evaluated how diatomite affects the performance of asphalt mixtures.

2. Materials

Base asphalt with 60–80 penetration was procured from Ezhou, China. Two SBS-modified bitumen samples were procured from Ezhou (EZ) and Inner Mongolia (IM), China, respectively. Their properties are presented in Tables 1 and 2, respectively. Basalt aggregate and limestone filler were also included in the asphalt mixtures. The fundamental properties of diatomite and limestone powder are listed in Table 3.

Table 1. Measured values of base asphalt.

Indexes	Measured Values	Specification
Specific gravity	1.034	N/A
Penetration at 25 °C (0.1 mm)	63	60–80
Ductility, 5 cm/min, 15 °C (cm)	>150	≥100
Softening point (°C)	48	≥46
Apparent viscosity, 135 °C (Pa·s)	0.48	≤1.5
Loss on heating (%)	+0.09	≤±0.8

Table 2. Measured values of styrene–butadiene–styrene (SBS)-modified bitumen.

Indexes	Measured Values		Specification
	EZ	IM	
Specific gravity	1.032	1.039	N/A
Penetration at 25 °C (0.1 mm)	55	68	30–80
Ductility, 5 cm/min, 5 °C (cm)	56	49	≥30
Softening point (°C)	69	52	≥50
Apparent viscosity, 135 °C (Pa·s)	0.95	1.23	≤3
Loss on heating (%)	+0.32	+0.56	≤±1

Table 3. Fundamental properties of limestone powder and diatomite.

Indexes	Diatomite	Limestone Powder
Color	light yellow	white
Apparent density (g/cm^3)	2.18	2.67
Water content (%)	1.81	0.55
Specific surface area (m^2/g)	29.35	1.47
Hydrophilic coefficient	0.5	0.6

3. Experimental Methods

3.1. Characteristic Methods for Diatomite and Asphalt Binder

The mineralogy, chemical composition and microscopic surface characteristic of diatomite were tested by D8 Advance X-ray Diffraction (XRD, Brooke AXS, Berlin, Germany), Axios X-ray Fluorescence (XRF, PANalytical B.V., Amsterdam, The Netherlands), and JSM-IT300 Scanning Electron Microscope (SEM, NEC Electronics Corporation, Tokyo, Japan), respectively. A Mastersizer 3000 laser particle analyzer (Malvern Instruments, Malvern, UK) was used to determine the particle size distributions of fillers.

A TriStarII3020 multi-channel ratio surface area and aperture analyzer (Micromeritics, Atlanta, GA, USA) was applied to investigate surface area and mesoporous distribution. The specific surface area was determined by the Brunauer–Emmett–Teller (BET) test method. Nitrogen was used as adsorbent, and helium or hydrogen was used as a carrier gas. The two gases were mixed at a certain proportion. When it achieved the specified relative pressure, the gas flowed through solid material. The sample was adsorbed physically by nitrogen. When the liquid nitrogen was taken away, the adsorbed nitrogen was desorbed, and a desorption peak appeared. Finally, calibration peaks were obtained by injecting pure nitrogen of known volume into the mixture. According to the peak area of the calibration peaks and desorption peaks, the adsorption amount under the relative pressure was calculated. By changing the mixing ratio of the nitrogen and carrier gas, the adsorption capacity of several nitrogen relative pressures could be determined. The specific surface area could be calculated according to the following formula:

$$\frac{p}{V(p_0 - p)} = \frac{1}{V_m C} + \frac{(C-1)}{V_m C}\frac{p}{p_0} \tag{1}$$

$$S_g = \frac{V_m N_A A_m}{2240W} \times 10^{-18} \qquad (2)$$

where p = partial pressure of nitrogen; p_0 = saturated vapor pressure of liquid nitrogen; V_m = amount of gas required to form a monolayer; V = total volume of adsorbed gas; C = constant; S_g = specific surface area; N_A = Avogadro constant; A_m = cross-sectional area of adsorbed gas; and W = sample quality.

The Nicolet6700 Fourier-Transform Infrared Spectrometer (FTIR, Thermo Electron Scientific Instruments, Columbia, IN, USA) was used to obtain the IR spectra of the base asphalt, diatomite-modified asphalt binder, and diatomite. Binder specimens were made with base asphalt and 12% weight-based diatomite after constant stirring at 120 °C for 0.5 h. The test procedure was as follows: The infrared light of a certain wavelength was irradiated to the measured substance. If the radiant energy was equal to the energy level difference between the ground state and the excited state of the molecular vibration, the molecule could absorb the infrared light energy. The vibration transitioned from the ground state to the excited state. The instrument recorded the degree of infrared light absorption with the wavelength of the change function to form the infrared spectrum. When detecting the asphalt, it was dissolved in CS_2 in a solution, and then, the KBr tablet method was used to prepare the sample. Because of the high volatility of CS_2, the solution should be equipped with the current test. The scanning wave number range was 500~4000 cm^{-1}, and the scanning frequency was 64 times.

3.2. Performance Evaluation of Mixtures

3.2.1. Preparation of Asphalt Mixture

Four mixtures—EZ-SBS-modified, IM-SBS-modified, diatomite-modified, and base asphalt—mixtures were studied. All of them were prepared with the same gradation at optimum asphalt content. The gradation was designed with 13.2-mm nominal maximum size. Figure 1 shows the gradation. Two blending methods, namely direct and indirect blending methods, were used in the preparation of the diatomite-modified asphalt mixture. When the direct blending method was used, the diatomite was added to the mixture of asphalt and aggregate with mineral powder. When the indirect mixing method was used, the diatomite-modified asphalt binder was prepared before the preparation of the mixture [22]. Yin's [23] research showed that the two blending methods led to approximately the same mix effect. Chen [20] determined that the optimum amount of compound diatomite modifier was 10%, while at this content, the Marshall specimens showed the best performances. Zhang [16] concluded that the optimum dosage of diatomite was 13% through the analysis of the low-temperature performance of binders and mixtures. Hence, 10–13% was a reasonable range of dosage. According to the literature review, the direct blending method was chosen, with 12% (mass ratio of diatomite and asphalt) diatomite content for the specimens' preparation.

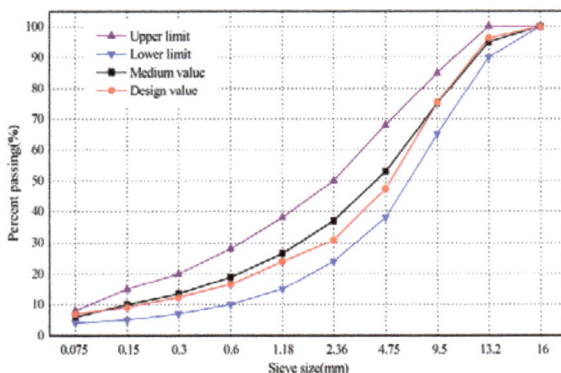

Figure 1. Asphalt concrete (AC)-13 gradation design used in this paper.

3.2.2. Low-Temperature Performance

A three-points bending test is the common approach for evaluating low-temperature cracking performance of asphalt mixtures. The test device is shown in Figure 2. Beam specimens with 250 ± 2.0 mm length, 30 ± 2.0 mm width, and 35 ± 2.0 mm height were used [24]. There were five parallel specimens in each type of mixture. The test was carried out on a Universal Testing Machine (UTM)-25 from Melbourne, Australia, and the experimental parameters were $-10\ ^\circ$C of temperature and a loading rate of 50 mm/min. The bending strain energy and bending strain energy density in this study were used to evaluate the four mixtures [25], and formulas for the calculations were as follows:

$$W = \int_s^{s_0} F ds \tag{3}$$

where W = bending strain energy; F = force; s = displacement; and s_0 = critical displacement.

$$\frac{dw}{dv} = \int_0^{\varepsilon_0} \sigma_{ij} d\varepsilon_{ij} \tag{4}$$

where dw/dv = bending strain energy density; σ_{ij} = stress component, ε_{ij} = strain component, and ε_0 = critical strain.

Figure 2. Three-points bending test set-up.

3.2.3. High-Temperature Performance

A rutting test is currently used to evaluate high-temperature stability. The size of the slab specimens was $300 \times 300 \times 50$ mm. The rolling speed of the wheel was 42 times/min, and the load was 0.7 MPa. The test time was 1 h, and the test temperature was $60\ ^\circ$C for a standard wheel tracking test.

3.2.4. Fatigue Performance

A four-points bending fatigue test was conducted by UTM-25 (IPC Global, Melbourne, Australia) as shown in Figure 3. The length, width, and height of the beam specimens were 380 ± 2.0 mm,

63.5 ± 2.0 mm, and 50 ± 2.0 mm, respectively. The test temperature was 15 °C. A haversine load pulse at 10 Hz was used. In the fatigue test, strain control loading mode was adopted to study the fatigue life of the asphalt mixture under a microstrain, such as 500 µε, 600 µε, 700 µε, and 800 µε.

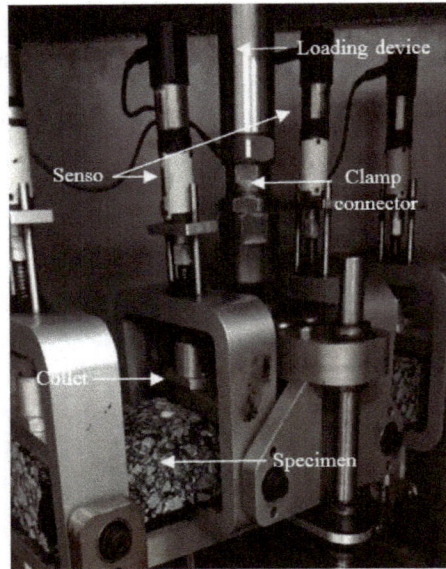

Figure 3. Four-points bending system for the fatigue test.

3.2.5. Water Stability

Water stability was used to evaluate the ability of asphalt to stripe from the aggregate surface when the asphalt mixture was subjected to water erosion. In this paper, the Marshall stability test and indirect tensile strength test were used to assess water stability.

4. Results and Discussion

4.1. Characteristics of Diatomite and Binder

4.1.1. Mineralogical Properties of Diatomite and Limestone Powder

Figure 4 shows the XRD patterns of diatomite and limestone powder. A strong diffraction peak appears at $2\theta = 26.66°$ in diatomite, which represents the mineral phase of SiO_2. From the retrieved mineral composition, it can be concluded that the group OH^- is contained in diatomite. It is an essential reason for the surface activity and absorptivity of diatomite [26]. For the XRD pattern for limestone, there is a very strong diffraction peak of $CaCO_3$ at $2\theta = 29.43°$. It implies the extremely high content of $CaCO_3$.

Table 4 shows the chemical components of diatomite. Silicon (Si) displays the highest contributions, followed by Al and Fe. The content of SiO_2 is one of the most important parameters by which to evaluate the quality of diatomite [27]. In particular, the surface of diatomite has very strong adhesion and adhesive strength due to the presence of these amorphous SiO_2. The inert nature of SiO_2 can also reduce the transmission speed in pavement and endow the pavement with heat insulation functions [6].

Table 4. Chemical components of diatomite and limestone powder.

Compound		SiO₂	CaO	Al₂O₃	Fe₂O₃	K₂O	MgO	TiO₂	Loss	Others
Content (wt %)	Diatomite	62.21	0.36	12.07	4.52	1.53	1.10	0.70	15.89	1.62
	Limestone powder	1.79	55.46	0.18	0.09	-	0.52	-	41.72	0.24

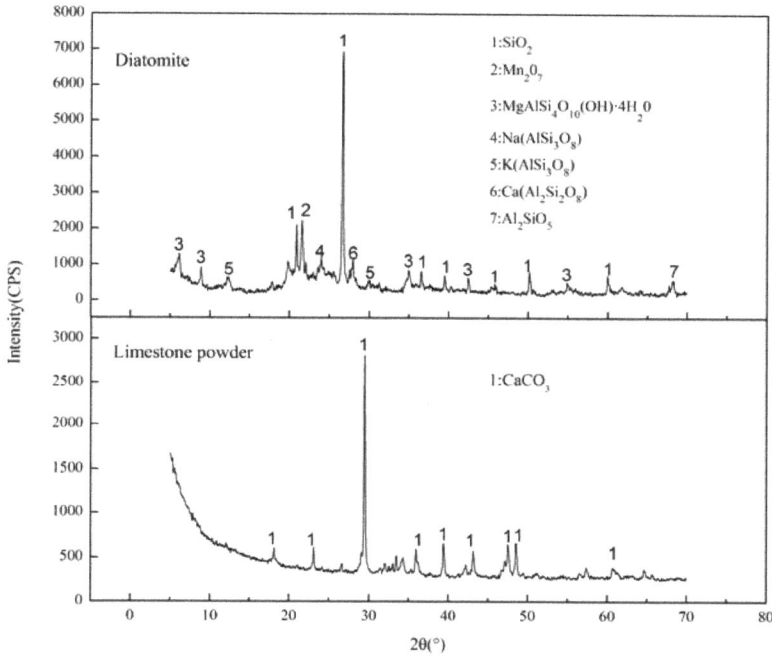

Figure 4. XRD pattern of the diatomite and limestone powder.

4.1.2. Particle Size Distribution of Diatomite

Figure 5 shows the particle size distributions of diatomite and limestone powder. From the frequency distribution curve, it can be seen that the average particle size of diatomite is slightly bigger than that of limestone powder. In addition, their particles mainly concentrate in between 5 µm and 50 µm. Cumulative distributions results show that the proportions of particle size of diatomite less than 14.48 µm and 36.52 µm reach 50% and 90%, respectively, and for limestone powder, it is 13.27 µm and 36.63 µm, respectively. The average particle size of diatomite and limestone powder is similar. Notably the average particle size is the main factor affecting the dispersion and compatibility of filler in asphalt. It is concluded that the use of diatomite contributes to the extension and filling of asphalt for its large specific surface area. It further lays a foundation for improving the performance of the mixture.

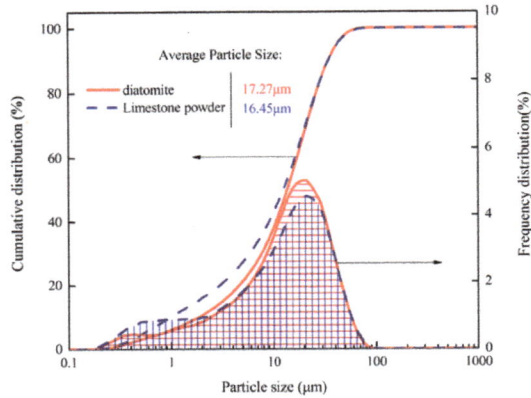

Figure 5. Particle size distributions of the diatomite and limestone powder.

4.1.3. Mesoporous Distribution of Diatomite

Figure 6 illustrates N_2 adsorption isotherms for pore size analysis. The quantity of adsorbed N_2 increases with the increase of relative pressure, where p and p_0 are equilibrium pressure and saturation pressure, respectively. Less adsorption in the low-pressure zone indicates that the force between the adsorbent and the adsorbate is quite weak. In the high p/p_0 range, with the rise of pressure, the adsorption rate increases significantly. It can be observed that the desorption isotherm is above the adsorption isotherm when the relative pressure is between 0.6 and 1.0. In this interval, the adsorbate condenses in capillary, resulting in desorption hysteresis. This result confirms the conclusion made by Garderen [28], who found that diatomite is a layered structure with a narrow number of mesopores in it. The mesoporous distribution diagram is highlighted in Figure 6 by using the adsorption branch data. It can be seen that the pore size of diatomite mainly concentrates from 1 nm to 8 nm. The average pore diameter is 5.4895 nm. The mesoporous structure can subsequently enhance the capability of the absorbing light components of asphalt, resulting in the improvement of viscosity and high-temperature performance of asphalt.

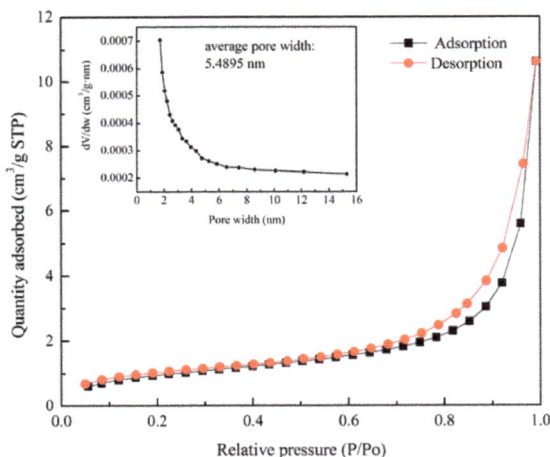

Figure 6. Adsorption isotherm and mesoporous distribution of diatomite.

4.1.4. SEM Results of Diatomite

It can be seen in the SEM images of Figure 7 that the shape of diatomite is a disc, believed to belong to cyclotella and stephanodiscus [21]. The average diameter of diatomite particles is about 20 μm. Further observation shows that there are a variety of small opening holes in the outer layer shell of diatomite. This specific structure of diatomite has certain influence on the asphalt mixture. It not only results in the large surface area of diatomite, but also facilitates the adsorption and wetting of asphalt.

Figure 7. Surface microstructure of diatomite: (**a**) 2000×; (**b**) 5000×; (**c**) 10,000×.

4.1.5. FTIR Test Results

Figure 8 shows the FTIR patterns for asphalt, diatomite, and asphalt binder specimens. The results indicate that the peak at 2954 cm^{-1} is the asymmetric stretching vibrations in CH_3. The two peaks at 2923 cm^{-1} and 2852 cm^{-1} are the asymmetric and symmetric stretching vibration in CH_2. The peak at 2728 cm^{-1} is the C–H stretching vibration in saturated alkyl. Peak at 1602 cm^{-1} is due to the stretching vibration of the benzene ring skeleton. The peaks at 1457 cm^{-1} and 1376 cm^{-1} stand for the symmetric and asymmetric flexural vibrations in CH_3. The absorption peaks of 873 cm^{-1} and 805 cm^{-1} are the outer flexural vibration absorption peaks of the hydrocarbon covalent bond of substituted benzene. The absorption peak at 747 cm^{-1} is the result of the alkyl flexural vibration.

There are three new peaks for diatomic modified asphalt, compared with base asphalt. They are the vibrational peaks of water at 3621 cm^{-1}, the stretching vibrations of Si–O bonds at 1033 cm^{-1}, and the vibrational bands of inorganic compounds near the 500 cm^{-1}, respectively. These peaks are the characteristic absorbed peaks for diatomite. It is observed that no new absorption peak appears on the spectrum of asphalt binder. Hence, it is proven that mixing of diatomite and asphalt is a simple physical blend. No new functional groups appear in the modified asphalt on account of the addition of diatomite.

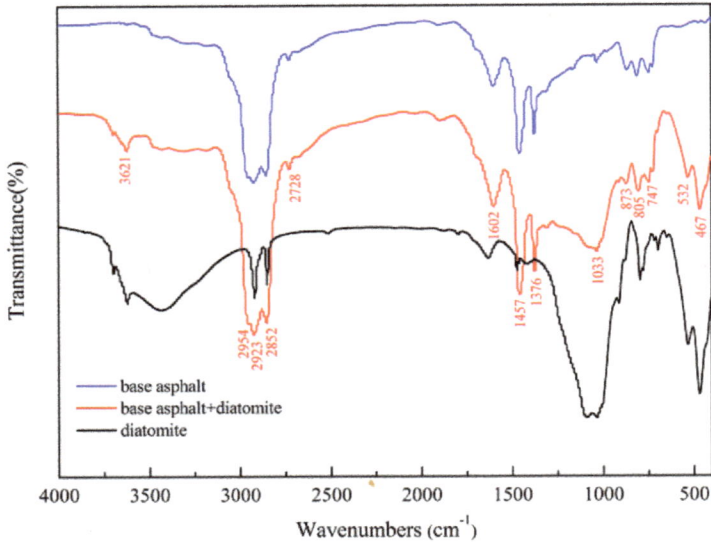

Figure 8. FTIR test results.

4.2. Performance of Asphalt Mixtures

4.2.1. Results of the Three-Point Bending Test

The low-temperature performance of an asphalt mixture is determined by the tensile strength of asphalt and its combination with the aggregate. Usually, inorganic filler can improve the rutting performance of an asphalt mixture but has little effect on the improvement of the low-temperature performance, because the addition of filler tends to enhance the hardness of asphalt and increase the possibility of brittle fracture at low temperatures. Therefore, the focus falls on the influence of diatomite on the anti-cracking performance of the asphalt mixture.

The test results are shown in Table 5 and Figure 9. It is seen that EZ-SBS-modified mixture shows the maximum tensile strain, strain energy density, and bending strain energy, followed by IM-SBS- and diatomite-modified mixtures. Compared with the base asphalt mixtures, the energy density and strain energy of mixtures with diatomite are improved to a certain extent. It may be ascribed to the hardening effects of asphalt adsorbed in the pores of diatomite at low temperatures, which enhanced the mechanical combination of asphalt and diatomite, further improving the low-temperature performance of diatomite asphalt mortar [29]. However, the extent of the improvement is not as great as that of SBS-modified asphalt.

Some discussions in the literature show that low-temperature performance of diatomite-modified asphalt mixture is comparable to or slightly lower than that of polymer-modified asphalt mixtures, such as SBS [30,31]. However, the test results showed that diatomite has little effect on the improvement of the low-temperature performance of an asphalt mixture. The reason is that an inorganic substance, such as diatomite, has no cross-linking and vulcanization with asphalt, unlike the modification mechanism of asphalt by SBS.

Table 5. Low-temperature bending test results of the asphalt mixtures.

Mixtures Types	Flexural Strength (MPa)	Tensile Strain (με)	Bending Stiffness Modulus (MPa)
EZ-SBS-modified	10.127	2077.16	4996.93
IM-SBS-modified	10.481	1576.36	7229.83
Diatomite-modified	8.411	1352.72	6570.01
Base	7.910	1130.84	7333.69

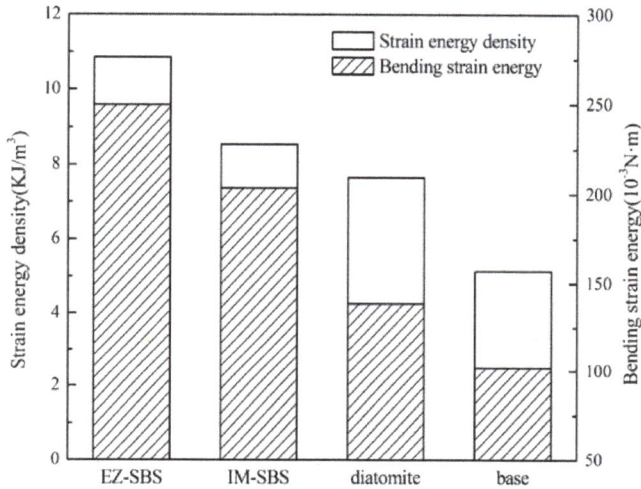

Figure 9. Strain energy density and bending strain energy of the four asphalt mixtures.

4.2.2. Results of the Rutting Test

It can be seen from the rutting tests results in Table 6 that the EZ-SBS-modified asphalt mixture has the highest dynamic stability. The dynamic stability of the diatomite-modified asphalt mixture is larger than that of the IM-SBS-modified asphalt mixture and about 3.4 times that of the base asphalt mixture. The reason is that diatomite, with its porous structure and large surface area, can absorb the light content of asphalt, which increases the overall complex shear modulus of the asphalt mortar and improves the rutting resistance of the mixture. In addition, diatomite is an inert substance with a high content of SiO_2, which is insensitive to the change of temperature [6]. Therefore, diatomite-modified asphalt pavement has the functions of thermal insulation and heat resistance.

Table 6. Rutting experiment results of asphalt mixtures.

Mixtures Types	45 min d_1 (mm)	60 min d_2 (mm)	Dynamic Stability (times/mm)
EZ-SBS-modified	1.229	1.295	9545
IM-SBS-modified	2.477	2.681	3088
Diatomite-modified	1.574	1.686	5625
Base	3.144	3.527	1645

4.2.3. Results of the Four-Point Bending Test

The fatigue lives of the three kinds of asphalt mixture beams specimens are shown in Figure 10. There are four microstrain levels including 500 με, 600 με, 700 με, and 800 με. It appears that the fatigue life of the IM-SBS-modified asphalt mixture is significantly greater than the other two kinds of mixtures. Meanwhile, compared with the base asphalt mixture, the fatigue performance of the asphalt mixture with diatomite is significantly improved. This is ascribed to the excellent compatibility between diatomite and asphalt, which can reduce mixing time, prevent aging of the modified asphalt, and directly improve the durability of asphalt mixture [27].

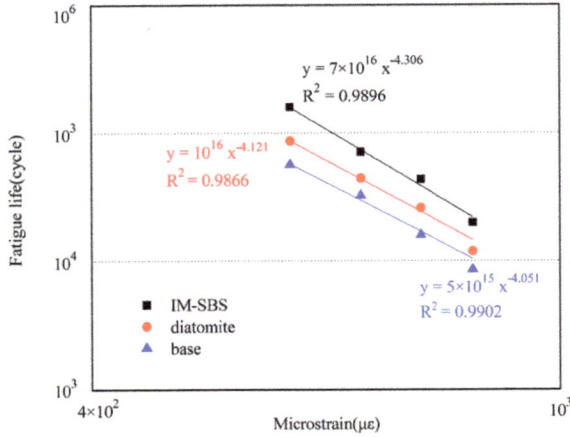

Figure 10. Fatigue curves of the three kinds of asphalt mixtures.

4.2.4. Results of Marshall Stability and Indirect Tensile Strength Test

The results of Marshall stability (MS), immersion Marshall stability (MS$_1$), and residual Marshall stability (RMS) tests are shown in Figure 11. The results of the indirect tensile strength of normal temperature group (RT$_1$), indirect tensile strength of freezing and thawing group (RT$_2$), and tensile strength ratio (TSR) were shown in Figure 12.

It can be seen that although the MS and MS$_1$ of the asphalt mixture with diatomite are lower than those of the IM-SBS-modified asphalt mixture, their RMSs are approximately equal. The TSR of the diatomite-modified asphalt mixture is slightly higher than that of the IM-SBS-modified asphalt mixture. With the addition of diatomite, the parameter value of the asphalt mixture is higher than that of base asphalt mixture, leading to the increase in the cohesive force between the asphalt and the aggregate. It contributes to the increase in shear resistance and stability [32].

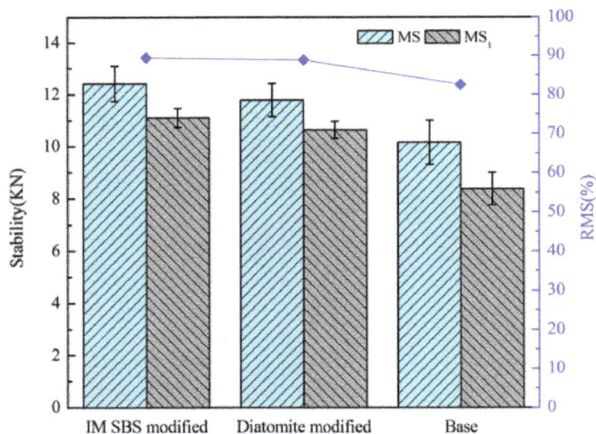

Figure 11. Results of the Marshall stability test.

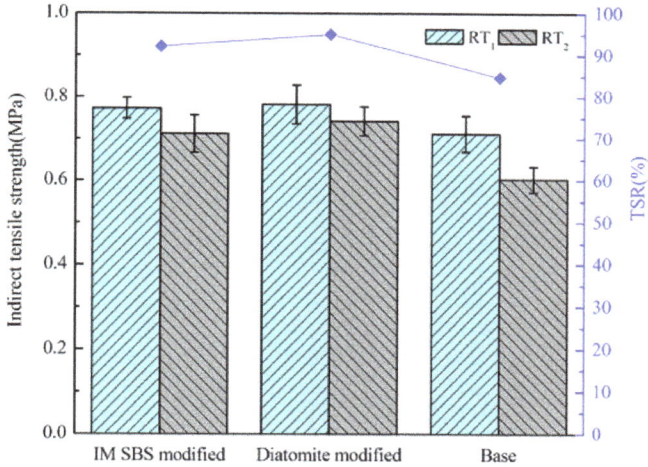

Figure 12. Results of the indirect tensile strength test.

5. Conclusions

This study investigated the characteristics of diatomite by various characterization methods. The influence of diatomite as modifier on asphalt mixture was also studied by comparing the pavement performance of SBS-modified, diatomite-modified, and base asphalt mixtures. According to the above results, the following items can be concluded:

(1) The group OH^- was contained in diatomite. It is an essential reason for the surface activity and absorptivity of diatomite. The porous structure of diatomite improves its adhesion and wet ability with asphalt. Small particle size, numerous mesopores, and large specific surface area enhance its adsordability for the light components of asphalt. The characteristics of diatomite contribute to its strong physical connection with asphalt. They provide a possible reason for its enhancement of asphalt mixture performance.

(2) The addition of diatomite resulted in an increase in the high-temperature performance of the asphalt mixture but resulted in little improvement of the low-temperature performance. Therefore, in terms of its practical engineering application, diatomite-modified asphalt mixture is not suitable for application in the upper layer of asphalt pavement in cold areas.

(3) Although it did not perform as well as the SBS-modified asphalt mixture, the asphalt mixture with diatomite showed better fatigue performance and water stability than the base asphalt mixture. In addition, due to its low cost and simple modification process, the economic benefits of the diatomite-modified asphalt mixture have great advantages compared with the traditional modified asphalt mixture.

Author Contributions: J.X., C.Y. conceived of and designed the experiments; C.Y. and X.Z. performed the experiments; C.Y. and Q.L. analyzed the data; L.P. contributed reagents, materials, and analysis tools; J.X., C.Y. and Q.L. wrote and edited the paper.

Funding: This research was funded by [National Natural Science Foundation of China] grant number [51708437 and 51778515], [Independent Innovation Foundation of Wuhan University of Technology of China] grant number [2018IVB039].

Acknowledgments: We gratefully acknowledge many important contributions from the researchers of all reports cited in our paper.

References

1. Xie, J.; Wu, S.; Lin, J.; Cai, J.; Chen, Z.; Wei, W. Recycling of basic oxygen furnace slag in asphalt mixture: Material characterization & moisture damage investigation. *Constr. Build. Mater.* **2012**, *36*, 467–474.

2. Chen, Z.; Wu, S.; Pang, L.; Xie, J. Function investigation of stone mastic asphalt (SMA) mixture partly containing basic oxygen furnace (BOF) slag. *J. Appl. Biomater. Funct. Mater.* **2016**, *14* (Suppl. 1), e68–e72. [CrossRef] [PubMed]

3. Long, Y.; Wu, S.; Xiao, Y.; Cui, P.; Zhou, H. VOCs reduction and inhibition mechanisms of using active carbon filler in bituminous materials. *J. Clean. Prod.* **2018**, *181*, 784–793. [CrossRef]

4. Arabani, M.; Tahami, S.A. Assessment of mechanical properties of rice husk ash modified asphalt mixture. *Constr. Build. Mater.* **2017**, *149*, 350–358. [CrossRef]

5. Pang, L.; Liu, K.; Wu, S.; Lei, M.; Chen, Z. Effect of LDHs on the aging resistance of crumb rubber modified asphalt. *Constr. Build. Mater.* **2014**, *67*, 239–243. [CrossRef]

6. Bao, Y.N. Study of Asphalt Mixture Modified by Diatomite. Master's Thesis, Chang'an University, Xi'an, China, 2005.

7. Modarres, A. Investigating the toughness and fatigue behavior of conventional and SBS modified asphalt mixes. *Constr. Build. Mater.* **2013**, *47*, 218–222. [CrossRef]

8. Baghaee Moghaddam, T.; Soltani, M.; Karim, M.R. Evaluation of permanent deformation characteristics of unmodified and Polyethylene Terephthalate modified asphalt mixtures using dynamic creep test. *Mater. Des.* **2014**, *53*, 317–324. [CrossRef]

9. Wulandari, P.S.; Tjandra, D. Use of Crumb Rubber as an Additive in Asphalt Concrete Mixture. *Procedia Eng.* **2017**, *171*, 1384–1389. [CrossRef]

10. Iskender, E. Evaluation of mechanical properties of nano-clay modified asphalt mixtures. *Measurement* **2016**, *93*, 359–371. [CrossRef]

11. Guo, Q.; Li, L.; Cheng, Y.; Jiao, Y.; Xu, C. Laboratory evaluation on performance of diatomite and glass fiber compound modified asphalt mixture. *Mater. Des. (1980–2015)* **2015**, *66*, 51–59. [CrossRef]

12. Cong, P.; Chen, S.; Chen, H. Effects of diatomite on the properties of asphalt binder. *Constr. Build. Mater.* **2012**, *30*, 495–499. [CrossRef]

13. Sun, Y.S.; Chen, X.L.; Han, Y.X.; Zhang, B. Research on Performance of the Modified Asphalt by Diatomite-Cellulose Composite. *Adv. Mater. Res.* **2010**, *158*, 211–218. [CrossRef]

14. Jiang, L.; Liu, Q.L. Application of Diatomite Modified Asphalt. *Appl. Mec. Mater.* **2013**, *477–478*, 959–963. [CrossRef]

15. Baldi-Sevilla, A.; Montero, M.L.; Aguiar, J.P.; Loría, L.G. Influence of nanosilica and diatomite on the physicochemical and mechanical properties of binder at unaged and oxidized conditions. *Constr. Build. Mater.* **2016**, *127*, 176–182. [CrossRef]

16. Zhang, Y.B.; Zhu, H.Z.; Wang, G.A.; Chen, T.J. Evaluation of Low Temperature Performance for Diatomite Modified Asphalt Mixture. *Adv. Mater. Rer.* **2011**, *413*, 246–251. [CrossRef]

17. Tan, Y.; Zhang, L.; Zhang, X. Investigation of low-temperature properties of diatomite-modified asphalt mixtures. *Constr. Build. Mater.* **2012**, *36*, 787–795.

18. Chen, Y.Z.; Li, Z.X. High Temperature Stability of Modified Asphalt Concrete. *Appl. Mec. Mater.* **2013**, *438–439*, 391–394. [CrossRef]

19. Wei, H.; He, Q.; Jiao, Y.; Chen, J.; Hu, M. Evaluation of anti-icing performance for crumb rubber and diatomite compound modified asphalt mixture. *Constr. Build. Mater.* **2016**, *107*, 109–116. [CrossRef]

20. Chen, W.F.; Gao, P.W.; Li, X.Y.; Le, J.; Jin, S. Effect of the New Type Diatomite Modifier on the Pavement Performance of Asphalt Mixture. *J. Mater. Sci. Eng.* **2007**, *25*, 578–581.

21. Bao, Y.N.; Jiang, X.H. Study on Laboratory Test of Road-performance of Diatomite-asphalt Mixture. *Highway Eng.* **2010**, *35*, 018.

22. Zhou, Z. Experimental Study on Mix Design of Diatomite Modified Asphalt Mixture. Master's Thesis, Jilin University, Changchun, China, 2008.

23. Yin, H.Y. Research on the Modification Mechanism of Diatomite and Dry Mixing of Diatomite-modified Asphalt Miture. Master's Thesis, Chongqing Jiaotong University, Chongqing, China, 2012.

24. *Standard Test Methods of Asphalt and Asphalt Mixtures for Highway Engineering*; Ministry of Transport: Beijing, China, 2011.

25. Ge, Z.S.; Huang, X.M.; Xu, G.G. Evaluation of asphalt-mixture's low-temperature anti-cracking performance by curvature strain energy method. *J. Southeast Univ. (Nat. Sci. Ed.)* **2002**, *32*, 653–655.

26. Xu, S.C. A study on the Specific Property and the Development of Diatomite. *J. Fuyang Teach. Coll.* **2002**, *17*, 29–31.

27. Li, Z.S. A Study on The Performance of Diatomite Modified Asphalt Mixture. Master's Thesis, Jilin University, Changchun, China, 2008.

28. Van Garderen, N.; Clemens, F.J.; Kaufmann, J.; Urbanek, M.; Binkowski, M.; Graule, T.; Aneziris, C.G. Pore analyses of highly porous diatomite and clay based materials for fluidized bed reactors. *Microporous Mesoporous Mater.* **2012**, *151*, 255–263. [CrossRef]

29. Zhang, Z.Q.; Zhang, D.L.; Yuan, J.A. The Influence of the Molecular Weight and the State Transition Characteristic on the Performance of Asphalt. *J. Xi'an Highway Univ.* **1998**, *18*, 207–211.

30. Wang, J.L.; Han, T. Study on temperature performance of diatomite modified asphalt mixture in Panxi, Sichuan. *Commun. Sci. Technol. Heilongjiang* **2014**, *11*, 26–28.

31. Zhang, X.K. Influence of Diatomite Material on Properties of Asphalt Mixture. Master's Thesis, Chongqing Jiaotong University, Chongqing, China, 2008.

32. Wang, Y.N. Study on Performance of Diatomite-Modified Asphalt Mixture. Master's Thesis, Jilin University, Changchun, China, 2007.

materials

MDPI

Article

Effects of Aluminum Hydroxide and Layered Double Hydroxide on Asphalt Fire Resistance

Menglin Li, Ling Pang *, Meizhu Chen, Jun Xie and Quantao Liu

Wuhan University of Technology, State Key Laboratory of Silicate Materials for Architectures, Luoshi Road 122, Wuhan 430070, China; limenglin@whut.edu.cn (M.L.); chenmzh@whut.edu.cn (M.C.); xiejun3970@whut.edu.cn (J.X.); liuqt@whut.edu.cn (Q.L.)
* Correspondence: lingpang@whut.edu.cn

Received: 27 August 2018; Accepted: 28 September 2018; Published: 11 October 2018

Abstract: When a fire occurs in a tunnel, the instantaneous high temperature and smoke cause great danger to people. Therefore, the asphalt pavement material in the tunnel must have sufficient fire resistance. In this study, the effects of aluminum hydroxide and layered double hydroxide on the fire resistance of styrene-butadiene-styrene (SBS) polymer-modified asphalt was investigated. The fire resistance of the asphalt was evaluated by using a limiting oxygen index (LOI). The impact of aluminum hydroxide (ATH), layered double hydroxide (LDHs), and mixed flame retardant (MFR) on LOI was studied. The synergistic fire resistance mechanism of ATH and LDHs in asphalt binder was analyzed by using an integrated thermal analyzer-mass spectrometry combined system (TG-DSC-MS) and Fourier transform infrared spectrometer (FTIR). The experimental results indicated that the main active temperature range of these flame retardants was 221–483 °C. The main components of smoke were methane, hydroxyl, water, carbon monoxide, aldehyde, carbon dioxide, etc. The addition of flame retardants could inhibit the production of methane, carbon monoxide, and aldehyde. Moreover, due to the good synergistic effects of ATH and LDHs, 20 wt % MFR had the best fire resistance.

Keywords: asphalt combustion; flame retardant; aluminum hydroxide; layered double hydroxide

1. Introduction

Asphalt pavement has been widely used in the construction of highway tunnels due to its excellent road performance and driving comfort [1,2]. Bitumen, as a petroleum product, is mainly composed of hydrocarbons, with an initial decomposition temperature of 200 °C [3]. This indicates that asphalt is flammable. Once a tunnel is ignited, asphalt in the pavement aggravates the spread of the fire. This could cause serious damage to a tunnel road. Therefore, it is important to improve the fire resistance of asphalt.

Adding flame retardants is a popular method of improving the fire resistance of asphalt. Based on previous research, the combustion process of asphalt can be divided into five stages: heating, decomposition, ignition, combustion, and propagation [4]. The addition of flame retardants destroys one or more of these stages [5]. Therefore, many studies have been done to improve the fire resistance of asphalt by adding flame retardants. The existing fire-resistant asphalts are made by adding various organic or inorganic flame retardants. Organic flame retardants have comparable efficiency. Nonetheless, halogen-based flame retardants release toxic smoke. This increases the concentration of poisonous and corrosive gases in the fire, which can easily lead to asphyxiation. Moreover, the amount of smoke released by fire-resistant asphalt mixtures prepared using halogen flame retardants is large, which could seriously affect the health of construction personnel [5–7]. For these reasons, inorganic flame retardants have received more attention [8]. As one kind of inorganic flame retardant, hydroxides have the merits of non-toxicity and environmental friendliness [9–12]. The initial decomposition temperature of both the light components in asphalt and aluminum hydroxide (ATH) is 200 °C [13–15]. Therefore, ATH can be used in asphalt to enhance its fire resistance.

LDHs are a family of hydrotalcite compounds that contain intercalated anions, metal cations, and water molecules. LDHs can also be used as a flame retardant in polymers, not only because of its unique chemical composition, similar to aluminum/magnesium hydroxide, but also because of its flake-like morphological structure, which can act as a barrier to heat and fuel transfer. The use of intercalated LDHs to enhance the fire resistance of polymers has been reported. The results showed that the LDHs had the effects of absorbing heat, releasing water, and forming a protective oxide layer that could prevent further degradation [16,17]. When Breu et al. [18,19] used modified LDHs as a flame retardant in polymers, the results showed that the fire resistance of polymers was improved greatly, and the total heat release value of the sample decreased a lot. Zhang et al. [20] combined PWA-LDH with an intumescent flame retardant (IFR) and the results showed that these two flame retardants had a good synergistic effect in promoting char formation and enhancing fire resistance. Xiao et al. [21–23] showed that the metal oxides produced by LDH had good inhibiting effects on the emissions of organic volatiles. However, the combination of ATH and LDHs was seldom employed to improve the fire resistance of asphalt.

In this study, a limiting oxygen index (LOI) test was conducted to evaluate the fire resistance of asphalt and flame-retardant-modified asphalt (FRMA). The combined TG-DSC-MS and FTIR techniques were employed to dynamically detect the amount of heat and smoke from asphalt and FRMA in the combustion process. The decomposition regularities of asphalt and FRMA were analyzed. Then the effects of flame retardants on the heat release and smoke suppression in the asphalt combustion were discussed. Finally, synergistic effects between ATH and LDHs were discovered during the combustion process.

2. Experimental

2.1. Materials

Styrene-butadiene-styrene (SBS) polymer-modified asphalt was obtained from Jiaotouzhiyuan New Material Industry in Hubei Province, China. Its properties include: SBS dosage of 4.3%, penetration of 5.8 mm at 25 °C, ductility of 57 cm at 5 °C, softening point of 70 °C, and viscosity of 0.95 Pa·s at 135 °C.

The flame retardant, namely ATH, was produced by KeXin Industry of Chemical Limited Company in Wuxi, Jiangsu Province, China. The nontoxic ATH flame retardant was a white powder with an average particle size of 1.5–2.0 μm, density of 2.4 g/cm^3, and purity of 99.0%.

The flame retardant, namely Mg-Al-LDHs, was produced by TaiKeLaiEr Chemical Industry in Beijing, China. The nontoxic Mg-Al LDHs flame retardant was also a white powder, with an average particle size of 2–20 μm, density of 0.33 g/cm^3, purity of 99.5%, and ratio of Mg to Al in LDHs of 2:1.

2.2. The Preparation of Flame-Retardant-Modified Asphalt

Flame-retardant powders (at a mass percentage of 20%) were added to SBS-modified asphalt at 165 ± 5 °C. First, the blends were stirred with a high-speed shear mixer (ESR-500, ELE, Shanghai, China) at 3000 rpm for 1 h to ensure the uniform distribution of flame retardants. Then, the FRMA was further stirred at a lower speed of 500 rpm for 15 min to reduce the amount of air bubbles. Finally, the FRMA was poured into a clean vessel with continuous hand stirring until it cooled to room temperature. SBS-modified asphalt was processed under the same conditions to ensure comparability with FRMA.

3. Methods

An LOI test is used to evaluate the relative combustion of plastics and other polymer materials so as to determine how easily the material can burn when exposed to fire in the air [6]. In order to prepare the specimens for LOI tests, a glass fiber surfacing mat with a density of 120 g/m^2 was cut into the samples with 120 mm length and 120 mm width; then we put a metal frame of 151 mm length and 151 mm width and 3 mm height on the middle of the glass fiber surfacing mat. After that, uniformly heated asphalt was carefully poured into the metal frame until the whole metal frame was filled up with asphalt. When cooled to room temperature, the asphalt stained with a glass fiber surfacing mat was cut into specimens of 115 mm length and 6.5 mm width for the LOI test. The fire resistance of asphalt was assessed by the limiting oxygen index according to ASTMD-2863-77. The LOI was defined according to Equation (1):

$$LOI = C_f + kd,$$ (1)

where LOI is the limiting oxygen index; C_f is the last oxygen concentration; k is a coefficient associated with oxygen concentration; and d is the change in oxygen concentration.

The integrated thermal analyzer-mass spectrometry combined system (TG-DSC-MS, ATA449F3) used in this study was produced by NETZSCH Group, Bayern, Germany. This technique was used to study the quality changes (TG), heat changes (DSC), and gas composition (MS) caused by chemical reactions and physical changes of samples. About 2.0 mg of the sample was used in each case and the respective peaks were recorded. The sample was heated from room temperature to 700 °C at a heating rate of 10 °C/min under air flowing at 10 mL/min.

The Fourier transform infrared spectrometer (FTIR, Nicolet 6700/Nicolet 6700) used in this study was produced by Thermo Fisher Scientific, Waltham, MA, USA. TG-FTIR experiments were performed using a thermal analyzer system coupled with a Fourier transform infrared spectrometer. The test was conducted under air flowing at 120 mL/min and a heating rate of 10 °C/min. About 10.0 mg of the sample were heated in the TG equipment from room temperature to 700 °C. The mass loss of the sample was recorded. At the same time, the gases released during combustion entered the gas cell for FTIR analysis, and the changes in the products with the temperature were monitored by FTIR spectrometry. The spectrum wavenumber ranged from 4 cm^{-1} to 4000 cm^{-1}.

The spectrum intensity was calculated according to Equation (2) or Equation (3):

$$Aborbance\ (\%) = \left(1 - \frac{I}{I_0}\right) \times 100\%$$ (2)

$$Transmittance\ (\%) = \frac{I}{I_0} \times 100\%,$$ (3)

where I_0 is the incident light intensity and I is the transmitted intensity.

4. Results and Discussion

4.1. Fire Resistance of Asphalt

The LOI values of SBS-modified asphalt and its composites are shown in Table 1. When the LOI value of the material was greater than 26%, it could be self-extinguished in air [6]. From Table 1, it can be observed that the LOI value of SBS-modified asphalt is just 19.3%, which means that SBS-modified asphalt has good flammability, since the concentration of oxygen in air is about 21% by volume. However, the LOI value of SBS-modified asphalt is obviously improved by the addition of flame retardants. When the content of ATH was 20%, the LOI value was 25%. To meet the requirement of self-extinguishing, the content of ATH in ATH-modified asphalt (AMA) was 25%. The LOI value of LDHs-modified asphalt (LMA) containing 5% LDHs was 20.7%, and the LOI value of 20% LDHs was just 22.5%, which indicated that LDHs had limited effects on the fire resistance of SBS-modified asphalt. The LOI value of mixed flame-retardant-modified asphalt (MMA) with 10% ATH and 7.5% LDHs was 26.0%, which met the requirement of self-extinguishment. In other words, MMA with 10% ATH, 7.5% LDHs, and AMA with 25% ATH had the same effects. It can be concluded that LDHs had good synergistic effects with ATH.

Table 1. LOI data of SBS-modified asphalt and its composites.

Name of Material	LOI (%)
SBS	19.3
SBS + 5% ATH	20.5
SBS + 10% ATH	21.7
SBS + 15% ATH	23.2
SBS + 20% ATH	25.0
SBS + 25% ATH	26.6
SBS + 30% ATH	28.4
SBS + 5% LHDs	20.7
SBS + 10% LDHs	21.3
SBS + 15% LDHs	21.6
SBS + 20% LDHs	22.5
SBS + 10% ATH + 2.5% LDHs	23.4
SBS + 10% ATH + 5% LDHs	24.6
SBS + 10% ATH + 7.5% LDHs	26.0
SBS + 10% ATH + 10% LDHs	27.3

4.2. TG Analysis

The TG curves of LDHs and ATH are shown in Figure 1. From Figure 1, the main decomposition temperature of ATH ranges from 231 °C to 314 °C. However, LDHs have three decomposition stages: 98~249 °C, 249~303 °C, and 303~498 °C. The stage with the most mass loss has the range 303~498 °C.

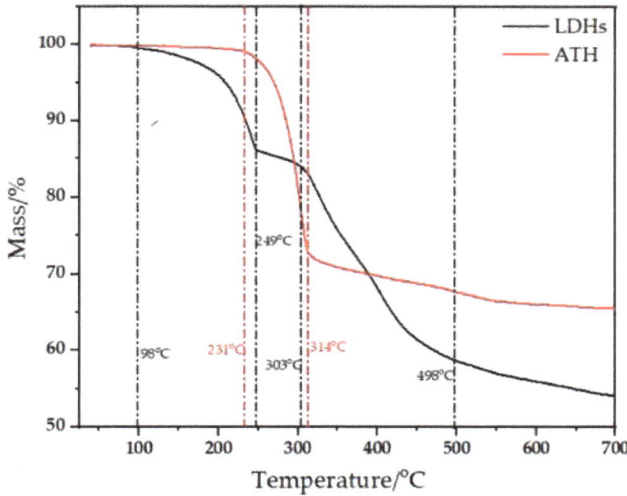

Figure 1. TG curves of LDHs and ATH.

The TG curves of ATH, LDHs, SBS-modified asphalt, AMA, LMA, and MMA are shown in Figure 2. The whole combustion process can be divided into several stages according to the peaks in the DTG curves.

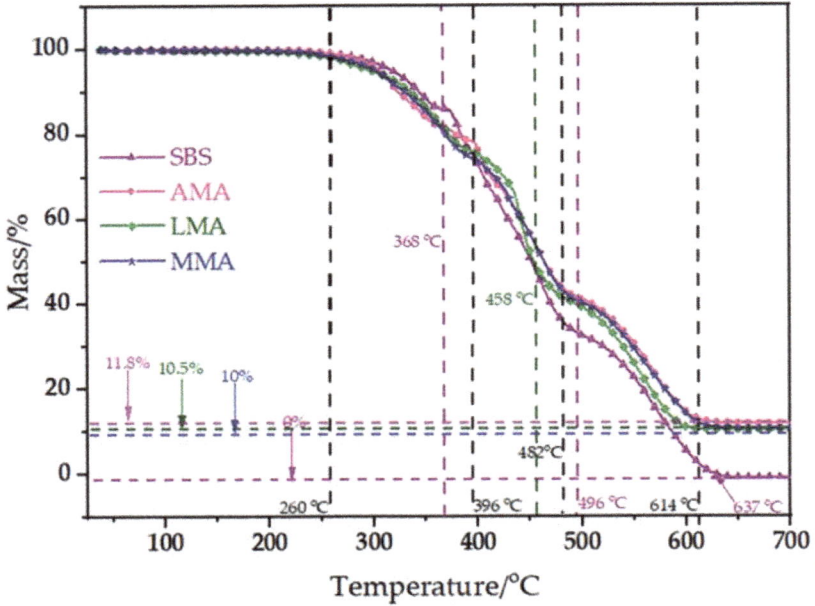

Figure 2. TG curves of SBS-modified asphalt, AMA, LMA, and MMA during pyrolysis.

Due to the complex composition of asphalt, it is hard to study the specific chemical composition. However, the composition can be approximately divided into four components—saturates, aromatics, resins, and asphaltenes—and each component has its own thermal properties.

From Figure 2, the combustion process of SBS-modified asphalt was divided into three stages with temperature ranges from 260 °C to 368 °C, 368 °C to 496 °C, and 496 °C to 637 °C. Stage 1 was from 260 °C to 368 °C. Most of the mass loss in this stage was due to the decomposition of the lightweight components, for example, saturates, aromatics, etc. [9]. In the meantime, due to the complexity of the lightweight components, a mass of volatiles was generated during combustion [24]. Under the combined action of thermal radiation and oxygen, the combustion of these flammable volatiles further promoted the decomposition of bitumen, and a carbon layer begun to form. Stage 2 was from 368 °C to 496 °C. This stage had the most mass loss of asphalt binder, with the most mass loss occurring at 423 °C. The three stages were due to the decomposition of aromatics and resins, and the chemical reactions were more complex [25,26]. In addition, a carbon layer was formed. Finally, Stage 3 was the temperature range from 496 °C to 637 °C. In this stage, the vast majority of bitumen had been decomposed. The most mass loss occurred at 570 °C. The main reason was the decomposition of asphaltenes [25,26].

From Figure 2, the decomposition of AMA could also be divided into three stages as the temperature ranged from 260 °C to 396 °C, 396 °C to 482 °C, and 482 °C to 614 °C, just as with SBS-modified asphalt. The general trend of the AMA curve was similar to SBS-modified asphalt, but there were some differences. Compared with SBS-modified asphalt, the temperature range of AMA was much larger. In addition, the first stage could be divided into two smaller stages by the remaining mass of asphalt. The first small stage was at the temperature from 260 °C to 388 °C, and the other one was at 388 °C to 396 °C. The most mass loss in the first small stage occurred at 316 °C, which was close to ATH. In this small stage, the mass loss of AMA was greater than that of SBS-modified asphalt. The main reason for this was the decomposition of ATH in AMA. Moreover, the decomposition of ATH was an endothermic reaction, and the decomposition product Al_2O_3 could also promote the formation of the carbon layer, which had an adverse effect on the decomposition and combustion of the lightweight components of asphalt in this stage. In the second stage, the undecomposed lightweight components were fully decomposed, which released a large amount of flammable volatiles. The burning of these volatiles further promoted the decomposition of bitumen, which also promoted the formation of a carbon layer. As a result, the char yield ratio of AMA was increased by 11.8 wt % to generate a tight carbon layer.

Figure 2 shows that decomposition of LMA also could be divided into three stages with temperature ranges from 260 °C to 396 °C, 396 °C to 458 °C, and 458 °C to 614 °C. In the first stage, the mass loss of LMA was greater than that of SBS-modified asphalt, but after 396 °C, the mass loss of LMA was smaller than that of SBS-modified asphalt, which was very similar to the trend of AMA. However, there were some notable differences between LMA and AMA. Firstly, the curve of AMA and SBS-modified asphalt only intersected at one point, while the curve of LMA and SBS asphalt intersected twice. Secondly, the second stage of LAM ranged from 396 °C to 458 °C, while that of AMA ranged from 396 °C to 482 °C. Thirdly, the first stage of LAM only had one process, but there were two small processes in the first stage of AMA. Finally, the char yield ratio of LMA was 10 wt %, while that of AMA was 11.8 wt %. Since the LDHs used for modification were first treated by sodium stearate, the first two differences might be caused by the presence of sodium stearate in the reaction. Moreover, the third difference indicated that LDHs had a greater temperature impact in the first stage than ATH, which meant that LDHs had a better inhibitory effect on the decomposition of lightweight components of asphalt.

From Figure 2, the decomposition of MMA could also be divided into three stages with temperature ranges from 260 °C to 396 °C, 396 °C to 482 °C, and 482 °C to 614 °C. Compared with LMA, the curve tendency of MMA was closer to that of AMA. There was only one process in the first stage of MMA, the curve of MMA and SBS-modified asphalt only intersected at one point, and the char yield ratio of MMA was 10.5%. As the MMA modifier contained ATH and LDHs, there were some interactions between ATH and LDHs during the whole combustion process, which eventually led to these results.

In conclusion, the combustion process of SBS-modified asphalt was mainly divided into three stages, and the flame retardants mainly affect the first two stages. The main functions of the flame retardants include: decomposing and absorbing heat, inhibiting the decomposition of light components in asphalt, and promoting the formation of a tight carbon layer.

4.3. DSC Analysis

DSC technology could be used for thermal analysis during combustion. The DSC test results of these asphalts are represented in Figure 3a-d. According to the TG results, the main active temperature range of flame retardants was 200–500 °C. Therefore, in the DSC analysis, heat released within the temperature range 221 °C to 483 °C was the main analysis target. The area of the curve within this temperature range was used to represent heat. The larger the area, the more heat was released in this temperature range. S_0 was the heat released by SBS-modified asphalt; S_1 was the heat released by AMA; S_2 was the heat released by LMA; and S_3 was the heat released by MMA.

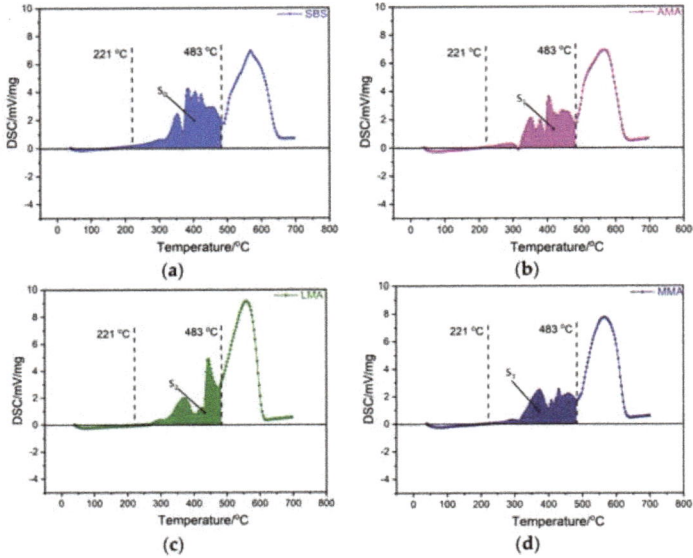

Figure 3. Results of DSC test of SBS-modified asphalt (a), AMA (b), LMA (c), and MMA (d).

The values of S_0 to S_3 are shown in Table 2.

Table 2. Values of DSC test results for 221–483 °C.

S_0	S_1	S_2	S_3
455.35	327.63	318.19	296.47

As seen from Table 2, $S_0 > S_1 > S_2 > S_3$, which meant that in th temperature range, SBS-modified asphalt releases the most heat, followed by AMA, then LMA, and finally MMA. This implied that the addition of flame retardants significantly reduced the heat released in this temperature range.

In order to better distinguish the effects of these three flame retardants, the flame retardant index (*HI*) is used in this study. The *HI* was defined as in Equation (5):

$$\Delta S_n = |S_n - S_0| \tag{4}$$

$$HI_n = \frac{\Delta S_n}{S_0}, \tag{5}$$

where HI_n is the flame retardant index of nth flame retardant; ΔS_n is the heat difference between nth flame-retardant-modified asphalt and SBS-modified asphalt; and S_0 is the heat released by SBS-modified asphalt.

Table 3 shows the values of HI_n and the heat differences.

Table 3. Values of flame retardant indexes.

HI_1	HI_2	HI_3
28.05%	30.12%	34.89%

Table 3 shows that $HI_3 > HI_2 > HI_1$, indicating that mixed flame retardants reduced heat the most when compared with the other two single flame retardants in this temperature range. Moreover, the mixed flame retardant was prepared by these two single flame retardants with a mass ratio of 1:1. So, ATH and LDHs had a good synergy effect.

4.4. FTIR Analysis

Figure 4 shows the infrared results of volatiles from SBS-modified asphalt. The components of the volatiles were characterized by the peak position in the infrared spectrogram. The peaks at 2357 cm^{-1} and 669 cm^{-1} were the absorption peaks of carbon dioxide. The peak at 2357 cm^{-1} was the asymmetrical stretching vibration absorption peak, and the peak at 669 cm^{-1} was the bending vibration absorption peak. The peak at 1737 cm^{-1} was the characteristic peak of THE carbonyl group, which meant aldehyde compounds might have been produced. The peaks at 2109 cm^{-1} were the absorption peaks of carbon monoxide. The peak at 2967 cm^{-1} was the asymmetric stretching vibration absorption of methyl. The peak at 2930 cm^{-1} was the asymmetric stretching vibration absorption of methylene, and the peak at 2860 cm^{-1} was the symmetric stretching vibration of methyl. The peak at 3015 cm^{-1} was the absorption peak of methane. All of these peaks showed that there were alkanes contained in flammable gases, and the gases were the main components for further combustion of asphalt. The peaks within the range from 3500 cm^{-1} to 3775 cm^{-1} were caused by the stretching vibration of the O-H bond, which indicated that the smoke contained water vapor.

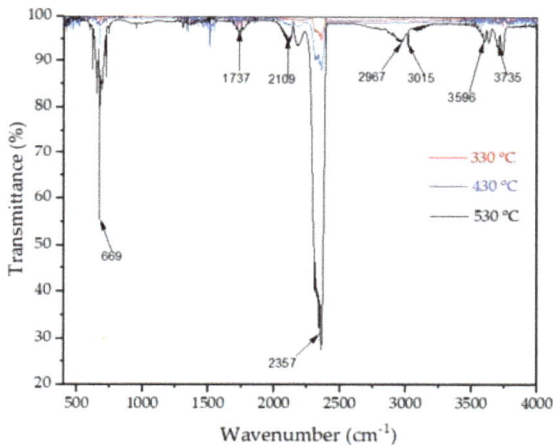

Figure 4. FTIR spectra of volatile products during the asphalt binder combustion process.

In conclusion, methane and carbon dioxide are the main components of greenhouse gases in smoke. Both carbon monoxide and aldehyde are toxic and are the main sources of toxicity in smoke. Hence, the products of the asphalt combustion process not only cause great harm to the environment, but can also seriously affect human health, especially in the case of a tunnel fire.

4.5. MS Analysis

MS technology was used to study the effect of flame retardants on the composition of bitumen smoke. Figure 4 shows that the main components of bituminous gas were carbon dioxide, aldehyde, carbon monoxide, alkanes, methane, and water vapor. Therefore, in the MS test, the ion strength of 16.13, 28.09, 29.09, and 44.06 relative molecular mass was mainly detected. These relative molecular masses represented methane, hydroxyl, water, carbon monoxide, aldehyde, and carbon dioxide, respectively. In order to better study the antismoking effect of flame retardants, the results of SBS-modified asphalt were standardized based on the research results. The relative ion strength of each component is shown in Figure 5.

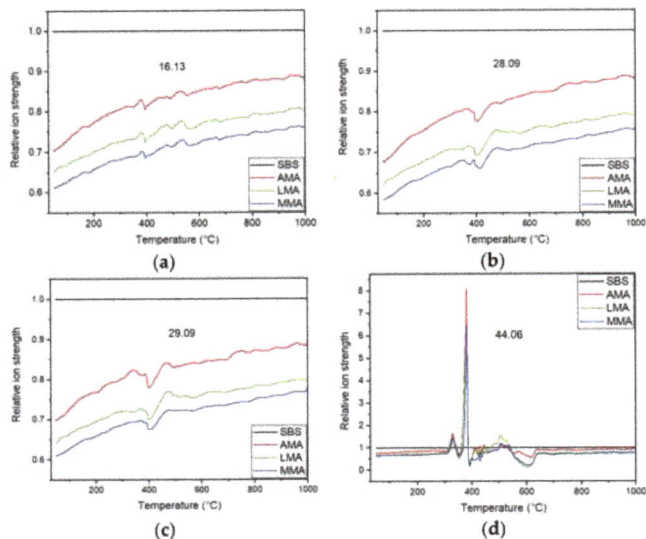

Figure 5. Relative ionic strength of four major components: 16.13 (**a**), 28.09 (**b**), 29.09 (**c**), and 44.06 (**d**).

It could be concluded from Figures 4 and 5a–c that the addition of these three flame retardants could effectively inhibit the production of methane, carbon monoxide, and aldehyde. The order of inhibition ability was: MIX > LDHs > ATH. Compared to ATH, LDHs has a good layered structure, which has a good promoting effect on the formation of carbon layer. Therefore, LDHs have a better antismoking effect than ATH. It could also be seen from the DSC analysis that ATH and LDHs had a good synergy effect. This synergy effect was also confirmed in the MS results. Reducing the content of methane in asphalt flue gas is of great importance for reducing greenhouse gas emission. The reduction of carbon monoxide and aldehyde can reduce the toxicity of asphalt flue gas, which is of great importance during evacuation and subsequent rescue. However, compared to the other three smoke components, these three flame retardants had little effect on carbon dioxide (Figure 5d). On the whole, flame retardants not only hindered the burning of asphalt, but also inhibited the production of smoke.

5. Conclusions

The following conclusion can be obtained from the analysis of the experimental results:

(1)   The combustion of SBS-modified asphalt could be roughly divided into three stages. The main active temperature range of these flame retardants was 221–483 °C, which covers the first two stages. The addition of flame retardants could significantly reduce the heat released within this temperature range, among which mixed flame retardants were the most effective, reducing the heat release by 34.89%.

(2)   The main components in smoke emitted were: carbon dioxide, aldehyde, carbon monoxide, alkanes, methane, and water vapor. Aldehyde and carbon monoxide were the main sources of smoke toxicity, and carbon dioxide and methane are the main greenhouse gases. The addition of flame retardants significantly reduced the content of aldehydes, carbon monoxide, and methane, which indicated that the flame retardant could also inhibit smoke emission.

(3)   The main role of the flame retardant in SBS-modified asphalt was to decompose and absorb heat, and to inhibit the decomposition of lightweight components in the initial stage of combustion.

Secondly, the decomposition products of flame retardants were inorganic metallic oxides, which could promote the formation of a carbon layer in the course of asphalt combustion.

(4) Due to the good synergistic effect of ATH and LDHs, the mixed flame retardant has the best flame-retardant and smoke-suppression effects.

Author Contributions: L.P. and M.L. conceived of and designed the experiments; M.L. performed the experiments; M.L. analyzed the data; L.P. and M.C. contributed reagents, materials, and analysis tools; M.L., J.X., Q.L., and L.P. wrote and edited the paper.

Funding: This research was funded by the National Key Research and Development Program of China (No. 2017YFE0111600), and Technological Innovation Major Project of Hubei Province (2016AAA023).

Acknowledgments: We gratefully acknowledge many important contributions from the researchers of all the reports cited in our paper.

Conflicts of Interest: The authors declare no conflict of interest.

References

1. Yang, Q.; Guo, Z.Y.; Lin, X.X. Research on fire-retarded open-graded friction course used in road tunnel. *J. Tongji Univ.* **2005**, *33*, 316–320.

2. Shu, B.; Wu, S.; Dong, L.; Wang, Q.; Liu, Q. Microfluidic Synthesis of Ca-Alginate Microcapsules for Self-Healing of Bituminous Binder. *Materials* **2018**, *11*, 630. [CrossRef] [PubMed]

3. Zhang, H.L.; Yu, J.Y.; Zhu, C.Z. Flame Retardants in Bitumens and Nanocomposites. *Eng. Mater.* **2015**, 167–186. [CrossRef]

4. Hull, T.R.; Stec, A.A.; Growth, F.; Processes, P.; Strategies, F.R.; Cinausero, N.; Laoutid, F.; Piechaczyk, A.; Leroy, E. *Fire Retardancy of Polymers: New Strategies and Mechanisms*; Hull, T.R., Kandola, B.K., Eds.; Royal Society of Chemistry: London, UK, 2009.

5. Wu, S.; Cong, P.; Yu, J.; Luo, X.; Mo, L. Experimental investigation of related properties of asphalt binders containing various flame retardants. *Fuel* **2006**, *85*, 1298–1304. [CrossRef]

6. Wu, S.; Mo, L.; Cong, P.; Yu, J.; Luo, X. Flammability and rheological behavior of mixed flame retardant modified asphalt binders. *Fuel* **2008**, *87*, 120–124. [CrossRef]

7. Cong, P.; Yu, J.; Wu, S.; Luo, X. Laboratory investigation of the properties of asphalt and its mixtures modified with flame retardant. *Constr. Build. Mater.* **2008**, *22*, 1037–1042. [CrossRef]

8. Xu, T.; Huang, X. A TG-FTIR investigation into smoke suppression mechanism of magnesium hydroxide in asphalt combustion process. *J. Anal. Appl. Pyrol.* **2010**, *87*, 217–223. [CrossRef]

9. Tao, X.; Huang, X. Study on combustion mechanism of asphalt binder by using TG–FTIR technique. *Fuel* **2010**, *89*, 2185–2190.

10. Xu, T.; Wang, H.; Li, G.; Huang, X. Inhibitory action of flame retardant on the dynamic evolution of asphalt;pyrolysis volatiles. *Fuel* **2013**, *105*, 757–763. [CrossRef]

11. Bonati, A.; Merusi, F.; Bochicchio, G.; Tessadri, B.; Polacco, G.; Filippi, S.; Giuliani, F. Effect of nanoclay and conventional flame retardants on asphalt mixtures fire reaction. *Constr. Build. Mater.* **2013**, *47*, 990–1000. [CrossRef]

12. Pei, J.; Wen, Y.; Li, Y.; Shi, X.; Zhang, J.; Li, R.; Du, Q. Flame-retarding effects and combustion properties of asphalt binder blended with organo montmorillonite and alumina trihydrate. *Constr. Build. Mater.* **2014**, *72*, 41–47. [CrossRef]

13. Levchik, S.V.; Weil, E.D. A Review of Recent Progress in Phosphorus-based Flame Retardants. *J. Fire Sci.* **2006**, *24*, 345–364. [CrossRef]

14. Zhang, C.; Xu, T.; Shi, H.; Wang, L. Physicochemical and pyrolysis properties of SARA fractions separated from asphalt binder. *J. Therm. Anal. Calorim.* **2015**, *122*, 241–249. [CrossRef]

15. Wu, K.; Zhu, K.; Kang, C.; Wu, B.; Huang, Z. An experimental investigation of flame retardant mechanism of hydrated lime in asphalt mastics. *Mater. Des.* **2016**, *103*, 223–229. [CrossRef]

16. Elbasuney, S. Surface engineering of layered double hydroxide (LDH) nanoparticles for polymer flame retardancy. *Powder Technol.* **2015**, *277*, 63–73. [CrossRef]

17. Shabanian, M.; Basaki, N.; Khonakdar, H.A.; Jafari, S.H.; Hedayati, K.; Wagenknecht, U. Novel nanocomposites consisting of a semi-crystalline polyamide and Mg–Al LDH: Morphology, thermal properties and flame retardancy. *Appl. Clay Sci.* **2014**, *90*, 101–108. [CrossRef]

18. Edenharter, A.; Breu, J. Applying the flame retardant LDH as a Trojan horse for molecular flame retardants. *Appl. Clay Sci.* **2015**, *114*, 603–608. [CrossRef]

19. Ding, P.; Kang, B.; Zhang, J.; Yang, J.; Song, N.; Tang, S.; Shi, L. Phosphorus-containing flame retardant modified layered double hydroxides and their applications on polylactide film with good transparency. *J. Colloid Interface Sci.* **2015**, *440*, 46–52. [CrossRef] [PubMed]

20. Cui, P.; Wu, S.; Xiao, Y.; Wan, M.; Cui, P. Inhibiting effect of Layered Double Hydroxides on the emissions of volatile organic compounds from bituminous materials. *J. Clean. Prod.* **2015**, *108*, 987–991. [CrossRef]

21. Zhang, S.; Yan, Y.; Wang, W.; Gu, X.; Li, H.; Li, J.; Sun, J. Intercalation of phosphotungstic acid into layered double hydroxides by reconstruction method and its application in intumescent flame retardant poly(lactic acid) composites. *Polym. Degrad. Stabil.* **2018**, *147*, 142–150. [CrossRef]

22. Xu, W.; Zhang, B.; Wang, X.; Wang, G.; Ding, D. The flame retardancy and smoke suppression effect of a hybrid containing $CuMoO_4$ modified reduced graphene oxide/layered double hydroxide on epoxy resin. *J. Hazard. Mater.* **2017**, *343*, 364–375. [CrossRef] [PubMed]

23. Xu, W.; Zhang, B.; Xu, B.; Li, A. The flame retardancy and smoke suppression effect of heptaheptamolybdate modified reduced graphene oxide/layered double hydroxide hybrids on polyurethane elastomer. *Compos. Part A* **2016**, *91*, 30–40. [CrossRef]

24. Kok, M.V.; Karacan, C.O. Behavior and Effect of SARA Fractions of Oil during Combustion. *SPE Reserv. Eval. Eng.* **1997**, *3*, 380–385.

25. Shi, H.; Xu, T.; Zhou, P.; Jiang, R. Combustion properties of saturates, aromatics, resins, and asphaltenes in asphalt binder. *Constr. Build. Mater.* **2017**, *136*, 515–523. [CrossRef]

26. Shi, H.; Xu, T.; Jiang, R. Combustion mechanism of four components separated from asphalt binder. *Fuel* **2017**, *192*, 18–26. [CrossRef]

materials MDPI

Article

Evaluating the Effect of Hydrophobic Nanosilica on the Viscoelasticity Property of Asphalt and Asphalt Mixture

Wei Guo [1], Xuedong Guo [1], Mengyuan Chang [2] and Wenting Dai [1,*]

[1]   School of Transportation, Jilin University, Changchun 130022, China; guowei17@mails.jlu.edu.cn (W.G.); guoxd@jlu.edu.cn (X.G.)
[2]   China Communications Second Highway Survey Design and Research Institute Co. Ltd., Wuhan 4300704, China; chenwx16@mails.jlu.edu.cn
*   Correspondence: daiwt@jlu.edu.cn; Tel.: +86-130-8681-4298

Received: 26 October 2018; Accepted: 16 November 2018; Published: 19 November 2018

Abstract: Viscoelasticity property of bitumen is closely related to the service life of bituminous pavement. This paper evaluated the impact of one of the most efficient and widely used nanomaterials in various industries called hydrophobic nanosilica on the viscoelasticity property of bitumen and asphalt mixture. In this paper, three hydrophobic nanosilica modified bitumens and asphalt mixtures were researched by conventional physical properties test, SEM test, FTIR test, DSC test, DSR test, static creep test and dynamic creep test. The results showed that the introduction of hydrophobic nanosilica could strengthen the viscosity of asphalt more effectively and had better dispersion than hydrophilic nanosilica in asphalt. From conventional physical properties test and rheological performance test, hydrophobic nanosilica could weaken the temperature susceptibility of bitumen observably. From DSR test, hydrophobic nanosilica modified asphalt had a lower sensitivity and dependence on temperature and frequency than hydrophilic nanosilica modified asphalt. The Cole–Cole diagrams indicated that hydrophobic nanosilica exhibited good compatibility with asphalt compared with hydrophilic nanosilica. Newly formed chemical bonds were found in the hydrophobic nanosilica modified asphalt and its mixture with stone according to SEM test, FTIR test, and DSC test, which is the biggest difference from the modification mechanism of hydrophilic nanosilica modified asphalt. Through static and dynamic creep test, it found that the addition of hydrophobic nanosilica can significantly reduce the creep strain at the same temperature.

Keywords: viscoelasticity; nanomaterial; hydrophobic nanosilica; hydrophilic nanosilica

---

1. Introduction

Asphalt material has been widely used in airfield and high-grade road pavement due to its numerous advantages such as smoothness, low-vibration, high-automated construction, and easy-maintenance [1–3]. As a typical viscoelastic material, the damage of asphalt pavement mainly depends on some external factors such as temperature, stress, oxygen, water and so forth. With the effect of temperature, vehicle, water, and oxygen, some damages including low-temperature cracking, high-temperature rutting, moisture-induced damage and deterioration of pavement structure would be generated easily [4]. He et al. investigated the rheological properties of lignosulfonate intercalated layered double hydroxides modified bitumen (LLMB) before and after UV aging by temperature susceptibility, dynamic viscoelastic properties and rheological model analyses. The results showed that the introduction of lignosulfonate intercalated layered double hydroxide (LS-LDHs) could strengthen the anti-UV aging ability of bitumen more effectively [5]. Liu et al characterized the effect of graphene oxide on viscosity, rheological properties, creep and recovery behavior, cracking resistance, and thermal properties of unaged/RTFO aged binders. The results demonstrated that the addition of trace amounts of graphene oxide (no more than 0.2 wt %) enhanced the paving temperature (120–135 °C) viscosity, high-temperature elasticity and rutting resistance of non-modified/SBS modified binders [6]. Gao et al. study the rheological behavior and the high-temperature performances of bio-asphalt binders. The results indicated the incorporation of bio-oil reduced the anti-rutting performance of asphalt [7]. In consequence, to lengthen the service life of the asphalt pavement, it is crucial to enhance overall performance and investigate rheological property of asphalt.

Because of the complexity of the pavement damage phenomenon, using proper additives and modifiers is considered the most cost-effective technique for prolong the pavement service life [8]. The modification of asphalt binder with polymeric materials and rubber powder is among the most common methods to improve the performance of the asphalt binder against most types of distresses [9–11]. The other method for prolonging the pavement service life is the addition of alternative environmental friendly materials. The use of industrial waste is a promising solution to minimize environmental pollution by sinking the accumulation of waste materials, which results in a reduction of the construction costs [12]. Golewski evaluated the effect of the low calcium fly ash (LCFA) addition on fracture processes in structural concretes. The results presented the properties of composites with the additive of LCFA depend on the age of the concrete tested, and mature concretes exhibit high fracture toughness at 20% additive of LCFA [13]. Woszuk analyzed the synthetic zeolites produced from fly ashes with 4 different types of crystalline structure. It was found that the viscosity of asphalt pastes with zeolitic materials increases with the increase in the amount of zeolite added. The potential usefulness of fly ash derived zeolites had also been proved [14]. Zhang et al. developed an innovative measure to solve this waste disposal problem by means of synthesizing zeolite A from the Sewage Sludge Ash (SSA) and using it as a Warm Mix Asphalt (WMA) additive, which can decrease the construction temperature of asphalt pavement, thus reducing the associated energy consumption and pollutant emission [15]. Another method for the improvement of asphalt binder characteristics that has recently gained attention is the modification of asphalt binder with nanomaterials [16]. Nano-material refers to a material with the size of less than 100 nmin at least one dimension. Due to the very tiny size and huge surface area, the properties of nanomaterials are much different from the normal-sized materials. The addition of these nano-sized particles to another material may overcome the limitations of monolithic, and asphalt binder is no exception [17]. Adding nanomaterials changes the rheological properties of the asphalt binder and might also lead to changes in the intermolecular forces within the asphalt binder structures. This is due to the fact that as dimensions reach the nanometer level, interactions at phase interfaces become largely increased. Nano silica is one of the nano-materials that have been extensively used in asphalt mixtures. Taherkhani et al. investigated the effects of nanosilica modification on some properties of a penetration grade asphalt cement and a typical asphalt concrete. The results showed that the addition of nanosilica results in the increase of stiffness, tensile strength, resilient modulus, fatigue life and resistance against permanent deformation and moisture damage [18]. Saltan et al. evaluated the performance of Hot Mix Asphalt (HMA) and bitumen by modifying bitumen with nanosilica according to Superpave mix design procedure. The results showed that nanosilica modified bitumen has the highest rutting performance, fatigue performance [19]. Firouzinia et al. examined the impact of nanosilica on the thermal properties of bitumen and asphalt mixture. The results showed that nanosilica can improve the temperature sensitivity of asphalt mixtures [20].

However, nanosilica as a kind of inorganic non-metallic nanomaterials is very prone to agglomeration. In addition, the asphalt binder is an organic cementitious material formed by high molecular hydrocarbons and non-metallic derivatives of these hydrocarbons, which makes nanosilica in asphalt binder have poor dispersibility and compatibility [21]. In order to improve the dispersion of nanosilica in organic solvents and enhance their interaction with the medium, and to broaden the application field of nanosilica, the commonly used method is to physically or chemically react the surface of the nanosilica with the surface modifier through a certain processing process [22].

The surface modification method can reduce the number of silanol structures on the surface of the nanosilica, change the structure of the surface functional groups and the atomic layer of the nanosilica particles. Thereby, the surface properties such as physicochemical adsorption of nanosilica are changed, the surface free energy of the nanoparticle and the phenomenon of agglomeration between particles is reduced.

The hydrophobic nanosilica obtained by modifying the surface of nanosilica has been applied in many fields, such as silicone rubber, thermosetting plastics, traditional coatings and many other product preparation [23,24]. Wang et al. incorporated hydrophobic nanosilica sol into polyvinylmethy-lsiloxane to prepare reinforced high-temperature vulcanized (HTV) silicone rubber. The resulted showed that HTV silicone rubber filled with 40 phr hydrophobic nanosilica sol has excellent mechanical and optical properties [25]. Ani et al. synthesized a series of nano-hybrid perfluoroalkyl oligosiloxane resins (FSi@SiO$_2$) using the hydrolysis and condensation of FVPS with tetraethylorthosilicate to improve the hydro-and oleo-phobic properties of anti-fingerprint coating [26]. Milionis et al. synthesized acrylonitrile butadiene styrene rubber (ABS) superhydrophobic nanocomposites containing various concentrations of hydrophobic silica nanoparticles by solution processing and spray coating on aluminum surfaces. The resulted showed that superhydrophobic nanocomposites has excellent wear abrasion resistance [27].

To the authors' knowledge, the open literature has no experimental studies of evaluating the effect of hydrophobic nanosilica on the viscoelasticity property of asphalt and asphalt mixture. In this study, the viscoelasticity of three hydrophobic nanosilica modified bitumens and asphalt mxitures were researched by laboratory test. Conventional physical properties test was used to prove the availability of hydrophobic nanosilica to bitumens. The SEM test, FTIR test, DSC test were carried to explore the modification mechanism of hydrophobic nanosilica. Moreover, the DSR test and creep test was employed to systematically investigated the effect of hydrophobic nanosilica on the viscoelasticity property of asphalt and asphalt mixture.

2. Materials and Sample Preparation

2.1. Asphalt

The asphalt used in this study was Donghai 70# asphalt produced by Qilu Branch of Sinopec Corp. (Qingdao, China). Laboratory tests were carried out in order to assess the conventional properties of asphalt according to Chinese standards. The test results of the asphalt used in this study are summarized in Table 1.

Table 1. Technical parameters of asphalt.

Technical Parameters	Penetration			25 °C Ductility	Softening Point	Wax Content	Flash Point	Solubility	Density
	15 °C	25 °C	30 °C						
Units		0.1 mm		cm	°C	%	°C	%	g·cm^{-3}
Test results	22.8	67.3	109.8	>130	48.8	1.8	328	≥99.9	1.036
Test procedure		GB/T0606-2011		GB/T0605 -2011	GB/T0606 -2011	GB/T0615 -2011	GB/T0611 -2011	GB/T0607 -2011	GB/T0603 -2011

2.2. Hydrophobic Nanosilica

The molecular formula of nanosilica is SiO$_2$·H$_2$O, which is an amorphous structure of SiO$_2$. The surface structure of nanosilica is shown in Figure 1. The hydroxyl groups on the surface can be classified into three types: isolated hydroxyl group, continuous hydroxyl group, and twin hydroxyl group. Isolated hydroxyl group is a free hydroxyl group that does not participate in the reaction; continuous hydroxyl group is formed by two hydrogen groups that generate hydrogen bonds and associate with each other; two hydroxyl groups attached to the same Si are called twin hydroxyl groups. Isolated hydroxyl group and continuous hydroxyl group have no hydrogen bond. The reinforcing effect of nanosilica as a modifier mainly comes from the active structure (–Si–OH) on the surface of the particles, and it is easy to bond the surrounding ions. The surface of nanosilica is uniformly distributed with a layer of silanol and siloxane group with strong water absorption. The water molecules can be physically covered on the surface of the particles or chemically bonded to the hydroxyl groups on the Si atoms. Therefore, nanosilica shows a strong affinity for water. In order to reduce the number of surface silanol structures and change the surface functional groups of the primary nanosilica particles, the surface modifier is used to modify the nanosilica to realize the transition from hydrophilic to hydrophobic.

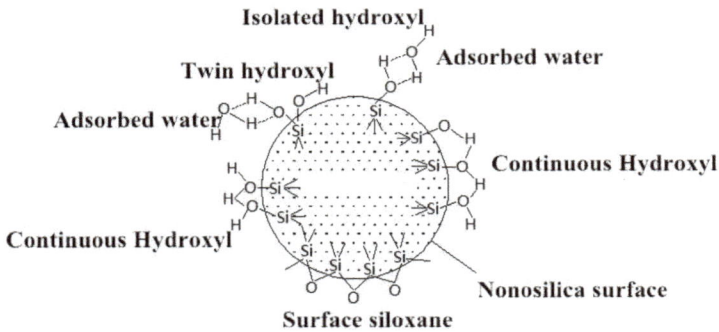

Figure 1. Surface structure of nanosilica.

The hydrophobic nanosilica used in this research included organochlorosilane surface modified nanosilica (OCS-Silica), silicone organic compound surface modified nanosilica (SOC-Silica) and silane coupling agent surface modified nanosilica (SCA-Silica). The nanosilica and three hydrophobic nanosilica materials are obtained from Changtai Weina Chemical Co., Ltd. (Shouguang, China). The technical properties of four nanosilica materials provided by Changtai Weina Chemical Co., Ltd. (Shouguang, China) have been presented in Table 2.

Table 2. Technical parameters of four nanosilica materials.

Type	Water Characteristics	BET (m²/g)	Average Particle Size (nm)	Loss on Drying (105 °C, 2 h, wt %)	Loss on Ignition (1000 °C, 2 h, wt%)	pH Value	SiO₂ Content (wt %)
Silica	Hydrophil-ia	130 ± 25	12	≤1.5	≤1	3.7–4.7	≥99.8
OCS-Silica	Hydropho-bicity	110 ± 20	16	≤0.5	≤2	3.6–4.3	≥99.8
SOC-Silica	Hydropho-bicity	100 ± 25	14	≤0.5	≤2	4.0–6.0	≥99.8
SCA-Silica	Hydropho-bicity	125 ± 20	12	≤0.5	1.5–2.5	5.0–8.0	≥99.8
Standard values	-	130 ± 30	≤20	≤3.0	≤10.0	3.7–6.5	≥99.8
Test procedure	GB/T 20020	GB/T 20020	GB/T 20020	GB/T 20020	GB/T 20020	GB/T 20020	GB/T 20020

2.3. Preparation of Samples

According to prior researches, mixing of nanoparticles into asphalt binder with a high shear mixer at the rate of more than 4000 RPM, at a temperature between 130 °C and 165 °C, and an approximate duration of 40 min to an hour is suggested [28]. It should be noted that the mentioned studies used various types of nanoparticles including surface modified ones which might need less mixing speed or mixing time to achieve proper dispersion.

Based on the modified asphalt preparation test, a mixture of nano Silica, nano OCS-Silica, nano SOC-Silica and nano SCA-Silica with asphalt binder is produced. 3 wt % nano Silica, nano OCS-Silica, nano SOC-Silica or nano SCA-Silica and the base asphalt were taken into a pot heated the mixture to 140 °C with oil-bath temperature control system. Then the mixture was stirred with the speed of 2000 rpm for 1 h to ensure homogeneous blending, to obtain nano Silica modified asphalt, OCS-Silica modified asphalt, SOC-Silica modified asphalt or SCA-Silica modified asphalt. The asphalt modified by nano Silica nano OCS-Silica, nano SOC-Silica and nano SCA-Silica were denoted by SBA, OCSSBA, SOCSBA, and SCASBA, respectively. Meanwhile, base asphalt was denoted by BA.

3. Experimental Procedures

3.1. Conventional Physical Properties Test

The conventional physical properties tests on BA, SBA, OCSSBA, SOCSBA, and SCASBA Such as softening point, ductility and penetration were performed according to Chinese standards GB/T0606-2011, GB/T0605-2011, and GB/T0604-2011, respectively. The viscosity test was carried out using a Brookfield rotational viscometer to measure the apparent viscosities of modified and unmodified asphalt binders according to Chinese standards GB/T0625-2011. In order to evaluate the aging resistance property of asphalt, the rolling thin film oven (RTFO) test was conducted to simulate the effect of construction aging according to Chinese standards GB/T0610-2011. After the rolling thin film oven tests, damaged specimens were collected for ductility test, penetration test and weighing test to identify the aging resistance property of modified asphalt.

3.2. SEM Test

SEM has been used to investigate the micromorphology of asphalt binders and mixtures. These microstructure further investigated into fracture morphology, combustion properties, rheological properties, modification mechanism interaction between modifier and aggregate, structural, viscoelastic and physicochemical properties. The microstructure of three hydrophobic nanosilica materials and nanosilica were examined by SU8000 electronic microscopy (Tianmei Co., Tokyo, Japan). The hydrophilic nanosilica and hydrophobic nanosilica particles were dispersed onto double-sided adhesive carbon tape, which was applied to the aluminum sample mounts. For the operation of SU8000 electronic microscopy, the accelerating voltage was 1–5 kV, the emission current was 9 A, and the working distance was 2.8–8 mm.

3.3. FTIR Test

Different molecular structures and functional groups have diverse absorptions of infrared radiation. Therefore the chemical composition can be identified by analyzing the infrared absorption spectra of samples. A Vertex 70 Fourier Transform Infrared Spectroscope (BRUKER OPTICS, Changchun, China) was used for the determination of the chemical compositions of both neat and modified binders. Spectrometry was performed at wavenumbers between 400 and 4000 cm^{-1} with a wavelength wavelength accuracy of 0.01 cm^{-1}, resolution of 4 cm^{-1}, the rate of 0.16 cm/s, and 32 scans. A small amount of each sample was placed on a diamond ATR prism and a certain amount of pressure was applied for all the samples.

3.4. DSC Test

Differential scanning calorimetry (DSC) analyzes energy differences between the neat and modified asphalt binders at different temperatures. DSC tests were conducted using a DSC-500B (Shanghai yinuo precision instrument Co. Ltd., Changchun, China) to get the thermographs of both neat and modified asphalt binders. Approximately 5 mg of each asphalt sample was used in the test. The range of the testing temperature was from −60 °C to 50 °C, and the heating rate was 5 °C/min under a nitrogen atmosphere. Two parameters, glass transition temperature T_g and endothermic energy value Δcp, were used to quantify the thermal properties of both neat and modified asphalt binders. Three specimens of each asphalt were measured to ensure reliability of the DSC results [29,30].

3.5. DSR Test

The viscosity and elasticity of asphalt binders can be evaluated using the DSR machine based on parameters such as complex modulus (G^*), phase angle (δ), Shenoy rutting parameter ($G^*/[1-(\sin\delta\tan\delta)^{-1}]$), storage modulus ($G^*\cdot\cos\delta$).

In this paper, temperature sweep test and frequency sweep test were performed using a Bohlin automatic dynamic shear (ADS) rheometer (DSRII, Malvern, UK), including the host, rheometer measuring system or measuring fixture, temperature control unit, and the instrument software.

3.5.1. Temperature Sweep Test

Temperature sweep tests were conducted on modified and unmodified asphalt binders samples under control deformation mode at strain equals to 12% and frequency equals to 1.59 Hz. This frequency of oscillation can simulate the shear stress corresponding to a traffic speed of approximately 100 Km/h. Complex shear modulus (G^*) and phase angle (δ) were measured at temperatures ranging from 28 °C to 76 °C at 2 °C increments for both asphalt binders. The diameter of the asphalt sample fixture was 25 mm, while the test spacing of asphalt sample was 1 mm.

3.5.2. Frequency Sweep Test

The frequency sweep tests were carried out to evaluate the shear-deformation performance and temperature sensitivity of the hydrophobic nanosilica modified binders. The frequency ranged from 0.1–200 rad/s and these tests were conducted at four different temperatures (i.e., 22 °C, 40 °C, 58 °C, and 76 °C). The parallel plates with diameter of 25 mm and gap of 1 mm were selected for the tests at the medium-high temperature range (e.g., 58 °C, 76 °C), whereas parallel plates with diameter of 8mm and gap of 2 mm were selected for the tests at the low-medium temperature range (e.g., 22 °C, and 40 °C).

From all of the two tests, the linear viscoelastic parameters such as G^* (complex shear modulus) and δ(phase angle) was obtained. Meanwhile, the Shenoy rutting parameter $G^*/[1-(\sin\delta\tan\delta)^{-1}]$ and storage modulus $G^*\cdot\cos\delta$ was calculated. The G^* represents the binders' ability of resistance to deformation and the δ reflects the binders' viscoelastic ratio. Consists of G^* and δ, the Shenoy rutting parameter $G^*/[1-(\sin\delta\tan\delta)^{-1}]$ is adopted as the index representing the asphalt binders' property (i.e., resistance to permanent deformation) and the storage modulus $G^*\cdot\cos\delta$ is adopted as the index representing the asphalt binders' elastic property [31].

3.5.3. Master Curve Generation Method

The master curve was used to describe the rheological properties of hydrophobic nanosilica modified asphalt in this study. The master curve of G^* was generated by the applying of the time-temperature superposition principle. To construct master curve, a reference temperature is selected and the dates collected from frequency sweeps at all other temperatures are shifted to one at reference temperature by shift factors. In this research, reference temperature was 40 °C.

3.5.4. Compatibility Analysis Method

The modifiers of nanomaterial can modify the asphalt from the microstructure, which is fundamentally different from other conventional modifiers, and whether the nanoparticles can be uniformly dispersed microscopically, and formed a stable structure with the asphalt component is the key to improve the asphalt. Cole–Cole plots were the most effective methods to identify compatibility and these plots have been used widely to study compatibility in polyblends. As a consequence, Cole–Cole diagrams are employed to evaluate the compatibility of hydrophobic nanosilica modified asphalt. Cole–Cole diagrams consist of representations of the complex viscosity components ($\eta^* = \eta' - \eta''$) in the complex plane (η', η''). Evolution of η'' with η' in conformity with symmetrical parabolas is considered as the proof of compatibility, while deviation from this symmetry is related to incompatibility for modified asphalt.

3.6. Viscoelastic Test of Asphalt Mixture

The viscoelasticity of modified asphalt mixtures can be evaluated by creep tests. SBA and SCASBA were selected to prepare asphalt concrete sample. The gradation of asphalt concrete is listed in Table 3. The Marshall mix design method is still widely used in the road laboratory, all around the world [32–34]. Optimal asphalt content of SBA-AC and SCASBA-AC mixture were ensured as 5.5% and 5.2% of dry aggregate using Marshall mix design method according to JTG E20-2011. The method of load application for creep testing in a laboratory environment can be classified into two types: static and dynamic loading. For both tests, the specimens were prepared according to Chinese standards GB/T0738-2011, with a diameter of 100 mm and height of approximately 63.5 mm. The specimens were kept in a climatic cabinet for 24 h.

Table 3. The gradation of AC-16.

Sieve (mm)	16	13.2	9.5	4.75	2.36	1.18	0.6	0.3	0.15	0.075
Gradation (wt %)	95	88	73	46	31	21.5	15.5	11.5	5.5	6

3.6.1. Static Creep Test

The test was performed as uniaxial static creep test, the specimen was subjected to a constant compressive load of 150 kPa at a test temperature of 30 °C, 40 °C, 50 °C, 60 °C. In this work, the test was performed for 3600 s.

3.6.2. Dynamic Creep Test

In a dynamic creep test, a repeated uniaxial stress is applied to an asphalt specimen for a number of load cycles while axial strain is measured in the same direction as the loading using linear variable differential transducers (LVDTs). The applied dynamic load used in this test was a sequence of rectangular pulses. The pulse duration was 1 s, and the rest period before the next pulse was 1 s. The test was repeated 1800 times. A static axial stress of 15 KPa was applied for 90 s to the top platen of the specimen for proper bedding, as in a static creep test. The deviator stress repeated loading was 150 kPa, and the testing temperature was set to 30 °C, 40 °C, 50 °C, and 60 °C.

4. Results and Discussion

4.1. Conventional Physical Properties Analysis

The results of empirical tests such as penetration, softening point, and 10 °C ductility for unmodified and modified asphalt are given in Table 4. It is readily seen by reference to this Table that the physical properties of asphalt were greatly affected following the addition of modifier. From penetration test, it was observed that addition of three hydrophobic nanosilica materials reduced the temperature sensibility of asphalt, but hydrophilic nanosilica is the opposite. The reason for the improvement in PI values of three hydrophobic nanosilica materials modified binder samples is the hydrophilic nanosilica is an inorganic modifier and is prone to agglomeration. Segregation and uneven dispersion may occur in the modified asphalt material, and the improvement of the temperature sensibility of the asphalt is not obvious. The hydrophobic nanosilica after surface modification is more compatible with asphalt, which is beneficial for adsorbing light oil (OCS-Silica, SOC-Silica) in the asphalt or bonding with some components in the asphalt (SCA-Silica), the original structural form of the asphalt is changed, so that the temperature sensitivity of the asphalt after the modification is lowered. From softening point test, it was observed that addition of hydrophobic nanosilica and hydrophilic nanosilica could improve the softening point. From 10 °C ductility test, it was observed that the ductility of modified asphalt binders was very close, and the ductility of hydrophobic nanosilica modified asphalt is slightly lower than that of hydrophilic nanosilica modified asphalt. Based on conventional tests, it can be concluded that addition of hydrophobic nanosilica results in relatively harder binder compared to SBA, which may be beneficial for rutting resistance.

Table 4. The conventional physical properties tests results of BA, SBA, OCSSBA, SOCSBA, and SCASBA.

Property	BA	SBA	OCSSBA	SOCSBA	SCASBA
15 °C Penetration (d-mm)	22.8	18.4	26.7	32.5	23.5
25 °C Penetration (d-mm)	67.3	56.7	64.1	66.1	58.9
30 °C Penetration (d-mm)	109.8	92.7	93	100.5	88.5
PI	−0.87	−1.06	0.64	1.45	0.24
Softening Point (°C)	48.8	50.1	54.4	53.7	53.7
10 °C Ductility (cm)	45.7	21.5	25.5	20.6	17.6

The high-temperature viscosity of asphalt was used to evaluate if the asphalt is fluid enough during pumping and mixing. Table 5 shows the viscosity of five asphalt binders at 115 °C, 135 °C and 155 °C. The viscosity of control BA binder was found to be 0.4503 pa.s at 135 °C, whereas SBA, OCSSBA, SOCSBA, and SCASBA showed viscosity of 0.7724 pa.s, 1.125 pa.s, 2.978 pa.s, and 1.987 pa.s. From the result it can be observed that all the binder combinations satisfied the Superpave specification (<3 pa.s). This shows that the satisfactory workability of asphalt mixes can be achieved by blending OCSSCA, SOCSBA, and SCASBA. The sequence of viscosity for the modified and unmodified asphalt binder were SOCSBA > SCASBA > OCSSBA > SBA > BA, respectively. This indicates that the addition of OCS-Silica, SOC-Silica, and SCA-Silica obviously increased the viscosity of asphalt compared with nanosilica, and SOC-Silica can improve the viscosity of asphalt most effectively among the four modifiers.

Table 5. The viscosity test results of BA, SBA, OCSSBA, SOCSBA, and SCASBA.

Property	BA	SBA	OCSSBA	SOCSBA	SCASBA
Rotational Viscosity at 115 °C (pa.s)	1.397	2.195	2.673	6.786	3.531
Rotational Viscosity at 135 °C (pa.s)	0.4503	0.7724	1.125	2.978	1.987
Rotational Viscosity at 155 °C (pa.s)	0.1927	0.2621	0.512	1.483	0.684

The paper used the Saal formula recommended by ASTMD 2493 to analyze the viscosity test results of asphalt binder samples. The Saal formula is as follows.

$$lglg\left(\eta \times 10^3\right) = n - m \times lg(T + 273.13),$$ (1)

Where η is the viscosity, T is the temperature, and the n, m is the regression coefficient. The regression coefficient m represents the temperature sensibility of asphalt.

The m value of BA, SBA, OCSSBA, SOCSBA, and SCASBA was −3.234, −3.253, −2.697, −1.906, and −2.252. It can be seen that the addition of three hydrophobic nanosilica materials reduced the temperature sensibility of asphalt obviously compared nanosilica. This conclusion is consistent with the results of penetration test.

The RTFO test was conducted for five asphalt binders. In the standard RTFO test, the binders were conditioned at 163 °C, for 85 min, with sufficient air. Then, the RTFO-aged samples were used for penetration test, 10 °C ductility test and weighing test. Table 6 shows the test results of five asphalt binders after RTFO test. The mass change of asphalt during RTFO tets is usually due to the volatilization of light oil and the oxygen gain in chemical reaction, and the volatile weight loss is much larger than oxygen gain. Thus, the mass loss rate can reflect the anti-aging ability of asphalt, and the smaller the mass loss rate, the better the anti-aging performance. After calculation, it is found that both hydrophilic nanosilca and hydrophobic nanosilca can effectively reduce the mass loss of BA after aging, but the mass loss of hydrophobic nanosilica modified asphalt is relatively smaller. The main reason may be that the hydrophobic nanosilica after surface modification has stronger lipophilicity and can adsorb more light oil in the asphalt. In the process of thermal oxygen aging, the hydrophobic nanosilica better reduces the volatilization of the light oil than the hydrophilic nanosilica, thereby improving its anti-aging property.

Table 6. Physical parameter of BA, SBA, OCSSBA, SOCSBA, and SCASBA before and after rolling thin film oven (RTFO) test.

Materials	Unaged Binder			RTFO Aged Binder		
	Weight (g)	25 °C Penetration (d-mm)	10 °C Ductility (cm)	Weight (g)	25 °C Penetration (d-mm)	10 °C Ductility (cm)
BA	35.21	67.3	45.7	35.0	42.6	12.2
SBA	35.52	56.7	21.5	35.4	38.54	9.2
OCSSBA	35.41	64.1	25.5	35.3	47.49	14.9
SOCSBA	35.25	66.1	20.6	35.2	44.58	13.7
SCASBA	35.31	58.9	17.0	35.2	40.74	8.6

The structural composition of asphalt after aging changes to show hardening. The 25 °C residual penetration ratio is the percentage of penetration at 25 °C of asphalt before and after aging. The greater the 25 °C residual penetration ratio, the less the degree of hardening of the asphalt, and the better the anti-aging ability. The 25 °C residual penetration ratio of five asphalt binders was 63.2%, 68%, 74.1%, 67.4%, and 69.1% respectively. This indicated that hydrophobic nanosilca is added into asphalt to form a new structure that can resist thermo-oxidative ageing and relieve asphalt hardening.

The 10 °C residual ductility is the percentage of ductility at 10 °C of asphalt before and after aging. The 10 °C residual ductility can reflect the low temperature performance of asphalt binders after RTFO test. The larger 10 °C residual ductility is, the better the anti-aging ability of asphalt. After RTFO, the ductility of BA decreased from 45.7 cm to 12.2 cm, and the 10 °C residual ductility of BA is 26.7%. The 10 °C residual ductility of SBA, OCSSBA, SOCSBA, and CASBA was 42.8%, 58.4%, 66.5% and 50.6%. This indicated hydrophobic nanosilica can effectively improve the low temperature mechanical properties of asphalt after aging compared with hydrophilic nanosilica.

4.2. SEM Analysis

The microstructure of three hydrophobic nanosilica materials and nanosilica were examined by SU8000 electronic microscopy (Tianmei Co., Tokyo, Japan). The scanning electron micrographs of nanosilica and three hydrophobic nanosilica materials observed at magnifying power of ×10,000 and ×60,000 are shown in Figures 2a–d and 3a–d.

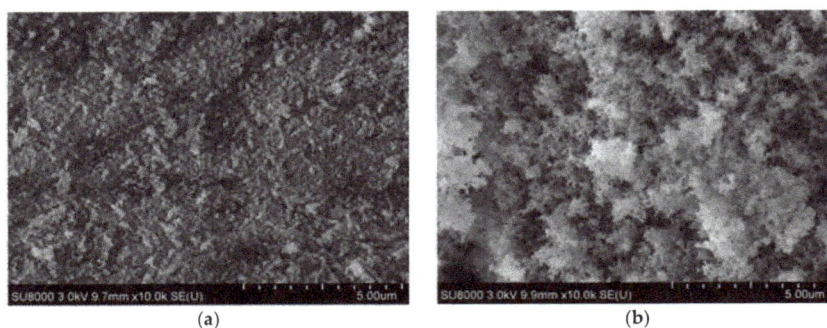

(a)

(b)

Figure 2. *Cont.*

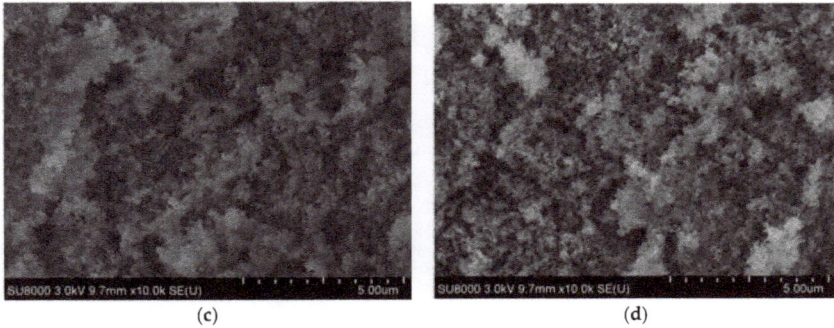

Figure 2. SEM images of nanosilica and three hydrophobic nanosilica materials at magnifications of ×10,000. (**a**) Silica; (**b**) OCS-Silica; (**c**) SOC-Silica; (**d**) SCA-Silica.

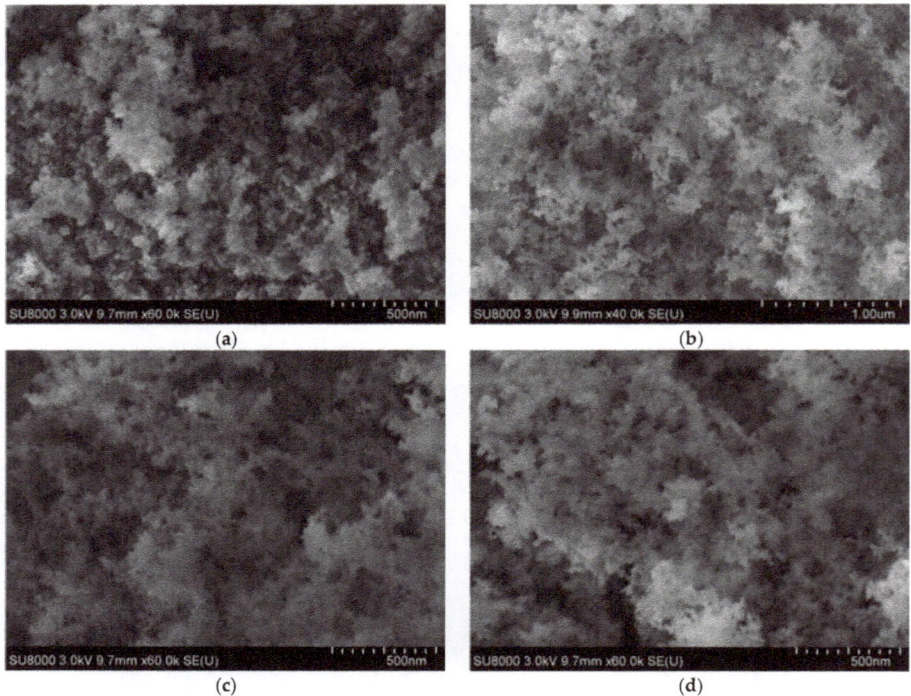

Figure 3. SEM images of nanosilica and three hydrophobic nanosilica materials at magnifications of ×60,000. (**a**) Silica; (**b**) OCS-Silica; (**c**) SOC-Silica; (**d**) SCA-Silica.

It can be seen from Figure 2 that the primary SiO_2 particles in the unmodified nannosilica are a zero-dimensional spherical nanomaterial, and the particles of the three hydrophobic nanosilica materials have a 3D continuous structure composed of many nanoparticles. These characteristics are highly valuable for hydrophobic nanosilica applications, serving excellent steric stabilization and ensuring efficient dispersion and good catalytic sites accessibility Further magnified SEM image of the four nanomaterials presented in Figure 3 indicates that a densely structured aggregate, agglomerate chain and a network structure formed by van der Waals force between spherical nanosilica particles in the hydrophilic silica, the structure can be easily destroyed by mechanical force, but it is easier to regenerate. The morphology of three hydrophobic nanosilica materials (OCS-Silica, SOC-Silica and SCA-Silica) is similar after surface modification. The main difference between the three hydrophobic nanosilica materials and the hydrophilic silica material is that the surface is grafted with organic groups. The high polymer forms an organic film on the surface of the nanosilica particles, and the organic film reduces the surface tension of particles,improves the agglomeration between particles and increases the dispersion in asphalt.

4.3. FTIR Analysis

The FTIR spectra of the nano silica and three hydrophobic nanosilica materials are given in Figure 4. Compared to the nanosilica, the FTIR spectra of three hydrophobic nanosilica materials has no absorption peak and wide absorption band in the region of 3700–3150 cm^{-1}, indicating that most of the hydroxyl groups on the surface of nanosilica have chemically reacted with the modifier, resulting in the change of the surface properties of nanosilica. The OCS-Silica, SOC-Silica and SCA-Silica produced –CH_3 symmetric deformation vibration peaks at 1273.63 cm^{-1}, 1265.25 cm^{-1}, and 1261.85 cm^{-1} respectively, indicating the organic structure of the modifier is grafted to the nanosilca. The new peaks around 2960.31 cm^{-1} corresponding to –CH_3 and –CH_2– stretching vibration from silicone modifier appeared in the FTIR spectra of the OCS-Silica, manifesting that the long chain of silicone organic compounds has been successfully grafted onto the surface of nanosilica. As can be seen from Figure 4d, the new peaks at 2920.61 cm^{-1} and 2851.53 cm^{-1} correspond to –CH_3 and =CH_2 antisymmetric stretching vibration from silane coupling agent. The observation suggested that the hydroxyl group on the surface of nanosilica reacted with the silane coupling agent, so that the nanosilica changed from hydrophilic to hydrophobic and lipophilic.

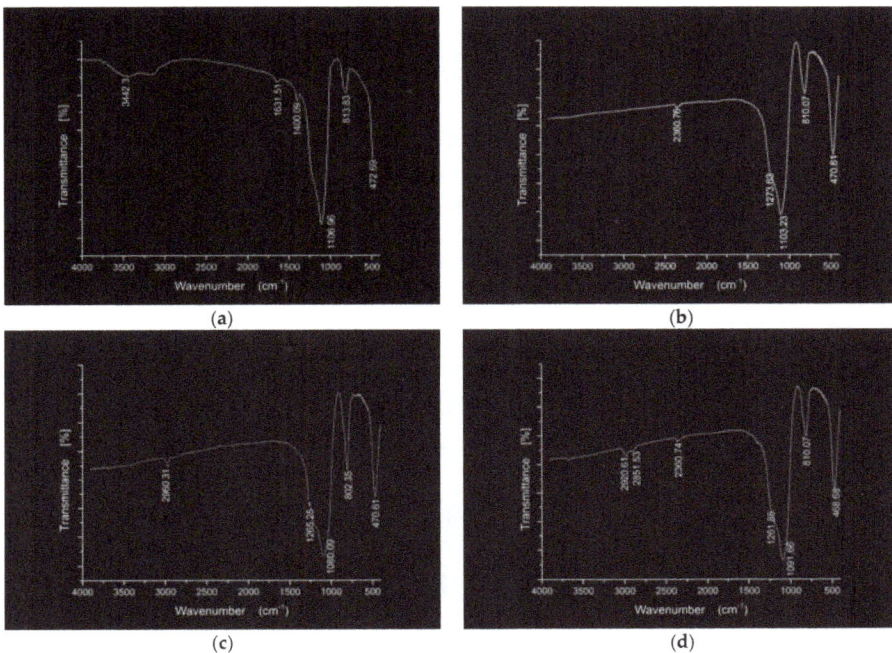

Figure 4. FTIR spectra of the nano silica and three three hydrophobic nanosilica materials. (**a**) Silica; (**b**) OCS-Silica; (**c**) SOC-Silica; (**d**) SCA-Silica.

In order to further explore the interaction between hydrophobic nanosilica materials and asphalt binder, five asphalt binders were tested by infrared spectroscopy. The FTIR spectra of the neat and modified asphalt binders are given in Figures 5 and 6.

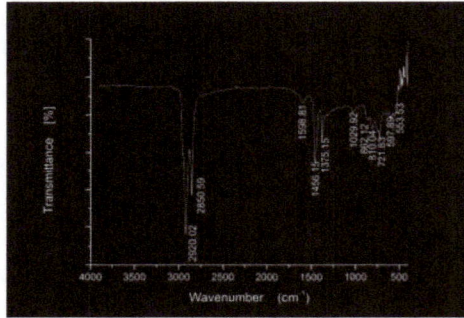

Figure 5. FTIR spectra of the BA materials.

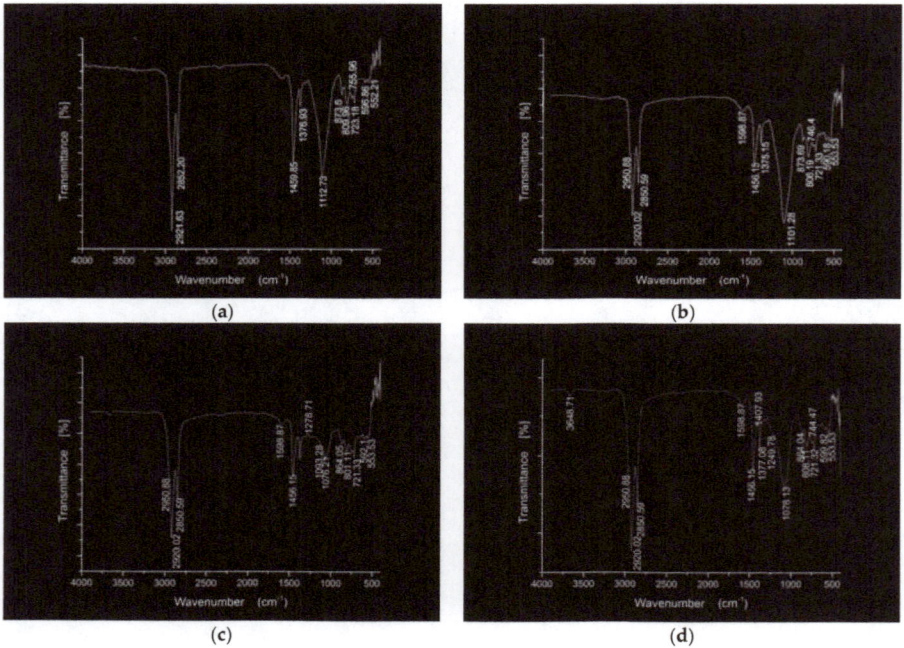

Figure 6. FTIR spectra of the SBA, OCSSBA, SOCSBA and SCASBA materials. (**a**) SBA; (**b**) OCSSBA; (**c**) SOCSBA; (**d**) SCASBA.

As shown in Figure 5, two peaks emerged at 2920.02 cm^{-1} and 2850.59 cm^{-1}, which correspond to –CH$_2$– antisymmetric stretching vibration and symmetrical stretching vibration, suggesting the presence of a certain saturated hydrocarbon in the BA sample. The peaks at 1598.80 cm^{-1} correspond to C=C stretching vibration from benzene ring skeleton in aromatic. The peaks at 1456.15 cm^{-1} and 1375.15 cm^{-1} correspond to –CH$_2$– flexural vibration and –CH$_3$ symmetric deformation vibration (Umbrella vibration) from aliphatic series, respectively. The peak at 1029.92 cm^{-1} are attributed to S=O vibration, which indicates that the thermo-oxidative aging occurred between base asphalt during production and transportation. The peaks at 862.12 cm^{-1} and 810.04 cm^{-1} correspond to =C–H out-of-plane deforming vibration from polycyclic aromatic hydrocarbon. The peak at 721.83 cm^{-1} are attributed to alkyl radical flexural vibration. The observations illustrate that asphalt is a complex mixture of aliphatic compounds, saturated hydrocarbons, aromatic hydrocarbons and some heteroatom derivatives.

Compared to BA, the FTIR spectra of SBA is similar to BA. Only one obvious difference is that SBA produced Si–O–Si symmetrical stretching vibration at 1112.73 cm^{-1} from nanosilica, indicating the nanosilica does not chemically react with the asphalt,but a simple physical blending in the nanosilica modified asphalt. Compared to BA, three hydrophobic nanosilica modified asphalt produced a new absorption peak at 2950.88 cm^{-1} which is due to –O–Si(CH$_3$) structure from OCS-Silica, the captain chain from SOC-Silica and methyl antisymmetric stretching vibration from SCA-Silica. Compared to BA, the SOCSBA produced O-Si stretching vibration at 1093.28 cm^{-1}, and the peaks at 864.05 cm^{-1} and 801.11 cm^{-1} is heightened due to the superposition of Si–(CH$_3$)$_2$ stretching vibration from the captain chain. Compared to BA, the FTIR spectrum of SCASBA produced continuous peaks at 1407 cm^{-1} and 1249.78 cm^{-1}. This may be due to the fact that the Q group on the surface of SCA-Silica reacts with the aromatic component in the modified asphalt, creating a new chemical bond. From the position of the peak, it can be inferred that the continuous peaks at 1407 cm^{-1} and 1249.78 cm^{-1} correspond to hydrogen bond stretching vibration from aromatic or aliphatic ether. Based on the above observations and analysis, it can be summarized that in addition to the physical blending reaction, there are certain chemical reactions in the three hydrophobic nanosilica modified asphalt.

The early diseases occurrence of asphalt pavement, such as low-temperature cracking, high-temperature rutting, moisture-induced damage and so forth, has a great relationship with the adhesion between asphalt and aggregates. This adhesion is an extremely complicated physical-chemical process. In order to clarify the connection between the hydrophobic nanosilica modified asphalt and the stone, base asphalt mortar, nanosilica modified asphalt mortar and SCA-Silica modified asphalt mortar mixed with limestone power were tested by infrared spectroscopy. Limestone is an alkaline stone with an alkaline active center on the surface. It is easy to react with acidic components in asphalt to form water-insoluble compounds. Therefore, limestone is selected as the test material. The base asphalt mortar, nanosilica modified asphalt mortar and SCA-Silica modified asphalt mortar mixed with 5 wt % limestone power with a particle size of less than 0.15 mm were denoted by BA-SP, SBA-SP and SCASBA-SP, respectively. The FTIR spectra of the neat and modified asphalt mortar are given in Figures 7 and 8.

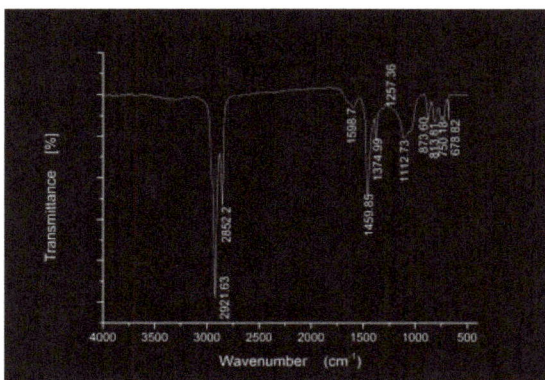

Figure 7. FTIR spectra of the BA-SP.

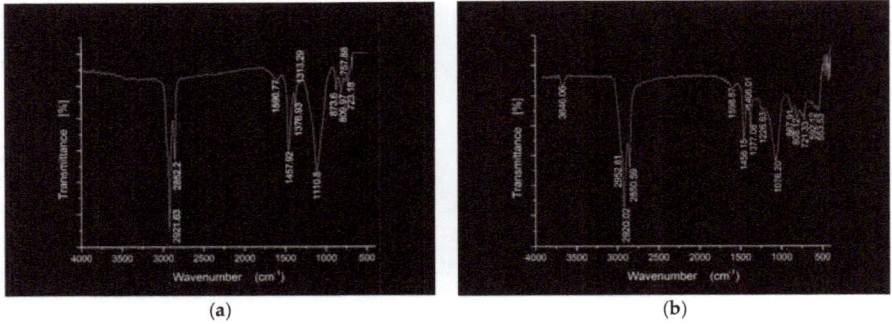

Figure 8. FTIR spectra of the SBA-SP and SCASBA-SP samples. (**a**) SBA-SP; (**b**) SCASBA-SP.

Compared to base asphalt (Figure 5), the FTIR spectra of BA-SP is similar to BA. Only one obvious difference is that BA-SP produced a new peak at 1112.73 cm^{-1} from CaCO$_3$, SiO$_2$ and AI$_2$O$_3$ of limestone. Qualitative assessment of spectrums of BA-SP and SBA-SP show no significant peak except for wavenumbers at 1110.8 cm^{-1} for SBA-SP. The marked peak at 1110.8 cm^{-1} for this sample is related to the superposition of Si-O-Si from nanosilica and limestone power vibration and indicates no obvious chemical reaction in the SBA-SP. Qualitative assessment of spectrums of SBA-SP and SCASBA-SP show no significant peak except for wavenumbers from 1600–1200 cm^{-1} for the SCASBA-SP. The new continuous tiny absorption peak indicates some chemical bonds between nano SCA-Silica modified asphalt and limestone powder were formed, but the number was less.

4.4. DSC Analysis

DSC was employed to evaluate the effect of nano silica and three hydrophobic nanosilica materials on the low-temperature thermal properties of the HMAs. Figure 9 presents the thermal properties of each type of studied asphalt binder, quantified by the glass transition temperature (T_g) and endothermic energy value (Δcp). The value of T_g indicates the temperature below which the asphalt binder changes from viscoelastic state to glassy state. The lower the T_g, the better the asphalt to resist low-temperature cracking. The value of Δcp represents the difference in heat capacity before and after the glass transition of asphalt binder, which reflects the amount of energy required to change the collective state of the unit mass of asphalt binder. Under the same system, the lower the Δcp value, the higher the crosslinking density of the components in the asphalt binder.

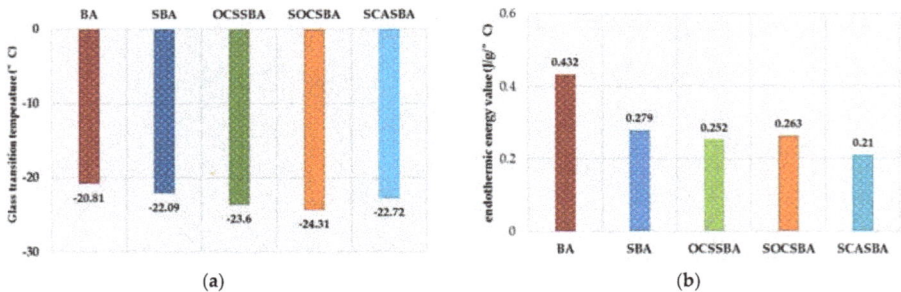

Figure 9. Differential scanning calorimetry (DSC) results of the neat and modified asphalt binders. (**a**) Glass transition temperature of the neat and modified asphalt binders; (**b**) Endothermic energy value of the neat and modified asphalt binders.

The incorporation of hydrophobic nanosilica materials greatly enhanced the low-temperature thermal properties of asphalt binder, as shown in Figure 9. The presence of 3 wt % nanosilica only resulted in a T_g 1.28 °C lower than the base asphalt without the nanosilica. In contrast, the presence of 3 wt % hydrophobic nanosilica (OCS, SOC and SCA) resulted in a T_g 2.79 °C, 3.5 °C 1.91 °C lower than the neat asphalt, respectively. This suggests that hydrophobic nanosilica is much more effective than nanosilica in reducing the glass transition temperature of the control asphalt. The addition of 3 wt % nanosilica in base asphalt decreased its Δcp value by 35.4%, and the presence of 3 wt % hydrophobic nanosilica (OCS, SOC and SCA) decreased the Δcp value of the base asphalt by 41.7%, 39.1% and 51.4%, respectively. These results demonstrate that the hydrophobic nanosilica improved the intermolecular crosslinking density in asphalt binders. According to the above FTIR spectra analysis, the hydrophobic nanosilica can chemically react with the control asphalt and thus restrict the movements of asphalt molecules, which may explain the observed increases in the crosslinking density of the asphalt binder.

In summary, adding hydrophobic nanosilica into asphalt will reduce the glass transition temperature of the binder (especially SOC modifier), and it will increase the crosslinking degree of the asphalt.

4.5. DSR Analysis

4.5.1. Temperature Sweep Test Results

The temperature sweep tests of BA, SBA, OCSSBA, SOCSBA, and SCASBA were carried out, and the complex shear modulus (G*), phase angle (δ), Shenoy rutting parameter $G^*/[1-(\sin\delta\tan\delta)^{-1}]$ and black diagram were analyzed.

As seen in Figure 10, for BA, SBA, OCSSBA, SOCSBA, and SCASBA, the complex shear modulus decreased with the increase of temperature, this indicated that, with the increase of temperature, the asphalt became soft, and its resistance to rutting declined. For the temperature range of 28–46 °C, the complex shear modulus of modified and unmodified asphalt binder is similar, the sequence of complex modulus for the modified and unmodified asphalt binder were G* (SCASBA) > G* (OCSSBA) > G* (SOCSBA) > G* (SBA) > G* (BA), respectively. The results showed that the addition of nanosilica and hydrophobic nanosilica provided a certain rheological resistance of asphalt. Moreover, the complex modulus of hydrophobic nanosilica modified asphalt was always greater than nanosilica modified asphalt, which is because the hydrophobic nanosilica particles have higher affinity with asphalt binder functional groups through surface attraction. For the temperature range of 46–76 °C, the complex shear modulus of five asphalt binders has a gap. The complex shear modulus of SOCSBA decreases faster than SBA after 68 °C, but still higher than BA. The OCSSBA and SCASBA have higher complex shear modulus, indicating that OCS-Silica or SCA-Silica can effectively improve the strength of asphalt and its ability to resist deformation. Moreover, a higher slope of the fitted line means that a higher temperature sensitivity of the asphalt is observed. As can be witnessed, the temperature sensitivity of hydrophobic nanosilica modified asphalt was lower than that of nanosilica modified asphalt and base asphalt, which is consistent with the conclusions of the conventional physical properties test.

As shown in Figure 11, for BA, SBA, OCSSBA, SOCSBA, and SCASBA, the phase angle gradually increased with the increase of temperature, this indicated that the viscous component increased with increasing temperature. The sequence of initial phase angle for the modified and unmodified asphalt binder were δ (SCASBA) > δ (SOCSBA) > δ (SBA) > δ (OCSSBA) > δ (BA), respectively. This indicated that the elastic components of modified asphalt increased with the addition of the nanosilica and hydrophobic nanosilica. For the temperature range of 52–76 °C, the increase of the phase angle of five asphalt binder becomes smaller with the increase of temperature. This indicated that five asphalt binder had become closer to viscous fluids in the temperature range. The addition of nanosilica does not change the viscoelastic nature of the asphalt material. It retains a small amount of elastic component at higher temperatures. While OCS-Silica, SOC-Silica and SCA-Silica exerts the advantages of nanosilica, it can form a more stable system with asphalt. With the increase of temperature, the decrease rate of complex shear modulus and the conversion rate of elastic components of hydrophobic nanosilica modified asphalt become slow compared with base asphalt and nanosilica modified asphalt, which proves the recoverable deformation increased in the hydrophobic nanosilica modified asphalt.

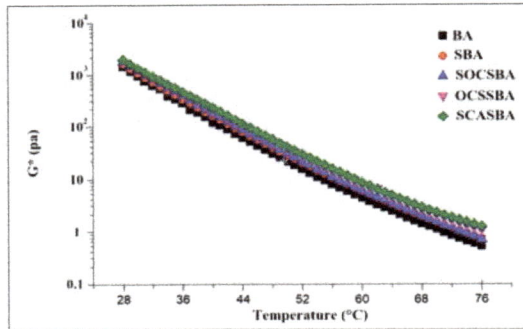

Figure 10. The relationship between complex shear modulus and temperature for modified and unmodified asphalt.

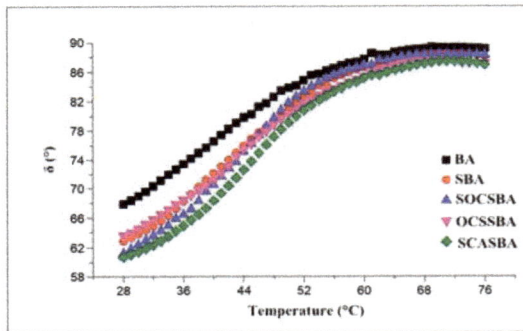

Figure 11. The relationship between complex shear modulus and temperature for modified and unmodified asphalt.

The research on permanent deformation of asphalt mixture is generally at 60 °C, but the fact is that the permanent deformation of asphalt mixture will produced within a certain temperature range. Shenoy proposed $G^*/[1-(\sin\delta\tan\delta)^{-1}]$ as a refinement to $G^*\cdot\sin\delta$. The parameter was derived through a semi-empirical approach. It represents the inverse of the non-recoverable compliance and is derived by linking the strain response in the creep experiment with the complex modulus G^* from oscillatory shear experiments at a matched timescale. This parameter is more sensitive to phase angle than the Superpave parameter; therefore, it better explains the changes in elastic properties when adding the polymeric modifier. In this paper, the Shenoy rutting parameter was adopted to evaluate the ability of modified asphalt and unmodified asphalt to resist permanent deformation.

Figure 12 shows the relationship between Shenoy rutting parameterShenoy rutting parameter and temperature. The changes of the Shenoy rutting parameters of five asphalt binders were relatively consistent with the changes of complex shear mudulus. The Shenoy rutting parameter of SBA has a certain improvement compared with BA, which manifested that the incorporation of nanosilica improves the high temperature performance of asphalt. The Shenoy rutting parameter of OCSSBA, SOCSBA, and SCASBA has a significant improvement compared with SBA, which indicated that incorporation of hydrophobic nanosilica can better improve the high temperature stability and resistance to high temperature permanent deformation of asphalt.

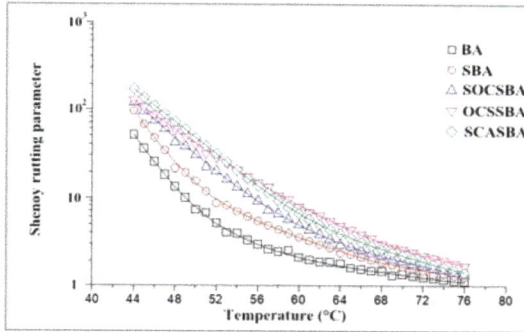

Figure 12. The relationship between Shenoy rutting parameter and temperature for modified and unmodified asphalt.

Complex shear modulus and phase angle obtained from temperature sweep tests can construct black diagram. Figure 13 shows the viscoelastic data of hydrophobic nanosilica modified asphalt using the black diagram. As shown in Figure 13, the modified asphalt has a smaller phase angle than the base asphalt in the case of the same complex shear modulus. The temperature of pavement in summer is generally around 60 °C, and the complex shear modulus of asphalt is about 5 KPa, the sequence of phase angle for the modified and unmodified asphalt binder were BA > SOCSBA > SBA > OCSSBA > SCASBA, respectively. This indicated that an increase in the elastic behavior and a better rutting resistance for hydrophobic nanosilica modified asphalt. From the curve trend, there is a certain gap between OCSSBA, SOCSBA SCASBA, and SBA, which also reflects the modification mechanism of hydrophobic nanosilica grafted with organic groups in asphalt is different from that of nanosilica.

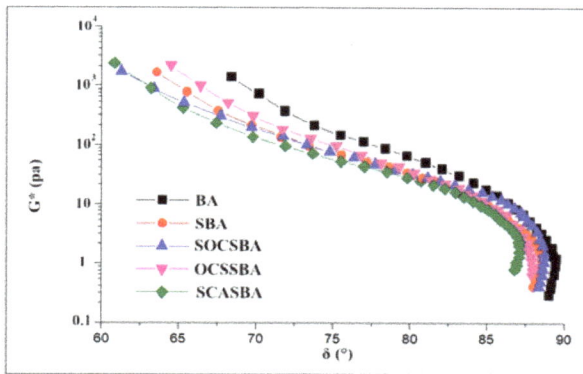

Figure 13. Black diagram for five modified asphalt binders.

4.5.2. Frequency Sweep Test Results

The frequency sweep tests of three hydrophobic nanosilica modified asphalt, nanosilica modified asphalt and base asphalt were carried out, and the complex shear modulus (G^*), phase angle (δ), storage modulus $G^* \cdot \cos\delta$ with frequency were analyzed.

The test results showed that the change trend of complex shear modulus of modified asphalt with frequency was consistent with that of base asphalt, which indicated that the addition of modifiers did not change the viscoelastic nature of the asphalt material. Figure 14 shows the change of complex shear modulus and phase angle of BA with frequency at 22 °C, 40 °C, 58 °C, and 76 °C. As can be seen in Figure 14, with the increase of frequency, the phase angle reduced gradually at the same temperature, while the complex shear modulus increased gradually. This indicated that the asphalt exhibited greater elastic properties as the frequency increased, while the phase angle was maintained at a plateau at 76 °C, which may be the reason that asphalt material has been approximated to viscous fluids viscous fluids caused by the increase in temperature, and the effect of load frequency on the elastic composition of the material is also greatly reduced.

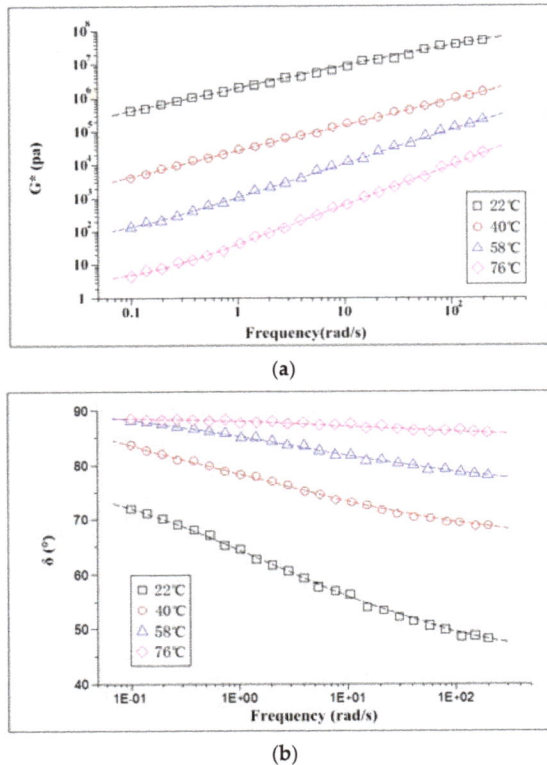

(a)

(b)

Figure 14. The changes of complex shear modulus and phase angle of base asphalt with frequency. (a) complex shear modulus (b) phase angle.

In the summer, the pavement is most likely to produce permanent deformation under continual load and the reference temperature is 58 °C. Figure 15 shows the changes of storage modulus and phase angle of five asphalt binder with frequency at 58 °C. As can be seen in Figure 15, the storage modulus of BA and modified asphalt increased with the increase of the frequency, and the storage modulus of hydrophobic nanosilica has a larger increase than hydrophilic nanosilica. Under the driving load, the vibration frequency of the pavement structure is about 15 Hz (94.2 rad/s). When the load frequency is 100 rad/s, the storage modulus of the BA is 0.508 KPa, and the storage modulus of SBA, SOCSBA, SCASBA, and OCSSBA is 0.589 KPa, 0.683 KPa, 0.745 KPa and 0.712 KPa. The improve of the storage modulus of SBA, SOCSBA, SCASBA and OCSSBA relative to BA is 15.9%, 34.4%, 40.2%, and 46.7%. Thus, the effect of hydrophobic white nanosilica on the storage modulus of asphalt is significant. This is because the nanosilica grafted with organic groups increases the complex shear modulus of asphalt on the one hand, and reduces the phase angle of the asphalt on the other hand, thereby increasing the storage modulus of the asphalt material. This can improve the deformation recovery of the asphalt and reduce the occurrence of permanent deformation during the hydrophobic nanosilica modified asphalt in application.

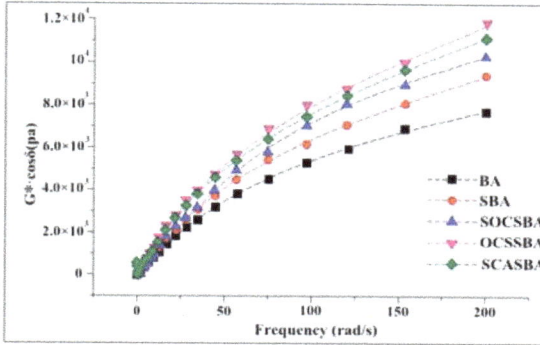

(a)

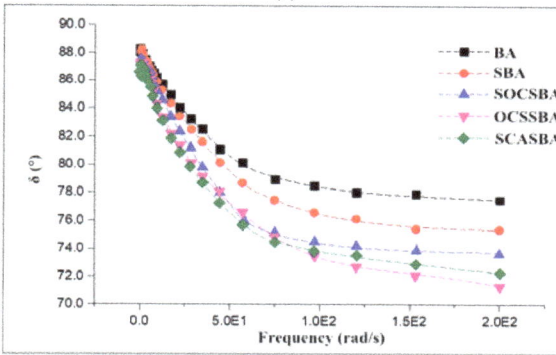

(b)

Figure 15. The changes of storage modulus and phase angle of five asphalt binder with frequency at 58 °C. (**a**) storage modulus (**b**) phase angle.

As shown in Figure 15, the phase angle of BA and modified asphalt decreased with the increase of the frequency, and the phase angle of five asphalt binders has a large drop before the frequency of 100 rad/s, and then the frequency has less influence on the phase angle of five asphalt binders. At a frequency of 100 rad/s, the phase angle of the hydrophobic nanosilica modified asphalt is reduced by 2–4° compared to hydrophilic nanosilica modified asphalt, which contributes greatly to the storage modulus of the asphalt at high temperatures.

4.5.3. Master Curve Generation

The complex shear modulus of modified asphalt in Figure 16 increased after mixing with modifiers at low and high frequency region compared with base asphalt. In the high frequency region, the complex modulus of five asphalts is relatively close, while in the low frequency region, the complex modulus of five asphalts has a large gap. Moreover, the hydrophobic nanosilica materials modified asphalt exhibits a higher complex modulus than BA and SBA, and the trend of the master curve of the hydrophobic nanosilica materials modified asphalt is relatively moderate, indicating that the hydrophobic nanosilica materials modified asphalt has a relatively low sensitivity and dependence on frequency and exhibits strong stability compared with BA and SBA.

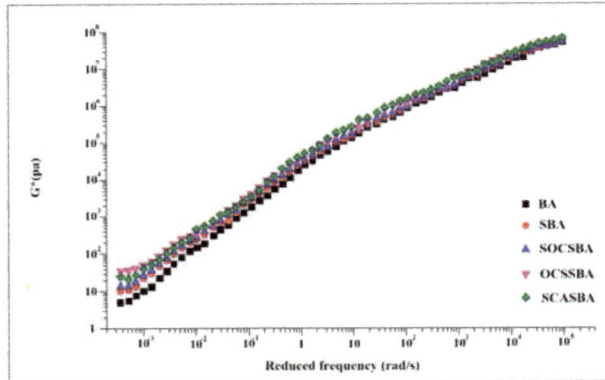

Figure 16. Complex shear modulus master curves for five asphalt binders.

The reason is that hydrophilic nanosilica acts as a physical cross-linking points in the asphalt. When the load acts on the modified asphalt, it can transmit a certain stress through the hydrophilic nanosilica, thereby the stability of the whole system is improved. However, a large amount of hydroxyl groups exist on the surface of the hydrophilic nanosilica, so that the nanoparticles are in a non-thermodynamic stable state, and the nanoparticles tend to agglomerate, forming a large-sized agglomerate with several weak interfaces, resulting in poor dispersion in asphalt.

The hydrophobic nanosilica successfully grafted with organic groups reduces the formation of agglomerates. In addition to the advantages of hydrophilic nanosilica, the hydrophobic nanosilica has better dispersion in asphalt due to the hydrophobic and oleophilic property. According to the previous analysis, it can be seen that $-Si(CH_3)_3$ on the surface of OCS-Silica, $-O-Si-CH=CH_2$ on the surface of SOC-Silica and the $-HN_2$, $-SH$ on the surface of SCA-Silica can be more closely combined with the organic components in the asphalt through the coupling effect, so that the hydrophobic nanosilica modified asphalt has more advantage against the load.

4.5.4. Compatibility Analysis Based on Cole–Cole Diagram

Figures 17 and 18 shows Cole–Cole diagrams of five asphalt binders at 58 °C and 76 °C.

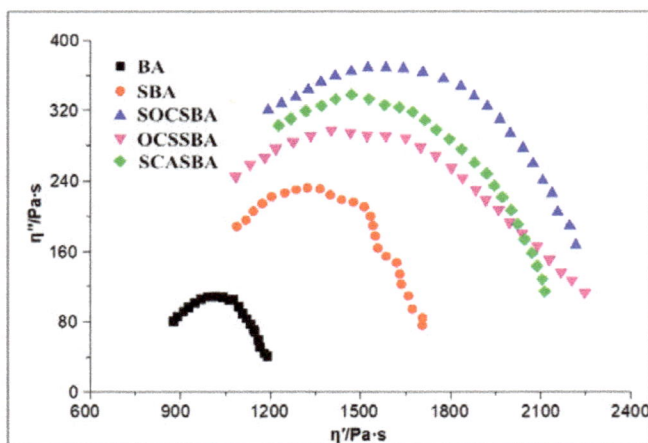

Figure 17. Cole–Cole diagrams of five asphalt binders at 58 °C.

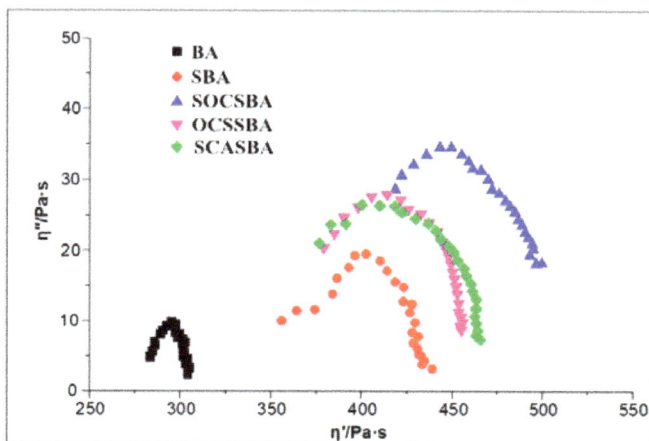

Figure 18. Cole–Cole diagrams of five asphalt binders at 76 °C.

As can be observed, η'' shows its descending following its ascending as the increase of η', where occurs peak value in the curves. What's more, the form of Cole–Cole plots differ from each other and base asphalt together with three kinds of hydrophobic nanosilica materials modified asphalt shows the best symmetrical parabolas, followed by nanosilica modifed asphalt (Figure 17). Data concentrated on the left side of the Cole–Cole curves mean the essential elastic properties at the tested temperature, whereas those shifted to the right show a transition from an elastic behavior at low temperature to a viscous behavior as the temperature increases. For all samples studied at 58 °C in this research, the curves occurred in the right side, predicting the dominant viscous behavior. The curves of goodness of fit for parabolas show that the compatibility of asphalt Table 7 shows the regression equation of the Cole–Cole curve of five asphalt binders at 58 °C and 76 °C. The R^2 index represents the correlation coefficient between the Cole–Cole curve and symmetrical parabolas. The higher the R^2 index, the higher the correlation of Cole–Cole curve and symmetrical parabolas, and the better the compatibility of the modified asphalt. The sequence of R^2 for the modified and unmodified asphalt binder at 58 °C were BA > SCASBA > OCSSBA > SOCSBA > SBA, respectively. This indicated that the compatibility of SBA is the worst and the surface modified nanosilica exhibits good compatibility with asphalt compared with untreated nanosilica. It is noteworthy that the form of Cole–Cole plots drastically changed and deviated from symmetrical parabolas as the higher temperatures (Figures 17 and 18), which shows the compatibility of asphalt becomes worse as the temperature increases. The sequence of R^2 for the modified and unmodified asphalt binder at 76 °C were BA > SOCSBA > OCSSBA > SCASBA > SBA, respectively.

In summary, the three hydrophobic nanosilica particles modified by different surface modification method have better compatibility with the asphalt compared with unmodifed nanosilica partilces at different temperatures, and form a more uniform blend system with asphalt.

There are three reasons to explain this phenomenon. First, the surface of the hydrophobic nanosilica particles is not easily affected by moisture to cause agglomeration, and physical mixing can produce more particles of original size. Second, according to the principle of similar compatibility, the three hydrophobic nanosilica particles shows lipophilic property. It is easier to move and homogenize in the organic phase system of asphalt subjected to the high-speed shearing. Third, the connection mechanism between the surface of the modified nanosilica particles and asphalt has changed.

Table 7. Regression equation of the Cole-Cole curve of five asphalt binders at 58 °C and 76 °C.

Temperature	Materials	Regression Equation	R^2
58 °C	BA	$y = -0.0016x^2 + 3.1951x - 1466.5$	0.9945
	SBA	$y = -0.001x^2 + 2.617x - 1483.7$	0.9756
	OCSSBA	$y = -0.0003x^2 + 0.896x - 361.66$	0.9904
	SCASBA	$y = -0.0005x^2 + 1.5916x - 848.81$	0.9924
	SOCSBA	$y = -0.0004x^2 + 1.3948x - 712.94$	0.9818
76 °C	BA	$y = -0.0138x^2 + 7.9311x - 1133$	0.9702
	SBA	$y = -0.0051x^2 + 4.0859x - 793.78$	0.8339
	OCSSBA	$y = -0.0024x^2 + 1.9758x - 380.09$	0.9323
	SCASBA	$y = -0.0048x^2 + 3.9745x - 798$	0.8874
	SOCSBA	$y = -0.0052x^2 + 4.6083x - 990.86$	0.9506

4.6. Viscoelastic Analysis of Asphalt Mixture

4.6.1. Static Creep Test Results

The corresponding static creep strain plotted against the loading time is shown in Figure 19. It can be seen that the static creep strain curves of all asphalt mixtures was basically the same, and the deformation can be divided into three parts: the instantaneous elastic part, which is generated at the initial stage of loading, can be regarded as the instantaneous loading of the initial stress; viscoelastic delay part, the deformation growth rate of the asphalt mixture gradually decreases with time; deformation stable part, the deformation tends to a stable growth rate (the creep failure stage is not reached in this experiment). As the temperature increases, the strain of the asphalt mixture increases. At the beginning, the creep strain increment is very large. After 100 s, the creep strain increment gradually slows down. After 500 s, the creep strain increment gradually stabilizes and decreases with the increase of temperature. The addition of SCA-Silica can significantly reduce the creep strain of asphalt mixture at the same temperature, and the reduction descreases as time increases. The creep strain difference of SBA-AC and SCASBA-AC gradually increases with increasing temperature.

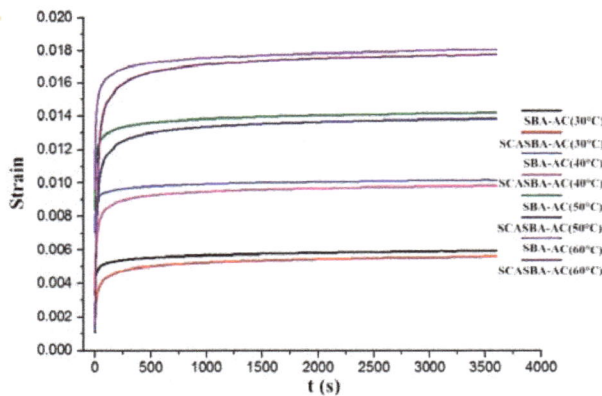

Figure 19. Static creep strain curves of SBA-AC and SCASBA-AC at different temperatures.

In order to further study the improvement of the viscoelasticity of asphalt concrete by adding SCA-Silica, the static creep Burger's model parameters of SBA-AC and SCASBA-AC based on Levenberg-Marquardt algorithm was calculated separately, as shown in Table 8. It should be taken in cosideration that by now, some more refined model have been developed in the literature. Pasetto et al. introduced a visco-elasto-plastic constitutive model for the characterisation of stress-strain behaviour in bituminous mixtures [35]. Li et al. obtained the Burger's viscoelastic parameters including E-1, eta (1) E-2, eta (2) and the steady-state creep rate K according to the creep curve, and analyzed the effects of test temperature, stress level and aggregate gradation on the viscoelastic parameters of asphalt mixture [36].

Table 8. Static creep Burger's model parameters of asphalt mixture under different temperatures.

Parameters	SBA-AC 30 °C	SCASBA-AC 30 °C	SBA-AC 40 °C	SCASBA-AC 40 °C	SBA-AC 50 °C	SCASBA-AC 50 °C	SBA-AC 60 °C	SCASBA-AC 60 °C
E_1	36.5	56.0	19.7	31.8	15.7	18.7	13.3	16.7
E_2	114.1	62.7	83.6	35.0	44.3	15.1	28.3	10.4
η_1	627,332.3	566,977.3	453,161.0	346,585.7	329,777.6	203,886.5	225,465.8	163,449.7
η_2	3632.8	3386.6	1271.1	959.2	860.7	356.4	525.9	273.0
R	0.973	0.987	0.982	0.991	0.975	0.991	0.979	0.994

It can be seen from Table 8 that the Burgers model parameters gradually decrease with increasing temperature, and the addition of SCA-Silica increases the elastic modulus (E_1) of asphalt mixture, and the elastic modulus (E_2) and viscosity coefficient (η_1, η_2) are correspondingly reduced. In addition, E_2 belongs to the Kelvin models, and its variation is affected by the parallel viscosity coefficient (η_2). Thus, it can be considered that, in a certain temperature range, the addition of SCA-Silica reduces the viscosity and enhances the elasticity of the asphalt mixture, so that the instantaneous elastic deformation is reduced, the viscous flow deformation is increased, and the delayed deformation of the viscoelasticity is increased.

4.6.2. Dynamic Creep Test Results

The characteristic change in dynamic creep strain can be interpreted as follows: as the number of cycles increases, the dynamic creep strain increases. It is interesting to note that the dynamic creep strain tends to increase with an increasing number of cycles only during the first 200 cycles; thereafter, the dynamic creep strain increment becomes negligibly small (Figure 20). Progressive increment in dynamic creep strain is obvious for all test specimens. The addition of SCA-Silica can reduce the creep strain of asphalt mixture at the same temperature, but it is not obvious at the beginning. When the number of cycles is the same, the reduction gradually decreases with the increase of temperature. At the same temperature, as the number of cycles increases, the difference of SBA-AC and SCASBA-AC gradually increases.

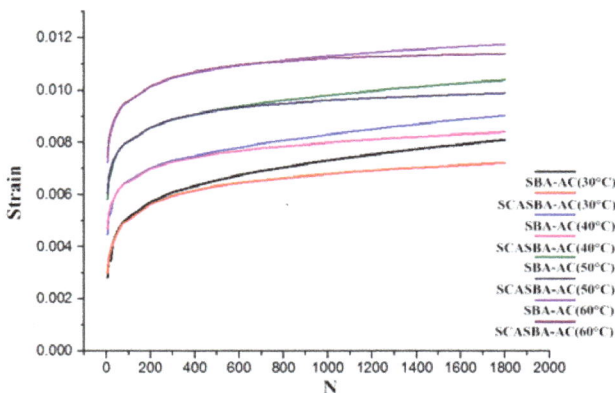

Figure 20. Dynamic creep strain curves of SBA-AC and SCASBA-AC at different temperatures.

265

In order to further study the improvement of the viscoelasticity of asphalt concrete by adding SCA-Silica, the dynamic creep Burgers model parameters of SBA-AC and SCASBA-AC based on Levenberg-Marquardt algorithm was calculated separately, as shown in Table 9.

Table 9. Dynamic creep Burger's model parameters of asphalt mixture under different temperatures.

Parameters	SBA-AC 30 °C	SCASBA-AC 30 °C	SBA-AC 40 °C	SCASBA-AC 40 °C	SBA-AC 50 °C	SCASBA-AC 50 °C	SBA-AC 60 °C	SCASBA-AC 60 °C
E_2	15.5	15.1	12.3	12.2	8.3	8.3	4.8	4.7
η_1	75,557.3	97,184.9	85,349.1	114,612.4	90,641.0	137,603.3	134,588.0	168,913.1
η_2	333.7	251.3	121.2	107.8	64.5	59.6	50.3	42.4
R	0.973	0.987	0.982	0.991	0.975	0.991	0.993	0.993

It can be seen from the above table that with the increase of temperature, the viscosity coefficient (η_1) of asphalt mixture increases under dynamic loading, and the elastic modulus (E_2) and viscosity coefficient (η_2) decrease gradually. When static load is applied, the incorporation of SCA-Silica increases the elastic modulus (E_1) of the asphalt mixture, the elastic modulus (E_2) and viscosity coefficient (η_1, η_2) decrease correspondingly, so that the instantaneous elastic deformation is reduced, the viscous flow deformation and the delayed deformation of the viscoelasticity is synchronously increased. When dynamic load is applied, the incorporation of SCA-Silica makes the elastic modulus (E_2) and viscosity coefficient (η_2) of asphalt mixture gradually decrease, the viscosity coefficient (η_1) increases, the viscous flow deformation decreases, and the delayed deformation of the viscoelasticity increases.

Overall, the hydrophobic nanosilica modifier can effectively improve the viscoelasticity property of asphalt and asphalt mixture. The price of normal asphalt is about 3000 rmb/t, and the price of hydrophobic nanosilica modified asphalt is about 3300 rmb/t. The comprehensive unit price analysis of mechanical paving hydrophobic nanosilica modified asphalt concrete of 7 cm thick is as follows. (a) Artificial cost: 2.1 rmb/cm^2 (b) Materials cost (including main material and auxiliary material): 97.7 rmb/cm^2 (c) Mechanical cost: 2.5 rmb/cm^2 (d) Other cost (including safe and civilized construction costs, fees and taxes): 7.4 rmb/cm^2. The total cost of hydrophobic nanosilica modified asphalt concrete is 109.7 rmb/cm^2, and the total cost of normal asphalt concrete is 102 rmb/cm^2. The cost of mechanical paving hydrophobic nanosilica modified asphalt concrete of 7 cm thick is 7.6% higher than normal asphalt concrete. Considering the ability of hydrophobic nanosilica modified asphalt to improve the viscoelasticity property of the asphalt concrete, an increase of 7.6% of the cost is still acceptable.

5. Conclusions

In this paper, the effect of hydrophobic nanosilica on the viscoelasticity property of asphalt and asphalt mixture was systematically investigated according to conventional physical properties test, SEM test, FTIR test, DSC test, DSR test, static creep test, and dynamic creep test. The main conclusions are as follows:

The conventional physical properties test results illustrated that addition of hydrophobic nanosilica results in relatively harder binder compared to hydrophilic nanosilica, which may be beneficial for rutting resistance. Moreover, the addition of three hydrophobic nanosilica materials could reduce the temperature sensibility of asphalt, and improve the viscosity and anti-aging property.

1. The SEM test, FTIR test, and DSC test were carried to explore the modification mechanism of hydrophobic nanosilica. The test results showed that the hydrophobic nanosilica had stronger adhesion property and can be better distributed in the asphalt without agglomeration. In addition, newly formed chemical bonds were found in the hydrophobic nanosilica modified and its mixture with stone, which is the biggest difference from the modification mechanism of hydrophilic nanosilica modified asphalt.

2. Temperature sweep test and frequency sweep test results showed that hydrophobic nanosilica modified asphalt had a lower sensitivity and dependence on temperature and frequency than hydrophilic nanosilica modified asphalt, showing strong stability. The Cole–Cole diagrams indicated that the compatibility of SBA was the worst and the surface modified nanosilica exhibited good compatibility with asphalt compared with untreated nanosilica.

3. Through static and dynamic creep test, it found that the addition of hydrophobic nanosilica can significantly reduce the creep strain at the same temperature. When static load is applied, the reduction descreases as time increases, and the creep strain difference of SBA-AC and SCASBA-AC gradually increases with increasing temperature. When dynamic load is applied, the reduction gradually decreases with the increase of temeprature. At the same temperature, as the number of cycles increases, the difference of SBA-AC and SCASBA-AC gradually increases.

Author Contributions: Data curation, W.G.; Formal analysis, W.G. and W.D.; Funding acquisition, X.G.; Investigation, X.G.; Project administration, X.G.; Writing—Original draft, W.G. and M.C.; Writing—Review & editing, X.G. and W.D.

Funding: This research was funded by the National Nature Science Foundation of China (NSFC) (Grant No. 51178204).

Conflicts of Interest: The authors declare no conflict of interest.

References

1. Yuan, J.; Wang, J.Y.; Xiao, F.P.; Amirkhanian, S.; Wang, J.; Xu, Z.Z. Impacts of multiple-polymer components on high temperature performance characteristics of airfield modified binders. *Constr. Build. Mater.* **2017**, *134*, 694–702. [CrossRef]

2. Özen, H.; Aksoy, A.; Tayfur, S.; Çelik, F. Laboratory performance comparison of the elastomer-modified asphalt mixtures. *Build. Environ.* **2008**, *43*, 1270–1277. [CrossRef]

3. Pasetto, M.; Baldo, N. Unified approach to fatigue study of high performance recycled asphalt concretes. *Mater. Struct.* **2017**, *50*, 113. [CrossRef]

4. Ahmedzade, P.; Gunay, T.; Grigoryeva, O.; Starostenko, O. Irradiated recycled high density polyethylene usage as a modifier for bitumen. *J. Mater. Civ. Eng.* **2017**, *29*, 04016233. [CrossRef]

5. He, B.Y.; Yu, J.Y.; Gu, Y.; Zhuang, R.H.; Sun, Y.B. Rheological properties of lignosulfonate intercalated layered double hydroxides modified bitumen before and after ultraviolet aging. *Constr. Build. Mater.* **2018**, *180*, 342–350. [CrossRef]

6. Liu, K.F.; Zhang, K.; Shi, X.M. Performance evaluation and modification mechanism analysis of asphalt binders modified by graphene oxide. *Constr. Build. Mater.* **2018**, *163*, 880–889. [CrossRef]

7. Gao, J.F.; Wang, H.N.; You, Z.P.; Hasan, M.R.M.; Lei, Y.; Irfan, M. Rheological behavior and sensitivity of wood-derived bio-oil modified asphalt binders. *Appl. Sci.* **2018**, *8*, 919. [CrossRef]

8. Sahebzamani, H.; Alavi, M.Z.; Farzaneh, O. Evaluating effectiveness of polymerized pellets mix additives on improving asphalt mix properties. *Constr. Build. Mater.* **2018**, *187*, 160–167. [CrossRef]

9. Zhao, Z.J.; Xu, S.; Wu, W.F.; Yu, J.Y.; Wu, S.P. The aging resistance of asphalt containing a compound of LDHs and antioxidant. *Petrol. Sci. Technol.* **2015**, *33*, 787–793. [CrossRef]

10. Kheradmand, B.; Muniandy, R.; Hua, L.T.; Solouki, A. A Laboratory investigation on the rheological properties of aged and unaged organic wax modified asphalt binders. *Petrol. Sci. Technol.* **2015**, *33*, 757–764. [CrossRef]

11. Liu, J.Y.; Li, P. Low temperature performance of sasobit-modified warm-mix asphalt. *J. Mater. Civ. Eng.* **2012**, *24*, 57–63. [CrossRef]

12. Woszuk, A.; Panek, R.; Madej, J.; Zofka, A.; Franus, W. Mesoporous silica material MCM-41: Novel additive for warm mix asphalts. *Constr. Build. Mater.* **2018**, *183*, 270–274. [CrossRef]

13. Golewski, G.L. Generalized fracture toughness and compressive strength of sustainable concrete including low calcium fly ash. *Materials* **2017**, *10*, 1393. [CrossRef] [PubMed]

14. Woszuk, A. Application of fly ash derived zeolites in warm-mix asphalt technology. *Materials* **2018**, *11*, 1542. [CrossRef] [PubMed]

15. Zhang, Y.; Leng, Z.; Zou, F.L.; Wang, L.; Chen, S.S.; Tsang, D.C.W. Synthesis of zeolite A using sewage sludge ash for application in warm mix asphalt. *J. Cleaner Prod.* **2018**, *172*, 686–695. [CrossRef]
16. Feldman, D. Polymer nanocomposites in building, construction. *J. Macromol. Sci. A Pure Appl. Chem.* **2014**, *51*, 203–209. [CrossRef]
17. Fang, C.Q.; Yu, R.E.; Liu, S.L.; Li, Y. Nanomaterials applied in asphalt modification: A review. *J. Mater.Sci. Technol.* **2013**, *29*, 589–594. [CrossRef]
18. Taherkhani, H.; Afroozi, S.; Javanmard, S. Comparative study of the effects of nanosilica and zyco-soil nanomaterials on the properties of asphalt concrete. *J. Mater. Civ. Eng.* **2017**, *29*, 04017054. [CrossRef]
19. Saltan, M.; Terzi, S.; Karahancer, S. Examination of hot mix asphalt and binder performance modified with nano silica. *Constr. Build. Mater.* **2017**, *156*, 976–984. [CrossRef]
20. Firouzinia, M.; Shafabakhsh, G. Investigation of the effect of nano-silica on thermal sensitivity of HMA using artificial neural network. *Constr. Build. Mater.* **2018**, *170*, 527–536. [CrossRef]
21. Jiangkongkho, P.; Arksornnukit, M.; Takahashi, H. The synthesis, modification, and application of nanosilica in polymethyl methacrylate denture base. *Dent. Mater. J.* **2018**, *37*, 582–591. [CrossRef] [PubMed]
22. Zhang, L.; Zhang, D.F. Study on the surface modification and characterization of nano-SiO_2. *Russ. J. Inorg. Chem.* **2005**, *50*, 925–930.
23. Ekin, N.; Icduygu, M.G.; Gultek, A. Investigation of long-term water absorption behavior of carbon fabric reinforced epoxy composites containing hydrophobic nanosilica. *Turk. J. Chem.* **2016**, *40*, 321–331. [CrossRef]
24. Suriyaprabha, R.; Karunakaran, G.; Kavitha, K.; Yuvakkumar, R.; Rajendran, V.; Kannan, N. Application of silica nanoparticles in maize to enhance fungal resistance. *IET Nanobiotechnol.* **2014**, *8*, 133–137. [CrossRef] [PubMed]
25. Wang, Q.; Zhang, Q.; Huang, Y.H.; Fu, Q.; Duan, X.J.; Wang, Y.L. Preparation of high-temperature vulcanized silicone rubber of excellent mechanical and optical properties using hydrophobic nano silica sol as reinforcement. *Chin. J. Polym. Sci.* **2008**, *26*, 495–500. [CrossRef]
26. Ani, Q.F.; Lyu, Z.J.; Shangguan, W.C.; Qiao, B.L.; Qin, P.W. The synthesis and morphology of a perfluoroalkyl oligosiloxane@SiO_2 resin and its performance in anti-fingerprint coating. *Coatings* **2018**, *8*, 100. [CrossRef]
27. Milionis, A.; Languasco, J.; Loth, E.; Bayer, I.S. Analysis of wear abrasion resistance of superhydrophobic acrylonitrile butadiene styrene rubber (ABS) nanocomposites. *Chem. Eng. J.* **2015**, *281*, 730–738. [CrossRef]
28. Fini, E.H.; Hajikarimi, P.; Rahi, M.; Nejad, F.M. Physiochemical, rheological, and oxidative aging characteristics of asphalt binder in the presence of mesoporous silica nanoparticles. *J. Mater. Civ. Eng.* **2016**, *28*, 04015133. [CrossRef]
29. Rasool, R.T.; Song, P.; Wang, S.F. Thermal analysis on the interactions among asphalt modified with SBS and different degraded tire rubber. *Constr. Build. Mater.* **2018**, *182*, 134–143. [CrossRef]
30. Liu, K.F.; Zhang, K.; Wu, J.L.; Muhunthan, B.; Shi, X.M. Evaluation of mechanical performance and modification mechanism of asphalt modified with graphene oxide and warm mix additives. *J. Cleaner Prod.* **2018**, *193*, 87–96. [CrossRef]
31. Shenoy, A. Model-fitting the master curves of the dynamic shear rheometer data to extract a rut-controlling term for asphalt pavements. *J. Test. Eval.* **2002**, *30*, 95–102. [CrossRef]
32. Xue, Y.; Wu, S.; Hou, H.; Zha, J. Experimental investigation of basic oxygen furnace slag used as aggregate in asphalt mixture. *J. Hazard. Mater.* **2006**, *138*, 261–268. [CrossRef] [PubMed]
33. Ahmedzade, P.; Sengoz, B. Evaluation of steel slag coarse aggregate in hot mix asphalt concrete. *J. Hazard. Mater.* **2009**, *165*, 300–305. [CrossRef] [PubMed]
34. Pasetto, M.; Baldo, N. Comparative performance analysis of bituminous mixtures with EAF steel slags: A laboratory evaluation. In Proceedings of the 2008 global symposium on recycling, waste treatment and clean technology, REWAS 2008, Cancun, Mexico, 12–15 October 2008.
35. Pasetto, M.; Baldo, N. Computational analysis of the creep behaviour of bituminous mixtures. *Constr. Build. Mater.* **2015**, *94*, 784–790. [CrossRef]
36. Li, P.; Jiang, X.; Guo, K.; Xue, Y.; Dong, H. Analysis of viscoelastic response and creep deformation mechanism of asphalt mixture. *Constr. Build. Mater.* **2018**, *171*, 22–32. [CrossRef]

MDPI

St. Alban-Anlage 66

4052 Basel

Switzerland

Tel. +41 61 683 77 34

Fax +41 61 302 89 18

www.mdpi.com

Materials Editorial Office

E-mail: materials@mdpi.com

www.mdpi.com/journal/materials